Advanced Structured Materials

Volume 55

More information about this series at http://www.springer.com/series/8611

Bilen Emek Abali

Computational Reality

Solving Nonlinear and Coupled Problems in Continuum Mechanics

Springer

Bilen Emek Abali
Berlin
Germany

ISSN 1869-8433 ISSN 1869-8441 (electronic)
Advanced Structured Materials
ISBN 978-981-10-2443-6 ISBN 978-981-10-2444-3 (eBook)
DOI 10.1007/978-981-10-2444-3

Library of Congress Control Number: 2016948788

Printed on acid-free paper

This Springer imprint is published by Springer Nature
The registered company is Springer Nature Singapore Pte Ltd.
The registered company address is: 152 Beach Road, #22-06/08 Gateway East, Singapore 189721, Singapore

"Herhangi bir kişinin, yaşadıkça mutlu, bahtiyar olması için gerekli olan şey, kendisi için değil, kendisinden sonra gelecekler için çalışmaktır."

Any individual shall not work themselves but the future generations into success, in order to gain happiness in their own life.

M.K. Atatürk (Yücel Dergisi, 02/1935)

Preface

Engineers may have different aims and abilities; however, they have a common task: modeling and solving a physical system. In reality, the systems are complex and can only be modeled by nonlinear and coupled equations. Nowadays, even laptops are capable of solving such equations numerically. Therefore, an engineer can model and solve such problems numerically, just by using a laptop.

The underlying work balances between two extremes: being a programmer without duty and being a theoretician without any useful results. The first one, let me call them a *pro*, is able to write an efficient code but pro lacks the knowledge of the governing equations. The second one, let me call them a *theo*, believes in the lengthy and complicated equations. Theo claims that the humanity cannot comprehend the utmost importance of the theory, but theo never performs a useful calculation. An engineer ought to be the fusion of pro and theo; trying to model and *compute* the *reality*.

This work aims for one single target: modeling and computing various engineering applications. The theory leading to nonlinear and coupled equations will be discussed and applied by simulating continuum mechanics problems. Open-source packages are utilized for creating a computational reality, where complex engineering problems are solved. Learning by doing is the key concept in this book; theory and practice are served on a silver platter!

Theory and the collection of engineering applications have been realized over the years with the aid of colleagues:[1] Wolfgang H. Müller, Christina Völlmecke, Andreas Brandmair, Holger Worrack, Arion Juritza, Guido Harneit, Bärbel Minx, Tabea Wilk, Paul Lofink, Felix Reich, Cheng-Chieh Wu, Robert Kersting, Wolfram Martens, Heino Henke, Ingo Müller, Dimitri V. Georgievskii, Maria Kashtalyan, Hans Walter, Wolfgang A. Wall, Volker Gravemeier, Ata Muğan, Holm Altenbach; with assistance of students: Jörg Christian Reiher, G. Gabin Noubissi M., Andre Klunker, Aditya Desai, Fanny Roziere, Wilhelm Hübner, Matthias Steinbach, Elias Büchner, Philipp Diercks, Vyacheslav Boyko, Mario Kierstein; and with support of

[1]No specific order has been used by noting the names.

invaluable friends: Çağrı Döner, Ata Iyiyazıcı, Çağrı Üzüm. Special gratitude is owed to Richard Murray, Chaitanya Raj Goyal, and Mark Searle for perusing different parts of the book and providing amendments to the text.

Moreover, I have been recharged by the motivation and love of my family, namely, Elisabeth Kindler-Abali, G. Ipek Abali, Lale Abalı, and A. Ertan Abalı.

Tons of thanks go to everyone helped me for putting science to work!

Berlin, Germany
May 2016

Bilen Emek Abali

Contents

Acronyms

A possibly incomplete list of symbols used in the book is given in the following. It has been a great effort to attain a unique use of every introduced symbol. There are no standards or rules, how to choose a symbol for a physical variable, however, conventions have led to many of the choices below.

Latin Symbols

Symbol	SI units	Description
A_i	Wb/m ≙ H A/m ≙ T m ≙ ≙ V s/m ≙ J/(A m)	Magnetic or vector potential
B_i	T ≙ Wb/m^2≙ V s/m^2≙ ≙ N/(A m) ≙ kg/(s^2 A)	Magnetic flux (area) density
B_{ij}	–	Left CAUCHY–GREEN deformation tensor
c	J/(kg K)	Specific heat capacity
	m/s^2	Speed of light in vacuum
C_{ijkl}	Pa ≙ N/m^2	Stiffness tensor
C_{ij}	–	Right CAUCHY–GREEN deformation tensor
da	m^2	Infinitesimal area element in current frame
dA	m^2	Infinitesimal area element in reference frame
dv	m^3	Infinitesimal volume element in current frame
dV	m^3	Infinitesimal volume element in reference frame
D_i	C/m^2	Charge potential (electric displacement)
d_{ij}	m/(m s)	Symmetric part of velocity gradient
e	J/kg	Specific (per mass) total energy
e_{ij}	m/m	EULER–ALMANSI strain tensor
E_i	N/C ≙ V/m ≙ kg m/(s^3 A)	Electric field

(continued)

(continued)

Symbol	SI units	Description
E_{ij}	m/m	GREEN–LAGRANGE strain tensor
f	Pa^2	Flow potential
f_i	N/kg	Specific supply of linear momentum, volumetric or body force
$f_i^{\text{Lor.}}$	N/kg	Specific LORENTZ force
F_i	W/m^2	Flux of total energy
F_{ij}	m/m	Deformation gradient
g_{ij}	–	Metric tensor
G_i	K/m	Temperature gradient
h	$W/(m^2\ K)$	Convective heat transfer coefficient
H_i	A/m	Current potential (magnetic field strength)
J	–	JACOBI determinant (of deformation gradient)
J_i	$A/m^2 \widehat{=} C/(s\ m^2)$	Electric current (area) density
$J_i^{\text{fr.}}$	$A/m^2 \widehat{=} C/(s\ m^2)$	Free electric current (area) density
M_i	A/m	Magnetic polarization (magnetization)
n_i	–	Plane normal in current frame
N_i	–	Plane normal in reference frame
p	$N/m^2 \widehat{=} Pa$	Pressure
P_i	C/m^2	Electric polarization (polarization)
P_{ij}	$Pa \widehat{=} N/m^2$	Nominal, engineering, PIOLA, or first PIOLA–KIRCHHOFF stress
q_i	W/m^2	Flux of internal energy or heat flux in current frame
Q_i	W/m^2	Flux of internal energy or heat flux in reference frame
r	W/kg	Specific supply of internal energy, internal heating, or radiant heating
Re	–	REYNOLD's number
s	W/kg	Specific supply of total energy
S_{ij}	$Pa \widehat{=} N/m^2$	Second PIOLA–KIRCHHOFF stress
T	K	Absolute temperature
u	J/kg	Specific internal energy
u_i	m	Displacement
v_i	m/s	Velocity of massive particles
v_i^{e}	m/s	Velocity of charged particles
w	J/m^3	Stored energy density
w_i	m/s	Domain velocity
x_i	m	Coordinates in a Cartesian system
z	C/kg	Specific (electric) charge
z_i	m	Coordinates in an arbitrary system

Greek Symbols

Symbol	SI units	Description
α_{ij}	1/K	Coefficient of thermal expansion
β_{ij}	Pa $\hat{=}$ N/m^2	Back stress
γ		Not always used as the same variable
Γ^i_{jk}	1/m	CHRISTOFFEL symbols
Γ	W/m^3 $\hat{=}$ J/(s m^3)	Production of internal energy
$\mathit{\Gamma}$	–	Gamma function
δ_{ij}	–	KRONECKER delta
δ	–	Variation symbol
ϵ_{ijk}	–	LEVI-CIVITA symbol
ε_0	F/m $\hat{=}$ C/(V m) $\hat{=}$ $\hat{=}$ A s/(V m)	Vacuum permittivity
$\varepsilon^{\mathrm{el.}}_{ij}$	F/m	Materials dielectric permittivity
$\bar{\varepsilon}^{\mathrm{el.}}_{ij}$	–	Materials relative permittivity
ζ		Not always used as the same variable
η	J/(K kg)	Specific entropy
θ		Not always used as the same variable
ι		Not used at all
κ	W/(m K)	Thermal conductivity
λ	Pa $\hat{=}$ N/m^2	LAME constant for solids
	Pa s $\hat{=}$ N s/m^2	Volume viscosity for fluids
$\Lambda^{\cdot}$	1/(Pa s)	Plastic multiplier
μ	Pa $\hat{=}$ N/m^2	LAME constant for solids
	Pa s $\hat{=}$ N s/m^2	Shear viscosity for fluids
μ_0	H/m $\hat{=}$ T m/A $\hat{=}$ V s/(A m)	Vacuum permeability
$\mu^{\mathrm{mag.}}_{ij}$	H/m	Materials magnetic permeability
$\bar{\mu}^{\mathrm{mag.}}$		Materials relative permeability
ν	–	POISSON's ratio
ξ		Not used at all
π	V/K	Thermoelectric coupling coefficient
π	–	Number pi
ρ	kg/m^3	Mass density in current frame
ρ_0	kg/m^3	Mass density in reference frame
σ_{ij}	N/m^2	CAUCHY's stress
ς	S/m $\hat{=}$ 1/(Ω m) $\hat{=}$ A/(V m)	Electrical conductivity
Σ	W/(K m^3)	Entropy production
τ_q, τ_T	s	Time-delay parameters for heat flux

(continued)

(continued)

Symbol	SI units	Description
υ		Not used at all
ϕ	$V \triangleq J/C$	Electric or scalar potential
Φ_i	$W/(K\ m^2)$	Entropy flux
$\chi^{el.}$	–	Electric susceptibility
$\chi^{mag.}$	–	Magnetic susceptibility
ψ	J/kg	Specific free energy
ω	m/s	Characteristic velocity

Script and Calligraphic Symbols

Symbol	SI units	Description
$\mathfrak{D}_i$	C/m^2	Free charge potential
$\mathcal{E}_i$	N/C	Electromotive intensity, objective electric field
$\mathcal{F}^\alpha$		Thermodynamic fluxes
$\mathfrak{H}_i$	A/m	Free current potential
$\mathcal{K}^\alpha$		Thermodynamic forces
$\mathcal{M}_i$	A/m	Objective magnetic polarization
$\mathcal{J}_i$	A/m^2	Material electric current (area) density
$\mathcal{J}_i^{fr.}$	A/m^2	Free (material) electric current (area) density
$\mathcal{L}$	J/m^3	LAGRANGEAN density
$\mathcal{S}$	J s	Action

Introduction

The author and the reader are simply denoted by "we" henceforth. In this book we will exploit the standard tensor calculus notation and rules of continuum mechanics in order to understand, describe, model, and compute engineering problems. Some hints and key explanations belonging to the tensor notation are given in the first sections, however, we skip a brief tensor calculus chapter and start directly with mechanics in Chap. 1, proceed with thermodynamics in Chap. 2, and finish with electromagnetism in Chap. 3.

The book consists of 20 sections gathered in the three chapters. We follow a bottom-up approach, therefore, we suggest to experience the sections in the written order. In each section we discuss and model another type of an engineering system, and compute its *primitive variables* by solving the corresponding *field equations*. A field equation is a differential equation, solution of which results in the primitive variable as a function in space and time. Different systems may have different primitive variables:

- A solid structure like a bridge, building, or a vehicle deforms under a mechanical loading. The sought-after primitive variable is *displacement*.
- A fluid flows in a pipe due to the pressure difference applied on both ends of the pipe. In this case *velocity* and *pressure* are the primitive variables.
- A laser welding on a steel plate produces heat leading to a temperature increase. *Temperature* is the primitive variable. If we also want to compute the deformation caused by the temperature distribution, then *temperature* and *displacement* are both primitive variables. The field equations for the primitive variables are coupled and nonlinear.
- A conductor creates electric and magnetic fields. *Electric* and *magnetic potentials* are the primitive variables to be computed. By conducting an electric current, a wire heats up leading to a temperature increase followed by a deformation. Then we need to compute in addition to the electric and magnetic potentials, *displacement* and *temperature* as primitive variables satisfying the coupled and nonlinear field equations.

Even many more engineering examples are theoretically discussed and numerically computed in the present book. We will analyze mechanical, fluid dynamical, thermodynamical, and electrodynamical systems. All is established by using the method known as *continuum mechanics*. The strength of the continuum mechanics is its abstraction in obtaining governing equations. For many different systems we can obtain field equations by following a general receipt in three steps:

- First we write a *balance* equation for each primitive variable.
- Second, we select adequate *constitutive* or *material* equations.
- Third, we insert the constitutive equations into the balance equations for acquiring the *field* equations. These are the governing equations of the underlying engineering system and their solutions result in the primitive variables that we search for.

This approach seems to be complicated at first, honestly, it is the simplest method enabling to cover so many different subjects in one book. The continuum mechanical framework is the strength of the *computational reality* created in this present book.

An engineering system is described by the primitive variables satisfying field equations. Different primitive variables like displacement, temperature, electric and magnetic potentials result in a multiphysics problem. The field equations of multiphysics problems are coupled and nonlinear, in other words, difficult to solve. In order to compute the coupled, nonlinear system of partial differential equations in space and time, we will exploit a novel collection of open-source packages developed under the FEniCS project [1] and start exploring FEniCS by reading Appendix A.1 on p. 293. All codes in this book are written in Python and tested in FEniCS version 1.6.0.[2]

Every section starts with a theoretical treatise leading to the necessary governing equations. We attain in each section a so-called *weak form* that is used in a code to solve an example on a simple geometry, like a beam, cube, or rectangle. The weak form is valid for any geometry, so the code can be used for other geometries, too. Indeed, in many real-life engineering problems the geometry is much more complicated than just a box. In such a case the complicated geometry can be made ready by preprocessing with the open-source program Salome. We have explained step-by-step how to transfer the complicated geometry into FEniCS in Appendix A.3 on p. 297 by using Salome version 7.5 and Gmsh version 2.8.[3]

Chapter 1 deals with mechanics for a continuum body. We start with deformable solids and observe linear and nonlinear elastostatics followed by hyperelasticity. By incorporating time we start elastodynamics in rheology and proceed with plasticity. Solid mechanics uses a LAGRANGEan frame, which is beneficial for material systems. Then we move on to open systems like a fluid flowing in a pipe described in an EULERian frame. Linear and nonlinear fluids are discussed and computed. We crown

[2]See release notes for newer versions in [1].

[3]See [2, 3].

this chapter by using both frames simultaneously for computing a fluid-structure interaction like a spoon stirring a coffee.

Chapter 2 amends the computational reality by involving thermodynamics. The applied thermodynamics aims at modeling the temperature distribution in a continuum body. We apply thermodynamics differently in macroscopic and microscopic length scales. The theoretical thermodynamics answers the question of how to select the constitutive equations. We introduce and use in every following section a methodology allowing a formal derivation of the appropriate constitutive equations. This method is presented in viscous fluids and utilized in viscoelastic and plastic solids.

Chapter 3 embodies electromagnetism in the computational reality. Electric current producing heat is discussed. Polarized materials are introduced by motivating MAXWELL's equations from balance laws. The coupling effects are discussed, for example, for the thermoelectric coupling in conductors the governing equations are deduced from the balance equations with the constitutive equations derived using thermodynamics. Deformation, temperature distribution, and electromagnetic potentials are solved monolithically. We further develop the approach for incorporating polarized materials in electrodynamics and acquire thermodynamically consistent constitutive equations for piezoelectricity as well as magnetohydrodynamics.

A range of applications is presented by using continuum mechanics for obtaining governing equations, by exploiting thermodynamics for deriving the constitutive equations, and by utilizing FEniCS project to compute engineering examples by solving nonlinear and coupled equations monolithically. Information contained in this book may be difficult to grasp and internalize at once, even for an expert in engineering. For the purpose of a deeper understanding, every step in the formulations is shown, as well as every line of code in the computations. Moreover, the reader is encouraged to try to accomplish the challenging tasks at the end of each section, since Albert Einstein convinced the author by his saying:

"Learning is experience. Everything else is just information."

References

1. FEniCS project: Development of tools for automated scientific computing, 2001–2016. http://fenicsproject.org (2016)
2. Geuzaine, C., Remacle, J.F.: Gmsh: A 3-d finite element mesh generator with built-in pre-and post-processing facilities. International Journal for Numerical Methods in Engineering **79**(11), 1309–1331 (2009)
3. Salome: The Open Source Integration Platform for Numerical Simulation, 1993–2016. http://salome-platform.org (2016)

Chapter 1
Mechanics

For a new engineering design, we have to perform various analyses. Many of these analyses belong to *mechanics*. As a consequence of a static or dynamic loading, deformation and stress occur in the continuum body. If the stress lies below the yield limit, the deformation is recoverable upon unloading. This behavior is called an *elastic* response. This elastic response is instantaneous, i.e., rate of loading does not matter. In order to bring in the effect of the loading rate, we need a *viscoelastic* response. This behavior is modeled by changing the constitutive (material) equation. The deformation is still recovered upon unloading. In case of remaining deformation after unloading, we need a constitutive equation modeling a *plastic* behavior. In all of the aforementioned phenomena, we ignore any change in temperature, thus the process is *isothermal*.

In this chapter we will discuss mechanical systems and compute the motion of particles belonging to a continuum body. We start with the linear elasticity in Sect. 1.1 and setup the *three* necessary steps for obtaining the so-called *weak form*. First, the necessary balance equations are derived. Second, the material equations are employed to close the system of equations. Third, the variational formulation is utilized to obtain the weak form. These three steps are going to be used in the whole book without being mentioned further. In Sect. 1.2 we present the solid body mechanics in a LAGRANGEan frame[1] and solve a problem with *geometric* nonlinearities. In Sect. 1.3 we introduce an abstraction of the method and discuss the *real* variational formulation, moreover, we employ and solve a problem with *material* nonlinearities. All of these computations belong to elasto*statics*. By including time rate in the equations, we start off with *dynamics* in Sect. 1.4. Examples of linear and fractional rheology are presented in Sects. 1.4 and 1.5, respectively. In Sect. 1.6 the plastic deformation is addressed, where the material starts *flowing* beyond the yield stress. We change the understanding of motion from a solid body to fluid by introducing a

[1] It is named after Joseph-Louis Lagrange.

B.E. Abali, *Computational Reality*, Advanced Structured Materials 55,
DOI 10.1007/978-981-10-2444-3_1

EULERian frame[2] in Sect. 1.7 and present the computation of flows of *linear* fluids. We discuss the linearization of coupled field equations by examining a fluid flow problem of a *nonlinear* fluid in Sect. 1.8. An amazing feature of the simultaneous use of the LAGRANGEan and EULERian frames results in the fluid-structure interaction presented in Sect. 1.9.

1.1 Linear Elastostatics

A design has to hold under known loading conditions. Suppose that we want to verify a design by using a simulation. The structure in the design can be a part of a vehicle (car, plane, train) or a part of a bridge construction. This structure is designed to hold on "forever." In other words, the design shall avoid any plastic deformation such that the deformation will be recovered upon unloading. For example, a truck crossing a bridge applies a loading such that the structure deforms. After the truck has crossed the bridge—after unloading—the bridge turns back to its designed shape, any deformation is recovered. In short, the structure shall be loaded below the *yield* limit. For most of the engineering materials, this limit is known as the yield stress measured with a one-axial tensile test. In the simulation we will obtain the so-called CAUCHY stress[3] tensor, σ_{ij}, which has 9 components in a three-dimensional continuum,

$$\sigma_{ij} = \begin{pmatrix} \sigma_{11} & \sigma_{12} & \sigma_{13} \\ \sigma_{21} & \sigma_{22} & \sigma_{23} \\ \sigma_{31} & \sigma_{32} & \sigma_{33} \end{pmatrix} . \tag{1.1}$$

Unfortunately, in a tensile test we determine a single value for the yield stress that sets the upper limit for our engineering design. Below this value only elastic (but not plastic) deformation occurs. For example, the engineering steel, AISI 1006 (cold drawn), possesses a yield stress[4] of $\sigma^{\text{yield}} = 285\,\text{MPa}$ (MPa $\hat{=}$ N/mm^2). The 9 components from the simulation shall be reduced to a scalar value in order to compare with σ^{yield} obtained from the one-axial test. We use the VON MISES equivalent stress, $\sigma^{\text{eq.}}$, for this purpose[5]

$$\sigma^{\text{eq.}} = \Big(\frac{3}{2}\sigma_{|ij|}\sigma_{|ij|}\Big)^{1/2} , \quad \sigma_{|ij|} = \sigma_{ij} - \frac{1}{3}\sigma_{kk}\delta_{ij} , \tag{1.2}$$

which has to be lower than the *allowed* upper limit. If we would design a lightweight structure—in a plane or race car design—then the allowed limit might be chosen

[2] The frame is named after Leonhard Euler.

[3] The stress is named after Augustin-Louis Cauchy.

[4] See [12] for materials data.

[5] KRONECKER delta, δ_{ij}, is simply the identity matrix having 1 on its diagonal and zero as the non-diagonal components. It is named after Leopold Kronecker.

near to the yield stress. In other words, the equivalent stress has to be lower than the yield stress, $\sigma^{\text{eq.}} < \sigma^{\text{yield}}$. Often the allowed limit is lower than the yield stress. The allowed stress is found by dividing the yield stress by a *safety factor*. This factor is proposed by "know-how" and it depends on the specific construction.

Throughout the book we will use a standard indicial notation. In order to explain the notation in Eq. (1.2), we write some terms explicitly:

$$\begin{aligned}
\sigma_{kk} &= \sum_{k=1}^{3} \sigma_{kk} = \sigma_{11} + \sigma_{22} + \sigma_{33} \,, \\
\sigma_{|ij|}\sigma_{|ij|} &= \sum_{i=1}^{3}\sum_{j=1}^{3} \sigma_{|ij|}\sigma_{|ij|} = \sigma_{|11|}\sigma_{|11|} + \sigma_{|12|}\sigma_{|12|} + \\
&+ \sigma_{|13|}\sigma_{|13|} + \sigma_{|21|}\sigma_{|21|} + \cdots + \sigma_{|33|}\sigma_{|33|} \,.
\end{aligned} \tag{1.3}$$

We are too lazy to write the summation symbols in continuum mechanics and henceforth we understand a summing up over doubly repeated indices, i.e., we apply EINSTEIN's summation convention.[6]

In this section we want to simulate a simple geometry and answer the following question: How can we obtain the stress tensor in elastostatics? For this aim, we briefly outline the *linear* elasticity and present the variational formulation leading to the *weak form*. The general procedure of obtaining the weak form consists of three steps. The step number one is to choose the appropriate *balance equations*. The second step is finding the *constitutive* or *material* equations capable of modeling the elasticity. The third step is to utilize both of them and to employ the *variational formulation* leading to the weak form. Out of that weak form, we obtain a numerical solution of the primitive variable. The primitive variable in elastostatics is the displacement in three-dimensional continuum that we want to compute numerically. As the numerical solution technique, we are going to apply the *finite element method*. In the end, we write a code and solve the displacement.

Step I: Balance equations

We aim at the computation of stress in three-dimensional space, which is calculated from the primitive variable: displacement. Without arguing, i.e., in an axiomatic manner, we simply assume that the primitive variable exists. We set our goal to compute the primitive variable (displacement), therefore, we need a differential equation, which will be deduced from the balance equations. The displacement, $u_i(\boldsymbol{x})$, in three-dimensional space is expressed in a Cartesian coordinate system $x_i = (x\,,\ y\,,\ z)$. A continuum body deforms under a known (given) external mechanical loading force. The connection between the deformation and the given forces is formulated by using the *balance of linear momentum*:

[6] The summation convention is named after Albert Einstein.

$$\rho v_i^{\bullet} - \frac{\partial \sigma_{ji}}{\partial x_j} = \rho f_i \ , \tag{1.4}$$

where the summation convention is utilized over the index "j." Velocity, v_i, is the rate[7] of displacement, $v_i = u_i^{\bullet}$. Mass density, ρ, is the weight per volume and the rate of velocity is simply the acceleration. The CAUCHY stress tensor, σ_{ij}, needs to be related to the displacement by means of a *constitutive* or *material* equation. A volumetric or body force, ρf_i, is a given quantity. A typical example in mechanics is the gravitational force. The specific force (per mass), f_i, in a coordinate system directed toward the Earth's center with the z-axis reads

$$f_i = \begin{pmatrix} 0 \ 0 \ 9.81 \end{pmatrix} \text{ N/kg} \ . \tag{1.5}$$

In general, the external loading on the surface of the continuum body deforms the body much more than the body force. Therefore, the deformation due to the weight is omitted. The balance of linear momentum reads

$$\rho v_i^{\bullet} - \frac{\partial \sigma_{ji}}{\partial x_j} = 0 \ , \tag{1.6}$$

An elastic response is instantaneous. Of course this is a simplification, we know that an instantaneous motion is impossible. In reality, every particle moves with a velocity, however, its motion is quicker than we can measure. The easiest way of visualizing this phenomenon is a tensile test in a hydraulic machine. Suppose we perform experiments at different loading rates and compare stress-strain plots. For an elastic material, e.g., for an engineering steel, all results are identical. Although the loading rates are different, the response is the same. The speed of loading is limited with the response time of the fluid used in the hydraulic system, which is greater than the response time of steel. In other words, the elastic response of steel is instantaneous with respect to the measurement system. Actually, it is only quicker than our measurement system such that the hydraulic system is not capable of detecting it. Since we fail to observe the rate of velocities, we simply neglect them, $v_i^{\bullet} = 0$, in statics. Now the balance of linear momentum reads

$$\begin{aligned} -\frac{\partial \sigma_{ji}}{\partial x_j} &= 0 \ , \\ -\sigma_{ji,j} &= 0 \ . \end{aligned} \tag{1.7}$$

We have introduced a shorthand notation for the space differentiation where again EINSTEIN's summation convention is applied,

[7] Rate means a change in time, given by the derivative in time.

$$\sigma_{ji,j} = \sigma_{1i,1} + \sigma_{2i,2} + \sigma_{3i,3} \,. \tag{1.8}$$

For the sake of clarity, we rewrite the three equations:

$$-\sigma_{j1,j} = 0 \,, \quad -\sigma_{j2,j} = 0 \,, \quad -\sigma_{j3,j} = 0 \,. \tag{1.9}$$

With three equations we will compute the three components of the primitive variable in space:

$$u_i = \begin{pmatrix} u_1(x_j) \\ u_2(x_j) \\ u_3(x_j) \end{pmatrix} , \tag{1.10}$$

since the deformation is instantaneous we cannot speak of time. In order to solve u_i we need differential equations in or *field* equations of u_i. For obtaining field equations, we need to discuss the second step and set σ_{ij} in relation to u_i.

Step II: Constitutive equations

We present a material or constitutive equation connecting CAUCHY stress, σ_{ij}, with the primitive variable, u_i. Displacement of a particle is intuitive. It is the motion from the initial position (before loading) to the current position (after loading). Displacement gradient, i.e., neighboring particles having different displacements, creates a tension in the matter, called stress. We need to relate the stress to displacement gradient by means of a constitutive equation. For the moment we just state that the stress tensor is symmetric, $\sigma_{ij} = \sigma_{ji}$. The reason behind this assumption will be discussed later. Since the stress tensor is symmetric we also want to use the symmetric part of the displacement gradient:

$$\varepsilon_{ij} = u_{(i,j)} = \frac{\partial u_{(i}}{\partial x_{j)}} = \frac{1}{2}\Big(\frac{\partial u_i}{\partial x_j} + \frac{\partial u_j}{\partial x_i}\Big) = \frac{1}{2}\big(u_{i,j} + u_{j,i}\big) \,, \tag{1.11}$$

which we call the *strain* tensor. It is also called the *kinematic condition*. This strain measure is linear and appropriate for small displacements. Linear elasticity is modeled by the constitutive relation between the stress and strain, referred to as HOOKE's law:[8]

$$\sigma_{ij} = C_{ijkl}\varepsilon_{kl} \,, \tag{1.12}$$

with the *elasticity* or *stiffness* tensor of rank four (four indices), C_{ijkl}. In three-dimensional space four indices make $3 \times 3 \times 3 \times 3 = 81$ components for C_{ijkl}. In one-dimensional case, for example in a one-axial tensile testing, the relation simply reads $\sigma_{11} = E\varepsilon_{11}$, where the elasticity or YOUNG's modulus, E, is the

[8]The law is named after Robert Hooke.

only component, C_{1111}, of the stiffness tensor. For a three-dimensional structure we need all components of the stiffness tensor—they shall be determined experimentally. It is quite difficult to construct 81 different experiments for determining the material parameters. Fortunately, this number gets reduced. The CAUCHY stress tensor is symmetric, $\sigma_{ij} = \sigma_{ji}$, and it is related to the symmetric strain tensor, $\varepsilon_{ij} = \varepsilon_{ji}$. Therefore, reduction of the number of experiments is possible. Since $\sigma_{ij} = C_{ijkl}\varepsilon_{kl} = \sigma_{ji} = C_{jikl}\varepsilon_{kl}$ and $\varepsilon_{kl} = (C_{ijkl})^{-1}\sigma_{ij} = \varepsilon_{lk} = (C_{ijlk})^{-1}\sigma_{ij}$ it is obvious that $C_{ijkl} = C_{jikl}$ and $C_{ijkl} = C_{ijlk}$. In each block of "ij" or "kl" we have 6 independent parameters. Hence, we can rewrite the stiffness tensor of rank four in a 6×6 matrix, known as the VOIGT notation:[9]

$$C_{IJ} = \begin{pmatrix} C_{1111} & C_{1122} & C_{1133} & C_{1123} & C_{1113} & C_{1112} \\ C_{2211} & C_{2222} & C_{2233} & C_{2223} & C_{2213} & C_{2212} \\ C_{3311} & C_{3322} & C_{3333} & C_{3323} & C_{3313} & C_{3312} \\ C_{2311} & C_{2322} & C_{2333} & C_{2323} & C_{2313} & C_{2312} \\ C_{1311} & C_{1322} & C_{1333} & C_{1323} & C_{1313} & C_{1312} \\ C_{1211} & C_{1222} & C_{1233} & C_{1223} & C_{1213} & C_{1212} \end{pmatrix} , \tag{1.13}$$

where I and J run from 1 to 6. This matrix is just an ordered list without any tensorial character. Tensors are objects with the same transformation properties as the coordinate frame transformations, thus they are reliable for constructing material relations. When the material response is defined with tensors and it holds in one coordinate system; then it holds in any coordinate system. Here we have introduced VOIGT notation just for an easier notation of a rank four tensor. In the computation we use the tensor notation. By employing symmetric stress and strain, we have reduced the number of different components to 36.

We will reduce the number furthermore by explaining the meaning of *linear* in linear elasticity. First we need to give the definition of work. As a consequence of a loading on the surface of the continuum body, $\partial\mathcal{B}$, there occurs a displacement, u_i, in the unit of length. By multiplying it with the force, F_i, directed toward the displacement direction gives the work done in the system,

$$\begin{gathered} W = F_i u_i = \int_{\partial\mathcal{B}} u_i \mathrm{d}F_i = \int_{\partial\mathcal{B}} u_i t_i \mathrm{d}A \, , \\ \dim(W) \hat{=} \mathrm{J(oule)} \, , \quad \dim(t_i) \hat{=} \mathrm{N(ewton)/m(eter)}^2 \, , \end{gathered} \tag{1.14}$$

where the traction vector t_i is an area density of the force. Using the CAUCHY tetrahedron argument, it is possible to change the traction vector to the stress tensor, $t_i = n_j \sigma_{ji}$, where n_j is the plane normal on the surface $\partial\mathcal{B}$ directed outward the body. By using the tetrahedron argument we can transform the surface integral into a volume integral with the help of GAUSS's law[10]

[9]It is named after Woldemar Voigt.

[10]The law is named after Carl Friedrich Gauß.

$$
\begin{aligned}
W &= \int_{\partial\mathcal{B}} u_i t_i \mathrm{d}A = \int_{\partial\mathcal{B}} u_i n_j \sigma_{ji} \mathrm{d}A = \int_{\mathcal{B}} \frac{\partial(\sigma_{ji} u_i)}{\partial x_j} \mathrm{d}V = \\
&= \int_{\mathcal{B}} \left(\frac{\partial \sigma_{ji}}{\partial x_j} u_i + \sigma_{ji} \frac{\partial u_i}{\partial x_j} \right) \mathrm{d}V = \int_{\mathcal{B}} w \mathrm{d}V \ , \quad \dim(w) \hat{=} \mathrm{J/m^3} \ .
\end{aligned} \tag{1.15}
$$

The first integrand vanishes owing to the balance of linear momentum, the work or energy density (energy per volume) reads

$$
w = \sigma_{ji} u_{i,j} \ . \tag{1.16}
$$

Any tensor of rank two, thus the displacement gradient can be decomposed into a symmetric and an antisymmetric part:

$$
\begin{aligned}
u_{(i,j)} = \frac{1}{2}\big(u_{i,j} + u_{j,i}\big) \ , \quad u_{[i,j]} = \frac{1}{2}\big(u_{i,j} - u_{j,i}\big) \ , \\
u_{i,j} = u_{(i,j)} + u_{[i,j]} \ .
\end{aligned} \tag{1.17}
$$

Since the CAUCHY stress is symmetric, $\sigma_{ji} = \sigma_{ij}$, its contraction with the antisymmetric part of the displacement gradient vanishes

$$
\sigma_{ji} u_{[i,j]} = \frac{1}{2}\big(\sigma_{ji} u_{i,j} - \sigma_{ji} u_{j,i}\big) = \frac{1}{2}\big(\sigma_{ij} u_{i,j} - \sigma_{ji} u_{j,i}\big) = 0 \ . \tag{1.18}
$$

This fact allows us to rewrite the energy density,

$$
\begin{aligned}
w = \sigma_{ji} u_{i,j} = \sigma_{ji}\big(u_{(i,j)} + u_{[i,j]}\big) = \sigma_{ji} u_{(i,j)} = \sigma_{ji} \varepsilon_{ij} \ , \\
\dim(\sigma_{ji}\varepsilon_{ij}) \hat{=} \mathrm{N/m^2} \times \mathrm{m/m} = \mathrm{N\,m/m^3} = \mathrm{J/m^3} \ .
\end{aligned} \tag{1.19}
$$

Suppose we have a monotonic strain, for example in a tensile test, the machine is steered by the displacement. Then the energy density, w, can be calculated $w = \int \sigma_{ji} \mathrm{d}\varepsilon_{ij}$. In a one-dimensional tensile test, this integral is the area underneath the curve on xy-plot with strain on abscissa and stress on ordinate. Now consider another tensile test where force is controlled such that the energy density is $w^* = \int \varepsilon_{ij} \mathrm{d}\sigma_{ji}$. This integral is the area underneath the curve on xy-plot with stress on abscissa and strain on ordinate. When the curve is a straight line, then $w = w^*$, and the material response is *linear*. The linear relation between stress and strain is given by a material parameter, called *modulus*. The components of stiffness tensor, moduli, are constant numbers, they are not functions of strain. Since C_{ijkl} consists of *constants*, we obtain

$$
\begin{aligned}
w = \int \sigma_{ji} \mathrm{d}\varepsilon_{ij} = \int C_{jikl} \mathrm{d}\varepsilon_{ij} = C_{jikl} \int \varepsilon_{kl} \mathrm{d}\varepsilon_{ij} \ , \\
w^* = \int \varepsilon_{ij} \mathrm{d}\sigma_{ji} = \int \varepsilon_{ij} \mathrm{d}\big(C_{jikl}\varepsilon_{kl}\big) = C_{jikl} \int \varepsilon_{ij} \mathrm{d}\varepsilon_{kl} = C_{klji} \int \varepsilon_{kl} \mathrm{d}\varepsilon_{ij} \ ,
\end{aligned} \tag{1.20}
$$

hence for linear elastic materials, the energy density is

$$w = w^* \Rightarrow C_{jikl} = C_{klji} \ . \tag{1.21}$$

In the VOIGT notation, this condition generates a symmetric stiffness matrix:

$$C_{IJ} = C_{JI} = \begin{pmatrix} C_{1111} & C_{1122} & C_{1133} & C_{1123} & C_{1113} & C_{1112} \\ & C_{2222} & C_{2233} & C_{2223} & C_{2213} & C_{2212} \\ & & C_{3333} & C_{3323} & C_{3313} & C_{3312} \\ & & & C_{2323} & C_{2313} & C_{2312} \\ & \text{symm.} & & & C_{1313} & C_{1312} \\ & & & & & C_{1212} \end{pmatrix} \ . \tag{1.22}$$

The number of different (independent) components, simply the number of material parameters in the stiffness tensor is reduced to 21 for the case of linear elasticity in three-dimensions.

Many engineering materials consist of a crystalline structure. This structure brings additional symmetries, which reduce the number of independent parameters further. For the case of no symmetries—the full group of anisotropic material—21 different moduli need to be determined. There are more than thirty different crystal classes and their symmetries are inspected under group theory calculations.[11] For the case of isotropic material, the number of different material parameters is only two, the LAME parameters,[12] λ, μ. The stiffness matrix of linear elastic isotropic material in the VOIGT notation:

$$C_{IJ} = C_{JI} = \begin{pmatrix} \lambda + 2\mu & \lambda & \lambda & 0 & 0 & 0 \\ & \lambda + 2\mu & \lambda & 0 & 0 & 0 \\ & & \lambda + 2\mu & 0 & 0 & 0 \\ & & & \mu & 0 & 0 \\ & \text{symm.} & & & \mu & 0 \\ & & & & & \mu \end{pmatrix} , \tag{1.23}$$

can also be written in the tensor notation (expressed in Cartesian coordinates)

$$C_{ijkl} = \lambda \delta_{ij} \delta_{kl} + \mu (\delta_{ik} \delta_{jl} + \delta_{il} \delta_{jk}) \ . \tag{1.24}$$

Hence the CAUCHY stress for linear elastic isotropic materials, like steel, aluminum, magnesium reads

$$\sigma_{ij} = C_{ijkl} \varepsilon_{kl} = \lambda \varepsilon_{kk} \delta_{ij} + 2\mu \varepsilon_{ij} \ , \tag{1.25}$$

---

[11] For a clear treatment of all the classes you can refer to the lessons of Bernhard J. Wuensch in MIT freely available under: http://www.academicearth.org/courses/symmetry-structure-and-tensor-properties-of-materials

[12] They are named after Gabriel Lamé.

where we have used

$$\varepsilon_{kl}\delta_{kl} = \varepsilon_{kk} = \varepsilon_{11} + \varepsilon_{22} + \varepsilon_{33} \,. \tag{1.26}$$

The constant parameters, λ and μ, can be expressed in terms of the widely known engineering parameters:

$$\begin{aligned} \lambda &= \frac{E\nu}{(1+\nu)(1-2\nu)} = \frac{2G\nu}{(1-2\nu)} = \frac{(E-2G)G}{3G-E} \,, \\ \mu &= \frac{E}{2(1+\nu)} = G \,, \end{aligned} \tag{1.27}$$

this notation is more useful since YOUNG's modulus[13] E, shear modulus G, POISSON's ratio[14] ν can all be determined by appropriate measurements.

Step III: Weak form

The balance of linear momentum in Eq. (1.7) is augmented by the constitutive (material) Eq. (1.25) for a linear elastic and isotropic material. We sum up the governing equations for linear elastostatics

$$-\sigma_{ji,j} = 0 \,, \quad \sigma_{ji} = \lambda\varepsilon_{kk}\delta_{ji} + 2\mu\varepsilon_{ji} \,, \quad \varepsilon_{ij} = u_{(i,j)} \,. \tag{1.28}$$

This field equation needs to be satisfied for all particles within the continuum body, $\forall x_k \in \mathcal{B}$. Additionally, for particles on boundaries, $\forall x_k \in \partial\mathcal{B}$, we need to define the *boundary conditions*:

$$\begin{aligned} u_i\big|_{x_k} &= \hat{u}_i \,, \quad \forall x_k \in \partial\mathcal{B}_{\mathrm{D}} \,, \\ n_j\sigma_{ji}\big|_{x_k} &= \hat{t}_i \,, \quad \forall x_k \in \partial\mathcal{B}_{\mathrm{N}} \,. \end{aligned} \tag{1.29}$$

For the so-called DIRICHLET boundary condition,[15] the displacement on $\partial\mathcal{B}_{\mathrm{D}}$ is given, $\hat{u}_i$. Analogously, for the so-called NEUMANN boundary condition,[16] the traction vector on $\partial\mathcal{B}_{\mathrm{N}}$ is given, $\hat{t}_i$. We have to define the whole boundary by using the boundary conditions. In this section we assume that the DIRICHLET and NEUMANN boundaries are disjunct, $\partial\mathcal{B}_{\mathrm{D}} \cap \partial\mathcal{B}_{\mathrm{N}} = \emptyset$, and they compile the whole boundary, $\partial\mathcal{B}_{\mathrm{D}} \cup \partial\mathcal{B}_{\mathrm{N}} = \partial\mathcal{B}$. In other words, a particle on the boundary belongs either to a DIRICHLET or to a NEUMANN condition. This condition is just necessary to understand the example better, it is by no means necessary for the variational formulation

---

[13] It is named after Thomas Young.

[14] It is named for Siméon Denis Poisson.

[15] It is named after Peter Gustav Lejeune Dirichlet.

[16] It is named for Carl Gottfried Neumann.

below. We will complement Eq. (1.28) with Eqs. (1.29) in order to obtain the weak form necessary for the *finite element method.* This method is a numerical approximation to the solution, where the residual equation is *weighted* by multiplying by a *test function*, δu_i. The variational notation δ will be explored later. For the moment, δu_i is an arbitrary function that can be *varied.* We multiply the residual by the test function globally (in the whole computational domain, i.e., the continuum body $\mathcal{B}$) as follows

$$-\int_{\mathcal{B}} \sigma_{ji,j}\, \delta u_i \mathrm{d}V = 0 \,, \tag{1.30}$$

in Cartesian coordinates. The volume element,[17] $\mathrm{d}V$, is

$$\mathrm{d}V = \mathrm{d}x_1 \mathrm{d}x_2 \mathrm{d}x_3 \,. \tag{1.31}$$

Since the solution on the DIRICHLET boundary is known, i.e., the exact values of the displacement is given on the DIRICHLET boundary, we skip the computation of the displacement on the DIRICHLET boundary. Usually, this fact is motivated mathematically that the test function vanishes on the DIRICHLET boundary. The DIRICHLET boundaries are implemented directly such that they are never presented in the variational formulation. The NEUMANN boundary condition has to be satisfied, too. We cannot apply it directly. In order to incorporate the NEUMANN boundary, we use integration by parts for tensors: First, we employ the product rule

$$\int_{\mathcal{B}} (\sigma_{ji}\, \delta u_i)_{,j} \mathrm{d}V = \int_{\mathcal{B}} \sigma_{ji,j}\, \delta u_i \mathrm{d}V + \int_{\mathcal{B}} \sigma_{ji}\, \delta u_{i,j} \mathrm{d}V \,, \tag{1.32}$$

secondly, we apply GAUSS's (or GAUSS–OSTROGRADSKIY's) theorem[18]

$$\int_{\mathcal{B}} (\sigma_{ji}\, \delta u_i)_{,j} \mathrm{d}V = \oint_{\partial\mathcal{B}} n_j \sigma_{ji}\, \delta u_i \mathrm{d}A \,, \tag{1.33}$$

where $\oint$ denotes the whole boundary. This notation is actually not necessary since we impose the integration over the whole boundary by writing the integration domain as $\partial\mathcal{B} = \partial\mathcal{B}_\mathrm{D} \cup \partial\mathcal{B}_\mathrm{N}$. We will mostly omit $\oint$ and use $\int$ instead. The following equation deduced from above is also called GREEN's identity[19]

$$-\int_{\mathcal{B}} \sigma_{ji,j}\, \delta u_i \mathrm{d}V = \int_{\mathcal{B}} \sigma_{ji}\, \delta u_{i,j} \mathrm{d}V - \int_{\partial\mathcal{B}} n_j \sigma_{ji}\, \delta u_i \mathrm{d}A \,. \tag{1.34}$$

---

[17]Mathematicians use the notation $\mathrm{d}x = \mathrm{d}x_1 \mathrm{d}x_2 \mathrm{d}x_3$ for a three-dimensional volume element in space $\boldsymbol{x} = (x_1, x_2, x_3)$, in order to generalize this to $\mathrm{d}x = \mathrm{d}x_1 \mathrm{d}x_2 \ldots \mathrm{d}x_n$ for an n-dimensional volume element in space $\boldsymbol{x} = (x_1, x_2, \ldots x_n)$. Also for the area element they use $\mathrm{d}s$ by referring to *surface* element. This notation is also used in the code that we will employ.

[18]The theorem is named after Carl Friedrich Gauß and Mikhail Vesilyevic Ostrogradskiy.

[19]It is named after George Green.

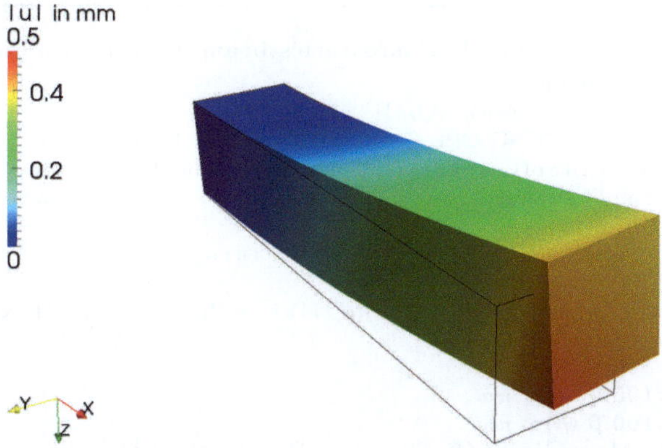

Fig. 1.1 Deformations, scaled 100 times, colors denote the magnitude of the displacement field

On the DIRICHLET boundary we know the displacement such that $\delta u_i\big|_{\partial\mathcal{B}_\text{D}} = 0$ results in

$$0 = -\int_{\mathcal{B}} \sigma_{ji,j}\, \delta u_i \mathrm{d}V = \int_{\mathcal{B}} \sigma_{ji}\, \delta u_{i,j} \mathrm{d}V - \int_{\partial\mathcal{B}_\text{N}} n_j \sigma_{ji}\, \delta u_i \mathrm{d}A\ . \tag{1.35}$$

Now by recalling the boundary condition in Eq. $(1.29)_2$, we obtain

$$\int_{\mathcal{B}} \sigma_{ji}\, \delta u_{i,j} \mathrm{d}V = \int_{\partial\mathcal{B}_\text{N}} \hat{t}_i\, \delta u_i \mathrm{d}A\ , \tag{1.36}$$

where the stress tensor is defined by the constitutive equation as given in Eq. (1.25). The latter integral form is called a weak form since we have reduced the necessary continuity of displacements by interchanging $\sigma_{ji,j}$ with σ_{ji} in the formulation. The weak form can be discretized by using GALERKIN's method,[20] where the test functions δu_i are chosen in the same vector space as the solution function, u_i. We apply GALERKIN's method with linear continuous elements and solve directly by employing the GAUSS elimination method[21] (LU for lower/upper decomposition). The code is written in Python by using *Unified Form Language* developed under the FEniCS project. After the code has run, we visualize the displacements with ParaView.[22] We use a simple geometry, a rectangular beam of length ℓ clamped on one side and loaded on the other side. The deformed structure can be depicted in Fig. 1.1 and its code is given below.

---

[20] The method is named after Boris Grigoryevich Galerkin.

[21] It is named after Carl Friedrich Gauß.

[22] See [17].

```
Computational reality 01, elastostatics beam deflection under
mechanical loading
__author__ = "B. Emek Abali"
__license__  = "GNU GPL Version 3.0 or later"
#This code underlies the GNU General Public License, http://www.gnu.org/licenses/gpl-3.0.en.html

# import all packages from fenics into cache
from fenics import *
# geometry is a 3D box, looks like a beam, with 15x15x75 elements in x,y,z
xlength=500.0 #in mm
ylength=100.0 #in mm
zlength=100.0 #in mm
mesh = BoxMesh(Point(0, 0, 0), Point(xlength, ylength, zlength), 50, 10, 10)
# vector space with polynomial degree 1
V = VectorFunctionSpace(mesh, 'P', 1)
# element surfaces, facets are 2D surfaces for 3D cells
# or they are 1D line boundaries for 2D elements (cells)
# or they are the edges of 1D elements (cells)
cells = CellFunction('size_t', mesh)
facets = FacetFunction('size_t', mesh)
dA = Measure('ds', domain=mesh, subdomain_data=facets)
dV = Measure('dx', domain=mesh, subdomain_data=cells)

# search for domains on the surface for clamping and loading
left = CompiledSubDomain('near(x[0], 0) && on_boundary')
right = CompiledSubDomain('near(x[0], length) && on_boundary', length=xlength)
# marking the parts of the surface
facets.set_all(0)
# integrating over dA(1) shall mean a surface integral
# over the domain 'right'
right.mark(facets, 1)
# traction vector
tr = Expression(('A+ B*x[1] + C*x[2]','0.0','0.0'), A=2.0, B=0.5, C=0.5) #in MPa
# a zero for the left clamped in the wall
null = Constant((0.0,0.0,0.0))
# stating the Dirichlet boundary values
# they will be applied after the assembly
bc = [DirichletBC(V, null, left)]
# definition for the variational formulation
u = TrialFunction(V)
del_u = TestFunction(V)

# material parameters of an AISI steel
nu = 0.3
E = 210000.0 #in MPa
G = E/(2.0*(1.0+nu))
# Lame parameters
```

```
la = 2.0*G*nu / (1.0 - 2.0*nu)
mu = G
# Kronecker delta in 3D
delta = Identity(3)
i,j,k = indices(3)
# strain tensor
eps = as_tensor(1.0/2.0*(u[i].dx(j)+u[j].dx(i)) , (i,j))
# Cauchy stress tensor
sigma = as_tensor(la*eps[k,k]*delta[i,j]+2.0*mu*eps[i,j] , (i
    ↪ ,j))
# Variational form
a = sigma[j,i]*del_u[i].dx(j)*dV
L = tr[i]*del_u[i]*dA(1)
disp = Function(V)
solve(a==L, disp, bcs=bc)
# write out
file_ = File('/calcul/CR01/CR01_elastostatics.pvd')
file_ << disp
# get the value of deflection on the tip
print 'tip deflection reads', disp(xlength,ylength/2.0,
    ↪ zlength/2.0)[2]
```

To-do

Get a pen and paper and redo the steps for integration by parts for tensors. Try to rewrite the code in Python and change the loading conditions in order to implement a tensile loading. Then apply a torque on the tip. Finally, apply a shearing force on the tip. Now get a technical handbook of mechanics and find for a simple beam, called a BERNOULLI beam, the analytical solution with the beam theory for the case of a shear loading on the tip. Try to find convenient answers for the following questions:

- Compute the deflection, u_3, on the tip, $x_1 = \ell$, and determine the accuracy of the numerical solution by comparing it to the analytical solution.
- Try to change the number of elements by increasing and decreasing them; how does the numerical solution change?
- Is the geometry chosen in the code appropriate for the beam theory? The BERNOULLI beam is *slender*, what is the ratio of length to thickness for a slender beam.
- We compute in a three-dimensional continuum; however, the beam theory is only one-dimensional. Which material parameter is excluded in the analytical solution?

Moreover, try to implement the following code and obtain the same results:

```
import numpy
def VoigtToTensor(A):
    A11, A12, A13, A14, A15, A16 = A[0,0], A[0,1], A[0,2], A
        ↪ [0,3], A[0,4], A[0,5]
    A22, A23, A24, A25, A26 = A[1,1], A[1,2], A[1,3], A[1,4],
        ↪ A[1,5]
    A33, A34, A35, A36 = A[2,2], A[2,3], A[2,4], A[2,5]
    A44, A45, A46 = A[3,3], A[3,4], A[3,5]
    A55, A56 = A[4,4], A[4,5]
    A66 = A[5,5]
    A21, A31, A41, A51, A61 = A12, A13, A14, A15, A16
    A32, A42, A52, A62 = A23, A24, A25, A26
    A43, A53, A63 = A34, A35, A36
    A54, A64 = A45, A46
    A65 = A56
    return as_tensor([ \
    [ \
    [ [A11,A16,A15], [A16,A12,A14], [A15,A14,A13]] , \
    [ [A61,A66,A65], [A66,A62,A64], [A65,A64,A63]] , \
    [ [A51,A56,A55], [A56,A52,A54], [A55,A54,A53]] \
    ] , [ \
    [ [A61,A66,A65], [A66,A62,A64], [A65,A64,A63]] , \
    [ [A21,A26,A25], [A26,A22,A24], [A25,A24,A23]] , \
    [ [A41,A46,A45], [A46,A42,A44], [A45,A44,A43]] \
    ] , [ \
    [ [A51,A56,A55], [A56,A52,A54], [A55,A54,A53]] , \
    [ [A41,A46,A45], [A46,A42,A44], [A45,A44,A43]] , \
    [ [A31,A36,A35], [A36,A32,A34], [A35,A34,A33]] ] \
    ])

C_voigt = numpy.array([ \
[la+2.*mu, la, la, 0, 0, 0],\
[la, la+2.*mu, la, 0, 0, 0],\
[la, la, la+2.*mu, 0, 0, 0],\
[0, 0, 0, mu, 0, 0],\
[0, 0, 0, 0, mu, 0],\
[0, 0, 0, 0, 0, mu] ])
C = VoigtToTensor(C_voigt)
eps = as_tensor(1.0/2.0*(u[i].dx(j)+u[j].dx(i)) , (i,j))
sigma = as_tensor(C[i,j,k,l]*eps[k,l] , (i,j))
```

Since the stress is coded by using the VOIGT notation, we can easily implement an anisotropic material. Copper is a cubic material with the following parameters in GPa:

$$C_{IJ} = \begin{pmatrix} 169.1 & 122.2 & 122.2 & 0 & 0 & 0 \\ & 169.1 & 122.2 & 0 & 0 & 0 \\ & & 169.1 & 0 & 0 & 0 \\ & & & 75.4 & 0 & 0 \\ & \text{symm.} & & & 75.4 & 0 \\ & & & & & 75.4 \end{pmatrix} . \tag{1.37}$$

Try to find out the crystal structure of a cubic material. The above parameters are valid when the coordinate system is along the material axis. Solve a bending beam out of a one-crystal copper material with the given material parameters.

1.2 Nonlinear Elastostatics

The linear elastostatics is accurate for small deformations. This assumption has been incorporated in Sect. 1.1 into "Step II" by using a *linear* strain measure and a *linear* constitutive equation. The first one; the linear strain measure simplification can be improved by choosing a nonlinear strain measure that is referred to as a *geometric* nonlinearity. The latter assumption; the linear constitutive equation can be changed to a nonlinear material equation that is called a *material* nonlinearity. In this section we will amend the simulation by generalizing our model for structures with a geometric nonlinearity. This generalization is necessary for deformations of the same order as geometric dimensions, in other words, for deformations that we can observe by a visual inspection. For structures like plates, which have a small size in one of dimensions, such *large* deformations occur. For their accurate computation we need to include the geometric nonlinearity in modeling.

The goal is to compute the deformation of each particle belonging to the continuum body. We can visualize the deformation as a change from a body of shape $\mathcal{B}_0$ at initial time, t_0, to a body of shape $\mathcal{B}$ at current time, t. The (initial) shape $\mathcal{B}_0$ is known such that we can identify the particles with their positions at t_0 expressed in Cartesian coordinates. The initial positions of particles, X_i, are used to define the motion of particles, i.e., the sought-after displacement is a function of particles and time, $u_i(X_j, t)$. The primitive variable is again the displacement field satisfying the balance of linear momentum in the *current* continuum body $\mathcal{B}$ at t, written in the global form

$$\Big(\int_{\mathcal{B}} \rho v_i \mathrm{d}v\Big)^{\bullet} = \int_{\partial\mathcal{B}} \sigma_{ji} n_j \mathrm{d}a + \int_{\mathcal{B}} \rho f_i \mathrm{d}v \,, \tag{1.38}$$

where ρ, σ_{ij}, and f_i stand for mass density, CAUCHY (symmetric) stress tensor, and body force, respectively. Caused by the displacement u_i, the material particles (identified by their initial positions) X_i move to their current positions x_i. For the same quantity (position) we use a capital letter X_i at t_0 and a small letter x_i at t. Therefore, we obtain

$$x_i = X_i + u_i \;. \tag{1.39}$$

Next we introduce a concept of *invariance*. In order to identify or observe a physical quantity, we need a coordinate system. We simply choose Cartesian coordinates.[23]

[23] In analytical mechanics the choice of the coordinate system is of importance. In a particular coordinate system the analytical solution might be much easier than in another coordinate system. In numerical mechanics this is very rarely the case, so we simply select Cartesian coordinates.

The orientation of axes is completely arbitrary. However, the coordinates of a particular position depend on the chosen axes, in other words, the description of a position depends on the coordinate system, we call that *frame* dependence. By using two coordinate systems, we can express the same position with respect to these two systems; the numerical values of position coordinates *vary* in different frames.[24] It is beneficial to use a quantity that is *invariant* under a frame change. Length is such a quantity. Suppose we have a deformable structure and a small rubber band of length $\mathrm{d}S$ attached to the system. It can be glued on the surface of the structure such that the deformation of the structure is mimicked by the rubber band—this idea is used in strain gauges. We measure the rubber band before the loading, $\mathrm{d}S$, and also afterwards, $\mathrm{d}s$. Indeed, this measurement is invariant and frame independent, thus, we have no necessity to introduce a coordinate system for the length measurement itself.

For the numerical treatise we have a (Cartesian) coordinate system, in which we realize a length measurement. Suppose that we mark two neighboring particles and draw an arrow between them. Before the deformation, at t_0, the arrow or vector is $\mathrm{d}X_i$. This vector changes to $\mathrm{d}x_i$ at t after the deformation. Both vectors, $\mathrm{d}X_i$ and $\mathrm{d}x_i$, are expressed in the same coordinate system. The length of this vector corresponds to the length of the rubber band since the rubber band is (infinitesimally) small by having chosen neighboring elements. In order to express this fact, we use the differential "d" in front of the length $\mathrm{d}S$ at t_0 and $\mathrm{d}s$ at t. The length can be calculated, either by means of the vector at the current time

$$\begin{aligned}(\mathrm{d}s)^2 &= \frac{\partial s}{\partial x_i}\mathrm{d}x_i \frac{\partial s}{\partial x_j}\mathrm{d}x_j = g_{ij}\mathrm{d}x_i\mathrm{d}x_j \ , \\ (\mathrm{d}S)^2 &= \frac{\partial S}{\partial x_i}\mathrm{d}x_i \frac{\partial S}{\partial x_j}\mathrm{d}x_j = c_{ij}\mathrm{d}x_i\mathrm{d}x_j \ ,\end{aligned} \tag{1.40}$$

or with respect to the vector at the initial time

$$\begin{aligned}(\mathrm{d}s)^2 &= \frac{\partial s}{\partial X_i}\mathrm{d}X_i \frac{\partial s}{\partial X_j}\mathrm{d}X_j = C_{ij}\mathrm{d}X_i\mathrm{d}X_j \ , \\ (\mathrm{d}S)^2 &= \frac{\partial S}{\partial X_i}\mathrm{d}X_i \frac{\partial S}{\partial X_j}\mathrm{d}X_j = G_{ij}\mathrm{d}X_i\mathrm{d}X_j \ .\end{aligned} \tag{1.41}$$

The difference between them denotes the length change and is the measure of stretching,

$$(\mathrm{d}l)^2 = (\mathrm{d}s)^2 - (\mathrm{d}S)^2 \ . \tag{1.42}$$

---

[24]There is a small difference between a coordinate system and a frame. A frame is only the *observer* in any coordinate system. Of course, in order to define physical quantities, the observer uses a coordinate system in its frame. We use them together in a sense that we immediately append a coordinate system to the frame.

We can express this measure either at the current time

$$(\mathrm{d}l)^2 = \Big(g_{ij} - c_{ij}\Big)\mathrm{d}x_i\mathrm{d}x_j \,, \tag{1.43}$$

or at the initial time

$$(\mathrm{d}l)^2 = \Big(C_{ij} - G_{ij}\Big)\mathrm{d}X_i\mathrm{d}X_j \,. \tag{1.44}$$

The deformation tensors, $g_{ij}, c_{ij}, C_{ij}, G_{ij}$, represent the state of the continuum body with respect to the initial vector, $\mathrm{d}X_i$, or with respect to the current vector, $\mathrm{d}x_i$. Between these vectors we find a mapping[25] as follows

$$\mathrm{d}x_i = \frac{\partial x_i}{\partial X_j}\mathrm{d}X_j \,, \quad F_{ij} = \frac{\partial x_i}{\partial X_j} \,, \tag{1.45}$$

where F_{ij} is called the *deformation gradient*. Analogously, we obtain

$$\mathrm{d}X_i = \frac{\partial X_i}{\partial x_j}\mathrm{d}x_j \,, \quad (\boldsymbol{F}^{-1})_{ij} = \frac{\partial X_i}{\partial x_j} \,. \tag{1.46}$$

Obviously, $\boldsymbol{F}$ and $\boldsymbol{F}^{-1}$ are inverse leading to

$$F_{ij}(\boldsymbol{F}^{-1})_{jk} = \frac{\partial x_i}{\partial X_j}\frac{\partial X_j}{\partial x_k} = \frac{\partial x_i}{\partial x_k} = \delta_{ik} \,. \tag{1.47}$$

We can comprehend the stretching or deformation in two different manifestations, as in Eq. (1.43) or in Eq. (1.44). Both have the same information but lead to different formulations. We demonstrate both in the following.

One possibility is to present the deformation with respect to the current vector, $\mathrm{d}x_i$. One of the marked particles possesses the coordinates, x_i, at the present time, t. Since the marked particles are infinitesimally close to each other, it is not important which particle's coordinates we choose. It is important to recall that we choose the coordinates at the present time. Thus, this manifestation is configured with the current coordinates such that it is referred to as a *current* or *present* frame. The distance vector after the deformation, $\mathrm{d}x_i$, is defined in the current frame. However, the distance vector before the deformation, $\mathrm{d}X_i$, needs to be rewritten. We utilize Eqs. (1.40) in the current frame,

$$\begin{aligned} (\mathrm{d}s)^2 &= \mathrm{d}x_i\mathrm{d}x_i = \delta_{ij}\mathrm{d}x_i\mathrm{d}x_j \\ &\Rightarrow g_{ij} = \delta_{ij} \,, \end{aligned} \tag{1.48}$$

---

[25]This mapping is according to the *tensor transformation rules*, as we will see it in Sect. 1.4 on p. 34.

as well as

$$
\begin{aligned}
(\mathrm{d}S)^2 &= \mathrm{d}X_k \mathrm{d}X_k = \frac{\partial X_k}{\partial x_i}\mathrm{d}x_i \frac{\partial X_k}{\partial x_j}\mathrm{d}x_j \\
\Rightarrow c_{ij} &= \frac{\partial X_k}{\partial x_i}\frac{\partial X_k}{\partial x_j} = (\boldsymbol{F}^{-1})_{ki}(\boldsymbol{F}^{-1})_{kj} \ .
\end{aligned}
\tag{1.49}
$$

Therefore, the length measure in Eq. (1.43) becomes

$$
(\mathrm{d}l)^2 = \Big(g_{ij} - c_{ij}\Big)\mathrm{d}x_i \mathrm{d}x_j = \Big(\delta_{ij} - (\boldsymbol{F}^{-1})_{ki}(\boldsymbol{F}^{-1})_{kj}\Big)\mathrm{d}x_i \mathrm{d}x_j \ . \tag{1.50}
$$

In the current frame the displacement gradient reads

$$
\begin{aligned}
u_i(x_k, t) &= x_i - X_i(x_k, t) \quad \Big| \frac{\partial}{\partial x_k} \ , \\
\frac{\partial u_i}{\partial x_k} &= \delta_{ik} - (\boldsymbol{F}^{-1})_{ik} \ .
\end{aligned}
\tag{1.51}
$$

We introduce the left CAUCHY–GREEN deformation tensor B_{ij} and its inverse $(\boldsymbol{B}^{-1})_{ji}$, which is also called FINGER–HENCKY deformation tensor,[26] as follows

$$
\begin{aligned}
B_{ij} &= \frac{\partial x_i}{\partial X_k}\frac{\partial x_j}{\partial X_k} = F_{ik}F_{jk} = \boldsymbol{F}\,\boldsymbol{F}^T \ , \\
(\boldsymbol{B}^{-1})_{ji} &= \frac{\partial X_k}{\partial x_i}\frac{\partial X_k}{\partial x_j} = (\boldsymbol{F}^{-1})_{ki}(\boldsymbol{F}^{-1})_{kj} = \boldsymbol{F}^{-T}\,\boldsymbol{F}^{-1} \ .
\end{aligned}
\tag{1.52}
$$

Now the EULER–ALMANSI or HAMEL's strain[27] tensor e_{ij} is introduced

$$
\begin{aligned}
(\boldsymbol{B}^{-1})_{ji} &= \frac{\partial X_k(x_l, t)}{\partial x_i}\frac{\partial X_k(x_l, t)}{\partial x_j} = \\
&= \frac{\partial}{\partial x_i}\Big(x_k - u_k(x_l, t)\Big)\frac{\partial}{\partial x_j}\Big(x_k - u_k(x_l, t)\Big) = \\
&= \Big(\delta_{ki} - \frac{\partial u_k}{\partial x_i}\Big)\Big(\delta_{kj} - \frac{\partial u_k}{\partial x_j}\Big) = \delta_{ji} - \frac{\partial u_j}{\partial x_i} - \frac{\partial u_i}{\partial x_j} + \frac{\partial u_k}{\partial x_i}\frac{\partial u_k}{\partial x_j} = \\
&= \delta_{ij} - 2e_{ij} \ ,
\end{aligned}
\tag{1.53}
$$

leading to

$$
e_{ij} = u_{(i,j)} - \frac{1}{2}u_{k,i}u_{k,j} \ , \tag{1.54}
$$

---

[26] It is named after Augustin Louis Cauchy, George Green, Josef Finger, and Heinrich Hencky.
[27] It is named after Leonhard Euler, Emilio Almansi, and Georg Karl Wilhelm Hamel.

that defines the length change in the current frame

$$(\mathrm{d}l)^2 = 2e_{ij}\mathrm{d}x_i\mathrm{d}x_j \ . \tag{1.55}$$

The EULER–LMANSI strain, e_{ij}, is defined by means of the current displacement vector and it is non-linear in displacement gradients, $u_{i,j}$.

The second possible choice relies on the description by means of the initial frame, where the description is based on the initial coordinates, X_k. This configuration is also called a LAGRANGEan frame. Initial coordinates of particles are known such that we *refer* to every single particle uniquely by using its coordinates. The configuration is also called a *reference* frame.[28] Moreover, it is a *material* system since all particles are included in the formulation. Before and after deformation, the totality of particles remains the same, i.e., a closed system. Within a closed or material system, no mass change applies. In the LAGRANGEan frame all functions have to depend on X_i. The distance vector, $\mathrm{d}x_i$, has to be transformed from the current to the LAGRANGEan frame. By employing Eqs. (1.41) we obtain

$$\begin{aligned}(\mathrm{d}s)^2 = \mathrm{d}x_k\mathrm{d}x_k = \frac{\partial x_k}{\partial X_i}\mathrm{d}X_i\frac{\partial x_k}{\partial X_j}\mathrm{d}X_j \Rightarrow C_{ij} = \frac{\partial x_k}{\partial X_i}\frac{\partial x_k}{\partial X_j} = F_{ki}F_{kj} \ , \\ (\mathrm{d}S)^2 = \mathrm{d}X_k\mathrm{d}X_k = \delta_{ij}\mathrm{d}X_i\mathrm{d}X_j \Rightarrow G_{ij} = \delta_{ij} \ .\end{aligned} \tag{1.56}$$

Hence, the length measure Eq. (1.43) becomes

$$(\mathrm{d}l)^2 = \Big(C_{ij} - G_{ij}\Big)\mathrm{d}X_i\mathrm{d}X_j = \Big(F_{ki}F_{kj} - \delta_{ij}\Big)\mathrm{d}X_i\mathrm{d}X_j \ . \tag{1.57}$$

In the LAGRANGEan frame the displacement gradient is expressed as

$$\begin{aligned}u_i(X_j,t) = x_i(X_j,t) - X_i \quad \Big| \frac{\partial}{\partial X_k} \ , \\ \frac{\partial u_i}{\partial X_k} = F_{ik} - \delta_{ik} \ .\end{aligned} \tag{1.58}$$

The deformation tensor in the LAGRANGEan frame is referred to as the right CAUCHY–GREEN deformation tensor:

$$C_{ij} = \frac{\partial x_k}{\partial X_i}\frac{\partial x_k}{\partial X_j} = F_{ki}F_{kj} = \boldsymbol{F}^T\boldsymbol{F} \ . \tag{1.59}$$

[28]The simplest choice is to refer to the particles at the initial time, since we know their initial positions. The LAGRANGEan frame has no restrictions of using the initial frame as the reference frame. Positions of particles at any moment can be chosen as the reference frame. In this work we will use the initial frame as the reference frame.

Therefore, the so-called GREEN–LAGRANGE strain tensor, E_{ij}, reads

$$
\begin{aligned}
C_{ij} &= \frac{\partial x_k}{\partial X_i}\frac{\partial x_k}{\partial X_j} = \frac{\partial}{\partial X_i}\Big(u_k(X_l,t)+X_k\Big)\frac{\partial}{\partial X_j}\Big(u_k(X_l,t)+X_k\Big) = \\
&= \Big(\frac{\partial u_k}{\partial X_i}+\delta_{ki}\Big)\Big(\frac{\partial u_k}{\partial X_j}+\delta_{kj}\Big) = \frac{\partial u_k}{\partial X_i}\frac{\partial u_k}{\partial X_j}+\frac{\partial u_j}{\partial X_i}+\frac{\partial u_i}{\partial X_j}+\delta_{ij} = 2E_{ij}+\delta_{ij} \,, \\
E_{ij} &= \frac{1}{2}u_{k,i}u_{k,j}+u_{(i,j)} \,.
\end{aligned}
\tag{1.60}
$$

The length change is now in the LAGRANGEan frame

$$
(\mathrm{d}l)^2 = 2E_{ij}\mathrm{d}X_i\mathrm{d}X_j \,. \tag{1.61}
$$

Strain tensor is quadratic in displacement gradients, $u_{k,i}$. Therefore, large displacements can be calculated more accurately than with the linear strain tensor used in Sect. 1.1.

We have seen two different configurations to setup the same problem. In the current frame we have obtained a strain by means of displacement derivatives with respect to x_i. In the initial, reference, or LAGRANGEan frame, however, strain is displacement gradients in X_i. In both configurations the shorthand comma notation is used with different meanings. The factors 2 in front of the strains have no particular meaning.

Although both formulations are correct, the LAGRANGEan frame is used in solid mechanics owing to its simplicity. The current frame uses the present positions of particles to define the functions; however, before calculating displacements, u_i, we lack the knowledge of the current positions of particles, x_i. Hence, the initial frame is more beneficial, especially for material systems. We can always switch between different strain measures by using

$$
E_{ij} = \frac{1}{2}(C_{ij}-\delta_{ij}) = \frac{1}{2}(F_{ki}F_{kj}-\delta_{ij}) \,, \tag{1.62}
$$

such that

$$
\begin{aligned}
E_{ij}(\boldsymbol{F}^{-1})_{jm}(\boldsymbol{F}^{-1})_{il} &= \frac{1}{2}\big(\delta_{kl}\delta_{km}-(\boldsymbol{F}^{-1})_{im}(\boldsymbol{F}^{-1})_{il}\big) = \frac{1}{2}\big(\delta_{lm}-(\boldsymbol{B}^{-1})_{lm}\big) \,, \\
E_{ij}(\boldsymbol{F}^{-1})_{jm}(\boldsymbol{F}^{-1})_{il} &= e_{lm} \,,
\end{aligned}
\tag{1.63}
$$

or equivalently,

$$
E_{nk} = e_{lm}F_{mk}F_{ln} \,. \tag{1.64}
$$

In this section we use the LAGRANGEan frame with the GREEN–LAGRANGE strain tensor and solve a deformable solid body in its initial frame.

We define a continuum body at initial time, $\mathcal{B}_0 = \mathcal{B}(t_0)$, with the coordinates expressed in a Cartesian coordinate system, X_i, and want to find its deformed shape $\mathcal{B}$ at t in the LAGRANGEan frame, i.e., we search for the displacement field, $u_i(X_j,t)$.

In the LAGRANGEan frame we use the GREEN–LAGRANGE strain tensor, E_{ij}. The balance equations are given in the current frame; we will transform them from the current to the initial frame in the following. We postpone the derivation of these identities[29] to Sect. 1.4 on p. 34 and transform the volume and area elements in the current frame, $\mathrm{d}v$, $\mathrm{d}a$, as follows

$$\begin{aligned} \mathrm{d}v &= \det\Big(\frac{\partial x_i}{\partial X_j}\Big)\mathrm{d}V = J\mathrm{d}V \,, \\ n_j\mathrm{d}a &= \Big(\frac{\partial x_j}{\partial X_k}\Big)^{-1} JN_k\mathrm{d}A = J(\boldsymbol{F}^{-1})_{kj}N_k\mathrm{d}A \,, \end{aligned} \tag{1.65}$$

where $\mathrm{d}V$ and $\mathrm{d}A$ are the volume and area elements in the initial frame. Plane normal in the current frame, n_i, and plane normal in the reference frame, N_i, are both of length 1, but they point to different directions.[30] The JACOBI determinant,[31] J, is the volume contraction; if the initial mass density ρ_0 is given, the current mass density can be determined by $\rho = \rho_0/J$, which is the balance of mass in the initial frame. We postpone its derivation to Sect. 1.4 on p. 34, by using $\rho = \rho_0 J$ we will satisfy the mass balance in the initial frame. The balance of momentum is given in the current frame

$$\Big(\int_{\mathcal{B}} \rho v_i \mathrm{d}v\Big)^{\boldsymbol{\cdot}} = \int_{\partial\mathcal{B}} \sigma_{ji} n_j \mathrm{d}a + \int_{\mathcal{B}} \rho f_i \mathrm{d}v \,, \tag{1.66}$$

so we transform it from the current to the initial frame by inserting the transformation of volume and area elements

$$\Big(\int_{\mathcal{B}_0} \rho v_i J\mathrm{d}V\Big)^{\boldsymbol{\cdot}} = \int_{\partial\mathcal{B}_0} \sigma_{ji}(\boldsymbol{F}^{-1})_{kj} JN_k\mathrm{d}A + \int_{\mathcal{B}_0} \rho f_i J\mathrm{d}V \,. \tag{1.67}$$

The numerical values of velocities v_i as functions in the current frame (x_k, t) are equivalent to the numerical values of v_i as functions in the initial frame (X_k, t). For the sake of a simplified notation we define a so-called *nominal stress*:[32]

$$P_{ki} = (\boldsymbol{F}^{-1})_{kj}\sigma_{ji}J \,, \tag{1.68}$$

---

[29] An identity is just an equation being true for any chosen arguments, here the left side is equal to the right side in every coordinates.

[30] Plane normals have no units, they just show the directions. Hence, it is also useful to write $n_i\mathrm{d}a = \mathrm{d}a_i$ or analogously $N_i\mathrm{d}A = \mathrm{d}A_i$.

[31] It is named after Carl Gustav Jacob Jacobi.

[32] There are many names for this stress in the literature: PIOLA's stress named after Gabrio Piola, the first PIOLA–KIRCHHOFF stress, named after Gabrio Piola and Gustav Robert Kirchhoff, nominal stress, and engineering stress are the most prominent names. Some textbooks distinguish between nominal and PIOLA's stresses and use PIOLA's stress as the transpose of the nominal stress as introduced as in this book.

which is *not* symmetric, while the deformation gradient, F_{ij}, does not have to be symmetric. However, the strain tensor is symmetric such that it is difficult to relate the nominal stress to the strain tensor.[33] This inconvenience can be rectified by introducing the second PIOLA–KIRCHHOFF stress tensor:

$$S_{kl} = (\boldsymbol{F}^{-1})_{li} P_{ki} = (\boldsymbol{F}^{-1})_{kj} (\boldsymbol{F}^{-1})_{li} \sigma_{ji} J \ , \tag{1.69}$$

which is obviously symmetric. The ST.VENANT–KIRCHHOFF constitutive model[34] for isotropic, linear elastic materials gives a tensor-linear relation between the symmetric stress and the symmetric strain

$$\begin{aligned} S_{ij} = C_{ijkl} E_{kl} \ , \quad C_{ijkl} = \lambda \delta_{ij} \delta_{kl} + \mu \delta_{ik} \delta_{jl} + \mu \delta_{il} \delta_{jk} \ , \\ S_{ij} = \lambda E_{kk} \delta_{ij} + 2\mu E_{ij} \ . \end{aligned} \tag{1.70}$$

The LAME parameters, λ, μ, have the same values as in Sect. 1.1. The linear constitutive equation is the reason, why the formulation is limited to applications with small strains—we will amend the formulation for material nonlinearities in Sect. 1.3 on p. 26. The strain measure is itself nonlinear, so the model is valid under large deformations, in other words, geometric nonlinearities are captured accurately.

Now we use the balance of mass, $\rho_0 = J\rho$, in Eq. (1.67)$_2$ and obtain the balance of linear momentum in the initial frame:

$$\begin{aligned} \Big(\int_{\mathcal{B}_0} \rho_0 v_i \mathrm{d}V \Big)^{\bullet} &= \int_{\partial \mathcal{B}_0} P_{ki} N_k \mathrm{d}A + \int_{\mathcal{B}_0} \rho_0 f_i \mathrm{d}V \ , \\ \int_{\mathcal{B}_0} \rho_0 v_i^{\bullet} \mathrm{d}V &= \int_{\partial \mathcal{B}_0} P_{ki} N_k \mathrm{d}A + \int_{\mathcal{B}_0} \rho_0 f_i \mathrm{d}V \ , \end{aligned} \tag{1.71}$$

where we have used the fact that the initial mass density as well as the initial volume element remains constant in time, $\rho_0^{\bullet} = 0$, $(\mathrm{d}V)^{\bullet} = 0$. After utilizing GAUSS's theorem we acquire the local form

$$\begin{aligned} \int_{\mathcal{B}_0} \rho_0 v_i^{\bullet} \mathrm{d}V &= \int_{\mathcal{B}_0} \frac{\partial P_{ki}}{\partial X_k} \mathrm{d}V + \int_{\mathcal{B}_0} \rho_0 f_i \mathrm{d}V \ , \\ \rho_0 v_i^{\bullet} &= \frac{\partial P_{ki}}{\partial X_k} + \rho_0 f_i \ , \\ \rho_0 v_i^{\bullet} - \frac{\partial P_{ki}}{\partial X_k} - \rho_0 f_i &= 0 \ . \end{aligned} \tag{1.72}$$

---

[33] Although we skip a discussion about the objectivity, the transformation properties of the strain and nominal stress are different. It is easier to construct constitutive equations between terms with equal transformation properties.

[34] It is named after Adhémar Jean Claude Barré de Saint-Venant and Gustav Robert Kirchhoff.

The rate of velocity, $v_i^{\bullet}$, is ignored in statics as aforementioned in Sect. 1.1. Moreover, we neglect the effect of gravity and obtain the balance of linear momentum in the LAGRANGEan frame:

$$-\frac{\partial P_{ki}}{\partial X_k} = 0 \,, \tag{1.73}$$

where this equation is also called the *equilibrium condition* in statics. Obviously, the differential equation is of same type as in Eq. (1.28). Once again we use the simplified notation

$$-P_{ki,k} = 0 \,, \tag{1.74}$$

where comma denotes a (partial) derivative with respect to $\boldsymbol{X}$ in the LAGRANGEan frame. For the variational formulation we multiply by a test function, δu_i, and obtain after an integration by parts the weak form:

$$\begin{aligned} -\int_{\mathcal{B}_0} P_{ki,k}\, \delta u_i \mathrm{d}V = 0 \,, \\ \int_{\mathcal{B}_0} P_{ki}\, \delta u_{i,k} \mathrm{d}V - \int_{\partial \mathcal{B}_0^{\mathrm{N}}} \hat{t}_i\, \delta u_i \mathrm{d}A = 0 \,, \end{aligned} \tag{1.75}$$

where the traction vector, $\hat{t}_i = N_j P_{ji}$, is given on the NEUMANN boundaries, denoted by $\partial \mathcal{B}_0^{\mathrm{N}}$. In order to sum up: the nominal stress reads

$$\begin{aligned} P_{ki} = F_{ij} S_{kj} \,, \quad S_{kj} = \lambda E_{ll} \delta_{kj} + 2\mu E_{kj} \,, \\ E_{kj} = \frac{1}{2}\big(C_{kj} - \delta_{kj}\big) \,, \quad C_{kj} = F_{ik} F_{ij} \,, \quad F_{ij} = u_{i,j} + \delta_{ij} \,. \end{aligned} \tag{1.76}$$

Since the strain tensor is quadratic in $u_{i,j}$ the weak form is nonlinear. Therefore, we need to linearize and solve the equation. We postpone the theoretical treatise to Sect. 1.8 on p. 86 and apply the NEWTON–RAPHSON linearization method.[35] It is fully automatized in FEniCS and we only need to utilize the code with the following nonlinear form:

$$\text{Form} = \int_{\mathcal{B}_0} P_{ki}\, \delta u_{i,k} \mathrm{d}V - \int_{\partial \mathcal{B}_0^{\mathrm{N}}} \hat{t}_i\, \delta u_i \mathrm{d}A \,. \tag{1.77}$$

Consider a thin plate of engineering steel with the dimensions of 100 × 50 × 11 mm. We model the plate as a three-dimensional continuum body. A mechanical loading is applied by using the traction vector, $\hat{t}_i$, on top of the plate within a circle of 10 mm radius. Its magnitude is 200 MPa and loads the plate in thickness direction. Therefore, the deformation is large and can be seen by the naked eye (without any scaling) in Fig. 1.2. This large deformation results in high equivalent stresses. The code below is used to compute the deformation with geometric nonlinearities.

[35] It is named for Isaac Newton and Joseph Raphson.

```
Computational reality 02, plate deformation with
geometric nonlinearities
__author__ = "B. Emek Abali"
__license__ = "GNU GPL Version 3.0 or later"
#This code underlies the GNU General Public License,
    ↪ http://www.gnu.org/licenses/gpl-3.0.en.html

from fenics import *
xlength=100.0 #in mm
ylength=50.0
zlength=1.0
#origin is in the middle of the plate
mesh = BoxMesh(Point(0,0,0), Point(xlength,ylength,zlength),
    ↪ 100, 50, 1)
V = VectorFunctionSpace(mesh, 'P', 1)
cells = CellFunction('size_t', mesh)
facets = FacetFunction('size_t', mesh)
dA = Measure('ds', domain=mesh, subdomain_data=facets)
dV = Measure('dx', domain=mesh, subdomain_data=cells)
left = CompiledSubDomain('near(x[0],0) && on_boundary')
right = CompiledSubDomain('near(x[0],l) && on_boundary',l=
    ↪ xlength)
top = CompiledSubDomain('pow(x[0]-X,2)+pow(x[1]-Y,2)<pow
    ↪ (10.0,2) && near(x[2],l) && on_boundary',X=xlength/4.,
    ↪ Y=ylength/3.,l=zlength)
facets.set_all(0)
top.mark(facets, 1)
tr = Constant((0.0,0.0,200.0)) # MPa
null = Constant((0.0,0.0,0.0))
bc1=DirichletBC(V, null, left)
bc2=DirichletBC(V, null, right)
bc = [bc1,bc2]
# definition for the variational formulation for a nonlinear
    ↪ form
du = TrialFunction(V)         # Incremental displacement
del_u  = TestFunction(V)      # Test function
u  = Function(V)              # Displacement
# material parameters of a stainless steel
nu = 0.3
E = 210000.0 #in MPa
G = E/(2.0*(1.0+nu))
# Lame parameters (lambda has another meaning in python,
# better it is not overwritten)
lambada = 2.0*G*nu / (1.0 - 2.0*nu)
mu = G
delta = Identity(3)
# index notation
i, j, k, l = indices(4)
# deformation gradient
F = as_tensor(u[i].dx(j) + delta[i,j] , (i,j))
J = det(F)
# right Cauchy-Green deformation tensor
```

```
C = as_tensor(F[i,k]*F[i,j] , (k,j))
# Green-Lagrange strain tensor
E = as_tensor(1./2.*( C[k,j] - delta[k,j] ) , (k,j))
# second Piola-Kirchhoff stress tensor
S = as_tensor(lambada*E[l,l]*delta[k,j] + 2.0*mu*E[k,j] , (k,
    j))
# nominal stress
P = as_tensor(F[i,j]*S[j,k] , (k,i))

Form = P[k,i]*del_u[i].dx(k)*dV - tr[i]*del_u[i]*dA(1)
Gain = derivative(Form, u, du)
solve(Form==0, u, bc, J=Gain, \
      solver_parameters={"newton_solver":{"linear_solver": "lu"
      , "relative_tolerance": 1e-3} }, \
      form_compiler_parameters={"cpp_optimize": True, "
      representation": "quadrature", "quadrature_degree"
      : 2} )

# write out
pwd='/calcul/CR02/'
file = File(pwd+'nonlin_elastostatics_deformations.pvd')
file << u
# Cauchy stress tensor
sigma = as_tensor(1./J*F[j,k]*P[k,i] , (j,i))
sigma_dev = as_tensor(sigma[i,j]-1.0/3.0*sigma[k,k]*delta[i,j
    ] , (i,j))
eqStress = as_tensor((3.0/2.0*sigma_dev[i,j]*sigma_dev[i,j])
    **0.5 , ())
# now we have displacements , u, so we can calculate
# the equivalent stress by projecting
eqS = project(eqStress , FunctionSpace(mesh, 'P', 1))
file = File(pwd+'nonlin_elastostatics_eqStress.pvd')
file << eqS
```

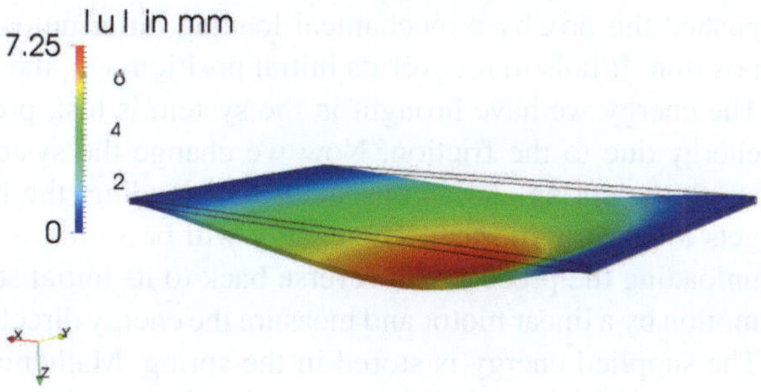

Fig. 1.2 Color indicates the magnitude of displacement field. The deformation is shown without scaling, therefore, the deformation is as big as the geometric dimensions

To-do

We have solved a nonlinear differential equation:

- Which term introduced the nonlinearity?
- Is there an analytical solution for this problem?
- Visualize the equivalent stresses and try to find a steel with such a high yield stress. How should we change the boundary conditions in order to reduce the equivalent stress?

1.3 Hyperelastic Materials in Statics

The balance of momentum is used to compute the deformation. By transforming the balance equation from the current to the reference frame, we have addressed the geometric nonlinearities and computed large deformations. The used material or constitutive equation was linear. In order to consider the material nonlinearities, we need to employ a nonlinear constitutive equation in this section. We may simply introduce a constitutive equation modeling hyperelasticity and implement it into the code given in the last section. Instead of doing so, we will present an alternative method for obtaining the weak form based on *energy*. This approach is by no means more or less beneficial than the method utilized in the last section, but it helps to comprehend the methodology used in the variational formulation better. The energy-based method sheds light on the real meaning of the *variational* formulation.

The first notion is *energy*. Any process evolves by utilizing energy. A box on a (horizontal, plain) desk will move, if one supplies mechanical energy into the system by simply pushing the box. We still solve problems in statics, therefore, we need to visualize a snapshot of the box before and after the motion. Suppose that we have pushed the box by a mechanical loading; after unloading the box rests in another position. It fails to recover its initial position, i.e., the process is not reversing back. The energy we have brought in the system is lost, probably it is dissipated as heat energy due to the friction. Now we change the system and connect the box with an elastic spring. As a consequence of loading, the box will move and the spring gets longer (or shorter). The energy will be *stored* in the (elastic) spring and upon unloading the process will reverse back to its initial state (position). We control the motion by a linear motor and measure the energy directly by counting watts in time.[36] The supplied energy is stored in the spring. Mathematically, we can calculate this energy (density) as already discussed in Sect. 1.1 from the first snapshot to the second (denoted by 1 and 2)

[36]W(att) is the unit of power named after James Watt. Power is the rate of energy. The energy used from the wall is measured in kWh (kilo-Watt-hours).

$$w = \int_1^2 \mathrm{d}w = \int_1^2 \sigma_{ji} \mathrm{d}\varepsilon_{ij} \, . \tag{1.78}$$

If the material model is linear

$$\sigma_{ji} = C_{jikl} \varepsilon_{kl} \, , \tag{1.79}$$

we obtain

$$w = C_{jikl} \int_0^{\varepsilon} \bar{\varepsilon}_{kl} \mathrm{d}\bar{\varepsilon}_{ij} = C_{jikl} \frac{1}{2} \varepsilon_{kl} \varepsilon_{ij} = \frac{1}{2} \sigma_{ji} \varepsilon_{ij} \, , \tag{1.80}$$

since $C_{jikl} = C_{klji}$ is constant in ε_{kl} and we start from a state without strain (zero strains in the first snapshot). This energy density is per mass, in other words, it is the energy for a material particle of the continuum body. The whole deformation energy in the current body, $\mathcal{B}$, reads

$$W = \int_{\mathcal{B}} w \mathrm{d}v \, . \tag{1.81}$$

If the geometric nonlinearity shall be included then we use EULER–ALMANSI strain tensor:

$$e_{ij} = \frac{1}{2} \big(\delta_{ij} - (\boldsymbol{B}^{-1})_{ij} \big) \, , \tag{1.82}$$

instead of the linearized strain tensor, ε_{ij}. In this case the energy reads

$$W = \int_{\mathcal{B}} \frac{1}{2} \sigma_{ml} e_{lm} \mathrm{d}v \, . \tag{1.83}$$

Energy is a scalar, in other words, its value remains the same for the current frame or for the reference frame. We may calculate the energy in the initial frame

$$W = \int_{\mathcal{B}_0} \frac{1}{2} \sigma_{ml} e_{lm} J \mathrm{d}V \, , \tag{1.84}$$

or even rewrite energy by using Eq. $(1.63)_3$ and obtain

$$W = \int_{\mathcal{B}_0} \frac{1}{2} \sigma_{ml} e_{lm} J \mathrm{d}V = \int_{\mathcal{B}_0} \frac{1}{2} \sigma_{ml} E_{ij} (\boldsymbol{F}^{-1})_{jm} (\boldsymbol{F}^{-1})_{il} J \mathrm{d}V \, . \tag{1.85}$$

Furthermore, by using Eq. (1.69) we acquire

$$W = \int_{\mathcal{B}_0} \frac{1}{2} S_{ji} E_{ij} \mathrm{d}V \, . \tag{1.86}$$

Therefore, the stress-strain pair, σ_{ji}, e_{ij}, in the current frame can be exchanged with the stress-strain pair, S_{ji}, E_{ij}, in the initial frame. Another way of seeing this relation relies on the power, i.e., the rate of energy,

$$w = \int_1^2 \mathrm{d}w = \int_{\bar{t}=t_0}^{\bar{t}=t} w^{\bullet}\mathrm{d}\bar{t} = \int_{\bar{t}=t_0}^{\bar{t}=t} \sigma_{ji} e^{\bullet}_{ij} \mathrm{d}\bar{t} \,, \tag{1.87}$$

since $\mathrm{d}w = \sigma_{ji}\mathrm{d}e_{ij}$. In order to discover another conjugate relation, we calculate the power in the whole continuum body:

$$\begin{aligned} W^{\bullet} &= \int_{\mathcal{B}} \sigma_{ji} e^{\bullet}_{ij} \mathrm{d}v = \int_{\mathcal{B}_0} S_{ji} E^{\bullet}_{ij} \mathrm{d}V = \int_{\mathcal{B}_0} S_{ji} \Big(\frac{1}{2} (F_{ni} F_{nj} - \delta_{ij}) \Big)^{\bullet} \mathrm{d}V = \\ &= \int_{\mathcal{B}_0} S_{ji} \frac{1}{2} (F^{\bullet}_{ni} F_{nj} + F_{ni} F^{\bullet}_{nj}) \mathrm{d}V \,, \end{aligned} \tag{1.88}$$

since $S_{ij} = S_{ji}$ we obtain

$$W^{\bullet} = \int_{\mathcal{B}_0} S_{ji} F^{\bullet}_{ni} F_{nj} \mathrm{d}V = \int_{\mathcal{B}_0} P_{in} F^{\bullet}_{ni} \mathrm{d}V \,. \tag{1.89}$$

Hence, the stress-deformation gradient pair, P_{ij}, F_{ji}, is also conjugate and can be used instead of S_{ji}, E_{ij}, or σ_{ji}, e_{ij}. Furthermore, owing to the relation:

$$w = \int_1^2 \mathrm{d}w = \int_1^2 P_{ji} \mathrm{d}F_{ij} \,, \tag{1.90}$$

we acquire CASTIGLIANO's theorem:[37]

$$P_{ji} = \frac{\partial w}{\partial F_{ij}} \,. \tag{1.91}$$

We need to have a constitutive relation in order to obtain P_{ji}. Now we can define an alternative way by using Eq. (1.91). Instead of defining a relation for P_{ji}, we can define the energy, w, and then by deriving it with respect to F_{ij}, we obtain P_{ji}. The so-called stored energy, w, is a *scalar potential field*. In other words, only the values in the states 1 and 2 are of interest. The values between the states are not important. Moreover, the change from the state 1 to the state 2 can happen in an arbitrary way; the amount of necessary energy remains the same. This property is the reason why we are allowed to write its so-called *first-integral*:

$$w = \int_1^2 \mathrm{d}w \,. \tag{1.92}$$

[37] Alberto Castigliano never wrote the equation in this way but he found out the variational method on a one-dimensional beam by using the relation: force is the energy differentiation with respect to the displacement. We can extend this theorem and motivate the relation that stress is the energy differentiation with respect to the strain.

The same phenomenon is known in the case of the gravitational potential energy, where the work is done by lowering or altering weights. The total work depends on the height difference; only the states 1 and 2 are important, not the states in between. Therefore, the stored energy may depend on the deformation gradient but not on the velocity.

We employ the neo-HOOKEan energy density used often for rubber like materials

$$w(F_{ij}) = \lambda \frac{1}{2} \ln^2(\det(F_{mn})) + \mu \left(\frac{F_{ji} F_{ji}}{2} - \frac{\delta_{jj}}{2} - \ln(\det(F_{kl})) \right) . \tag{1.93}$$

The LAME parameters, λ, μ, are the material parameters to be determined by the experiments. As aforementioned we can recall the weak form in Eq. (1.77) (multiplied by minus one and the gravitational force is incorporated)

$$\text{Form} = \int_{\mathcal{B}_0} \left(- P_{ki}\, \delta u_{i,k} + \rho_0 f_i\, \delta u_i \right) \mathrm{d}V + \int_{\partial \mathcal{B}_0^{\mathrm{N}}} \hat{t}_i\, \delta u_i \mathrm{d}A . \tag{1.94}$$

By using Eq. (1.91) and the stored energy in Eq. (1.93) we can setup the problem as follows

$$\begin{aligned} P_{ji} &= \frac{\partial w}{\partial F_{ij}} = \frac{\lambda}{2} 2 \ln(J) \frac{1}{J} \frac{\partial J}{\partial F_{ij}} + \mu \left(F_{kl} \delta_{ki} \delta_{lj} - \frac{1}{J} \frac{\partial J}{\partial F_{ij}} \right) , \\ P_{ji} &= \frac{1}{J} \frac{\partial J}{\partial F_{ij}} (\lambda \ln(J) - \mu) + \mu F_{ij} . \end{aligned} \tag{1.95}$$

The derivative of the determinant can be calculated as follows

$$\begin{aligned} J &= \epsilon_{ijk} F_{1i} F_{2j} F_{3k} , \\ \frac{\partial J}{\partial F_{1j}} F_{1j} = 0 , \quad & \frac{\partial J}{\partial F_{2j}} F_{2j} = J , \quad \frac{\partial J}{\partial F_{3j}} F_{3j} = 0 , \\ \frac{\partial J}{\partial F_{ij}} F_{ij} &= J + 0 + 0 = J , \end{aligned} \tag{1.96}$$

since the latter has to hold for arbitrary F_{ij} we obtain

$$\frac{\partial J}{\partial F_{ij}} = (\boldsymbol{F}^{-1})_{ji} J . \tag{1.97}$$

Therefore, the nominal stress reads

$$P_{ji} = (\lambda \ln(J) - \mu)(\boldsymbol{F}^{-1})_{ji} + \mu F_{ij} . \tag{1.98}$$

We can change the code in Sect. 1.2 and get the material nonlinearity modeled with this neo-HOOKEan material equation. In the literature, for the neo-HOOKEan material equation, the second PIOLA–KIRCHHOFF stress is given by

$$S_{jl} = (\boldsymbol{F}^{-1})_{li} P_{ji} = (\lambda \ln(J) - \mu)(\boldsymbol{F}^{-1})_{li}(\boldsymbol{F}^{-1})_{ji} + \mu(\boldsymbol{F}^{-1})_{li} F_{ij} = \\ = (\lambda \ln(J) - \mu)(\boldsymbol{C}^{-1})_{jl} + \mu\delta_{jl} \ . \tag{1.99}$$

The derivative of the stored energy is cumbersome and this can be handled by an automatized method. Moreover, we can introduce an abstraction to the formulation as an alternative method, which explains the real meaning of the variational formulation. This abstraction starts with the definition of the so-called *action*:

$$\mathcal{S} = \int_{\mathcal{T}} \int_{\mathcal{B}_0} \mathcal{L} \mathrm{d}V \mathrm{d}t + \int_{\mathcal{T}} \int_{\partial\mathcal{B}_0} W_\mathrm{s} \mathrm{d}A \mathrm{d}t \ , \tag{1.100}$$

which is an energy integral over time with a LAGRANGEan density, $\mathcal{L}$, to be defined and a potential work on surfaces, W_s, to be given on the boundaries. Since the time definition is redundant in statics, we reformulate the action

$$\mathcal{S}^* = \int_{\mathcal{B}_0} \mathcal{L} \mathrm{d}V + \int_{\partial\mathcal{B}_0} W_\mathrm{s} \mathrm{d}A \ . \tag{1.101}$$

The variational formulation is based on the action definition and can be applied by using the *principle of least action* stating that the variation of action:

$$\delta\mathcal{S}^* = \int_{\mathcal{B}_0} \delta\mathcal{L} \mathrm{d}V + \int_{\partial\mathcal{B}_0} \delta W_\mathrm{s} \mathrm{d}A \ , \tag{1.102}$$

vanishes

$$\delta\mathcal{S}^* = 0 \ , \tag{1.103}$$

such that the action is minimum by assuming that action is always positive valued. The principle of least action is based on the formulations made by Pierre Louis Maupertuis, Leonhard Euler, and Joseph Louis Lagrange. In the way that we use herein, the variational principle is applied by William Rowan Hamilton and Lord Rayleigh[38] in the mid-19th century.

We aim at calculating the displacement field that is the primitive variable. The LAGRANGEan density depends on the primitive variables and their derivatives. For hyperelasticity in homogeneous materials the LAGRANGEan density depends on displacement and its first derivative:

$$\mathcal{L} = \mathcal{L}(u_i, u_{i,j}) \ . \tag{1.104}$$

The potential on the NEUMANN surfaces depend only on displacement:

$$W_\mathrm{s} = W_\mathrm{s}(u_i) \ . \tag{1.105}$$

[38] John William Strutt, 3rd Baron Rayleigh.

Now we can calculate the variation of action in statics:

$$\text{Form} = \delta\mathcal{S}^* = \int_{\mathcal{B}_0} \Big(\frac{\partial \mathcal{L}}{\partial u_i} \delta u_i + \frac{\partial \mathcal{L}}{\partial u_{i,j}} \delta u_{i,j} \Big) \mathrm{d}V + \int_{\partial\mathcal{B}_0} \frac{\partial W_\mathrm{s}}{\partial u_i} \delta u_i \mathrm{d}A \ . \quad (1.106)$$

An integration by parts would lead to the EULER–LAGRANGE equations leading to the balance equations. Since we only need to obtain the weak form, we have accomplished the variational formulation. The weak form in Eq. (1.106) has to be identical with the weak form in Eq. (1.94). If we define the LAGRANGEan density and the potential as

$$\mathcal{L} = -w + \rho_0 f_i u_i \ , \quad W_\mathrm{s} = \hat{t}_i u_i \ , \quad (1.107)$$

then the weak form reads

$$\text{Form} = \int_{\mathcal{B}_0} \Big(\rho_0 f_i \, \delta u_i - \frac{\partial w}{\partial u_{i,j}} \delta u_{i,j} \Big) \mathrm{d}V + \int_{\partial\mathcal{B}_0} \hat{t}_i \, \delta u_i \mathrm{d}A \ , \quad (1.108)$$

since the stored energy depends only on the deformation gradient. By using

$$\frac{\partial w}{\partial u_{i,j}} = \frac{\partial w}{\partial F_{kl}} \frac{\partial F_{kl}}{\partial u_{i,j}} = P_{lk} \frac{\partial (u_{k,l} + \delta_{kl})}{\partial u_{i,j}} = P_{lk} \delta_{ki} \delta_{lj} = P_{ji} \ , \quad (1.109)$$

we can convince ourselves that the chosen $\mathcal{L}$ leads to the same weak form as in Eq. (1.94). The action formulation is quite abstract, however, really useful. We define the action and the balance of linear momentum comes out with the help of the variational formulation. It is an ongoing discussion between researchers, which axiom is less restrictive: postulating an action or postulating a balance equation. An energy-based method is beneficial for detecting the material response, since measuring an energy is much easier than measuring a stress. Consider a tensile testing. The force can be measured by an accelerometer, simultaneously and independently, the displacement can be tracked by an optic sensor. Force times displacement is the energy. More easier is to measure the energy directly on the motor, since the power supplied to the motor as well as the standard losses of the motor are known a priori. Basically, the measurement of energy is quite natural. However, if we want to calculate the stress from the force we need to know the cross-sectional area. We cannot measure the area correctly since it is changing throughout the experiment; mostly this change is neglected. Especially for hyperelastic materials, we have to take account the cross-sectional change, hence, the energy function, w, is a more accurate observable than the stress.

We simulate $40 \times 100 \times 100$ cm sample of a silicone gel loaded with a line force. This force has been applied by using a GAUSSian distribution[39] on yz-plane at $y = 500$ and along z in the direction of plane normal:

[39] It is named after Carl Friedrich Gauß.

$$\hat{t}_i = \Big(a \exp\big(b(y - 500.0) + c(y - 500.0)^2\big)\Big) N_i \,. \tag{1.110}$$

A sketch of the GAUSSian distribution can be seen in Fig. 1.3 by using $a = -50$, $b = 0$, $c = -0.001$ as in code. The deformation field is shown in Fig. 1.4, consider the large deformation with respect to the geometric sizes. The computation is realized by using the symbolic differentiation capabilities in FEniCS. We implement the weak form as in Eq. (1.108). The stored energy function is differentiated symbolically. Thus, one can use the same code even for more complicated energy formulations. The geometric and material nonlinearities are captured accurately by using the linearization at the partial differential level, again by using the symbolic differentiation. For modeling the line force accurately we have used a fine meshing leading to a high number of freedoms. Therefore, an iterative solver with preconditioning has been used for the computation. The code is given below.

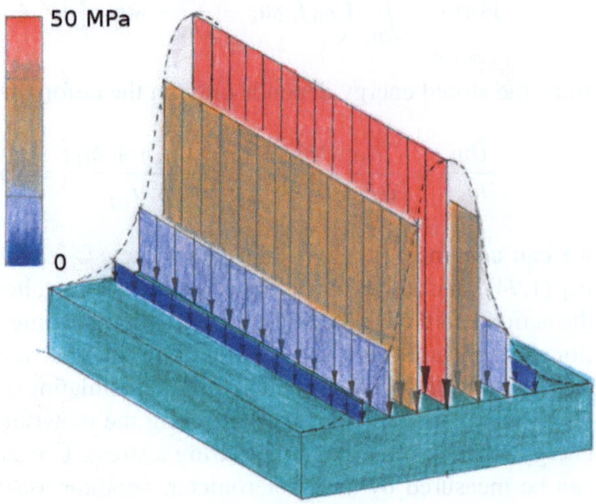

Fig. 1.3 The GAUSSian distribution for $\hat{t}_i$ is also called the *bell-curve* for obvious reasons

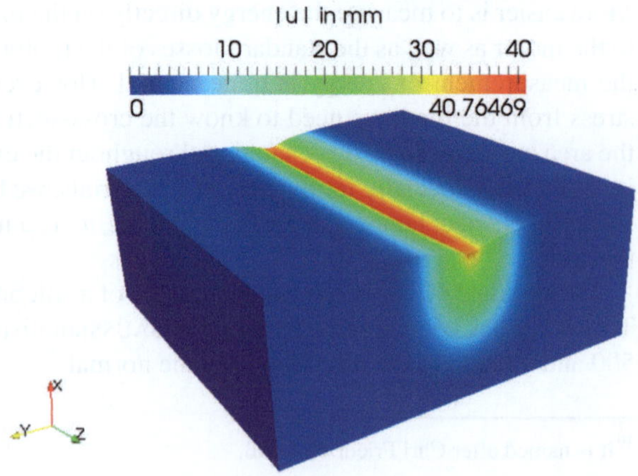

Fig. 1.4 Deformation as in reality (scale factor = 1), colors indicate the magnitude of the displacement field

```
Computational reality 03, example of a nonlinear
hyperelastic material for silicon gel.
__author__ = "B. Emek Abali"
__license__  = "GNU LGPL Version 3.0 or later"
#This code underlies the GNU General Public License, http://www.gnu.org/licenses/gpl-3.0.en.html

from fenics import *
xlength=400.0 #[mm]
ylength=1000.0 #[mm]
zlength=1000.0 #[mm]
mesh = BoxMesh(Point(0, 0, 0), Point(xlength, ylength, zlength), 15, 45, 25)
V = VectorFunctionSpace(mesh, 'P', 1)
cells = CellFunction('size_t', mesh)
facets = FacetFunction('size_t', mesh)
dA = Measure('ds', domain=mesh, subdomain_data=facets)
dV = Measure('dx', domain=mesh, subdomain_data=cells)
left = CompiledSubDomain('near(x[0],0) && on_boundary')
right = CompiledSubDomain('near(x[0], length) && on_boundary', length=xlength)
bottom = CompiledSubDomain('near(x[1],0) && on_boundary')
top = CompiledSubDomain('near(x[1],length) && on_boundary', length=ylength)
back = CompiledSubDomain('near(x[2],0) && on_boundary')
front= CompiledSubDomain('near(x[2],length) && on_boundary', length=zlength)
facets.set_all(0)
right.mark(facets, 1)
# like a gel filled in a box
bc1 = DirichletBC(V, (0.0,0.0,0.0), left)
bc2 = DirichletBC(V.sub(1), (0.0), top)
bc3 = DirichletBC(V.sub(2), (0.0), top)
bc4 = DirichletBC(V.sub(1), (0.0), bottom)
bc5 = DirichletBC(V.sub(2), (0.0), bottom)
bc6 = DirichletBC(V.sub(1), (0.0), back)
bc7 = DirichletBC(V.sub(2), (0.0), back)
bc8 = DirichletBC(V.sub(1), (0.0), front)
bc9 = DirichletBC(V.sub(2), (0.0), front)
bc=[bc1,bc2,bc3,bc4,bc5,bc6,bc7,bc8,bc9]
# definition for the variational formulation
du = TrialFunction(V)       # Incremental displacement
del_u = TestFunction(V)     # Test function
u  = Function(V)            # Displacement from previous iteration
# material parameters of a silicone gel TSE3062
rho_0 = 1.1E-9 #in ton/mm3
nu = 0.4
E = 89.0 #in MPa
lambada = Constant(E*nu/(1.0+nu)/(1.0-2.0*nu))
mu = Constant(E/(2.0*(1.0+nu)))
i, j, k, l = indices(4)
```

```
delta = Identity(3)

def stored_energy(g_u):
    F = delta + g_u
    return as_tensor(lambada*1.0/2.0*ln(det(F))**2 + mu*(F[j,i]*F[j,i]/2.0 - delta[i,i]/2.0 - ln(det(F)) ), [])

grad_u = variable(grad(u))
stored = stored_energy(grad_u)
# traction vector
distr = Expression('a*exp(b*(x[1]-500.0) + c*(x[1]-500.0)*(x[1]-500.0) )',a=-50.0,b=0.0,c=-0.001)
N = FacetNormal(mesh)
tr = distr*N
f = Constant((0.0,0.0,0.0)) # body force
Form = (rho_0*f[i]*del_u[i] - diff(stored,grad_u)[i,j]*grad(del_u)[i,j] )*dV + tr[i]*del_u[i]*dA(1)
Gain = derivative(Form,u,du)
solve(Form==0, u, bc, J=Gain, \
    solver_parameters={"newton_solver":{"linear_solver": "cg", "preconditioner": "ilu", "relative_tolerance": 1e-3} }, \
    form_compiler_parameters={"cpp_optimize": True, "representation": "quadrature", "quadrature_degree": 2} )
pwd = '/calcul/CR03/'
file = File(pwd+'displacement.pvd')
file << u
```

To-do

We have solved a silicon gel by using a hyperelastic material model. The variational form has been defined in an abstract manner by using the principle of least action.

- Identify the symbolic differentiation of the stored energy and the symbolic differentiation for linearization in the code.
- We use another solver from Trilinos packages. Conjugate gradients is an iterative solver used for large problems with symmetric matrix. Is it possible to print out the number of degrees of freedom in the computation?
- Find the energy function for material models named as MOONEY–RIVLIN and BLATZ–KO for rubber materials. Try to apply them into the code.
- Derive the stored energy for the linear model called ST. VENANT's law.

1.4 Linear Rheology

A continuum body with boundary conditions forms a system. Due to the applied force (input) the response of this system (output) is computed. All aforementioned material models respond the same under different loading rates. In order to include the effect of the loading rate we need to use the so-called *rheological* material models. The structure responds stiffer upon quicker loading—this phenomenon is called a viscous behavior. It can be realized easily on a bicycle pump that is a dashpot filled with a fluid (air). If we want to pump quicker, the necessary force is higher. In a solid body this viscous behavior occurs and it can be modeled by incorporating stress rate and strain rate into the constitutive equation. In this section we will apply such models and simulate a *nanoindentation* experiment for a viscoelastic material.

The primitive variable is the displacement field in space and time. In a LAGRANGEan frame, space denotes the initial positions of particles, X_i. Time can be seen as subsequent snapshots: $t = [0, \Delta t, 2\Delta t, 3\Delta t, \ldots n\Delta t]$. We use the same time step, Δt. There is no restriction about using a variable time step, in some cases it may even shorten the computation time by improving the convergence rate. In reality, or as we try to understand our observations, we do assume that time is continuous by utilizing an infinitesimal time step, $\Delta t \to 0$. The time discretization becomes more accurate by applying a smaller time step. A natural way of discretizing in time reads

$$
\begin{aligned}
v_i^{\bullet} &= \frac{\partial v_i}{\partial t} = \frac{v_i - v_i^0}{t - t^0} = \frac{v_i - v_i^0}{\Delta t} \,, \\
v_i(X, t) = u_i^{\bullet}(X, t) &= \frac{\partial u_i}{\partial t} = \frac{u_i - u_i^0}{\Delta t} \,,
\end{aligned}
\tag{1.111}
$$

where this method relies on the (finite) difference schema and is an *implicit* method, often called the backward EULER method—we refer to Appendix A.5 on p. 304 for the meaning of implicit. Basically, we exchange the partial derivative with a difference function.

The balance of linear momentum is defined in the current frame for an arbitrary coordinate system[40] as follows

$$
\Big(\int_{\mathcal{B}} \rho v_i \mathrm{d}v \Big)^{\bullet} = \int_{\partial\mathcal{B}} \sigma_{ji} \mathrm{d}a^j + \int_{\mathcal{B}} \rho f_i \mathrm{d}v \,,
\tag{1.112}
$$

where $\mathrm{d}a^j = n^j \mathrm{d}a$. We have not chosen a coordinate system, yet. The latter equation holds for any coordinate system since we employ the co- and contravariant notation. The lower suffix denotes the covariant components and the upper suffix identifies the contravariant components. The summation convention is between upper and lower indices. In an arbitrary coordinate system, z, the covariant components, z_i, are the orthogonal projections to the base vectors, whereas the contravariant components, z^i, are the parallel projections to the base vectors.[41] The relation of the coordinates

[40] A cylindrical and a spherical coordinate system are curvilinear orthogonal coordinate systems, where the base vectors are orthogonal to each other. If the base vectors are not orthogonal then we use the term *oblique*. An oblique and curvilinear coordinate system is called an arbitrary coordinate system. We use coordinate systems fixed in time (they do not move according to the observer).

[41] For tensor calculus, see [7, 15], or [26].

in the coordinate system, z, to the coordinates in the Cartesian coordinate system, $\boldsymbol{x}$, is given by the metric tensor:

$$g_{ij} = \frac{\partial x_k}{\partial z^i}\frac{\partial x_k}{\partial z^j} \; . \tag{1.113}$$

We recall that in Cartesian coordinates the parallel/orthogonal projections are identical, $x^i = x_i$, such that we utilize only lower suffix. In the case of an arbitrary coordinate system, the distinction is of importance since $z^i \neq z_i$. The covariant (lower suffix) and the contravariant (upper suffix) can be interchanged by using the metric tensor,

$$v^i g_{ij} = v_j \; , \tag{1.114}$$

or its inverse,

$$g^{ij} = \frac{\partial z^i}{\partial x_k}\frac{\partial z^j}{\partial x_k} = (\boldsymbol{g}^{-1})^{ij} \; , \quad g^{ij} g_{jk} = g^i_k = \delta^i_k \; , \tag{1.115}$$

as follows

$$v_i g^{ij} = v^j \; . \tag{1.116}$$

An often used notation for the determinant of metric reads

$$g = \det(g_{ij}) \; , \quad \frac{1}{g} = \det(g^{ij}) \; . \tag{1.117}$$

The permutation symbol $e^{ijk} = e_{ijk} = \pm 1$ if $[ijk]$ is cyclic/anticyclic else zero. For example, $e_{123} = 1$, and $e_{321} = -1$, but $e_{112} = 0$. The permutation symbol leads to the LEVI-CIVITA symbol (in three-dimensions):

$$\epsilon_{ijk} = \frac{1}{g} e_{ijk} \; , \quad \epsilon^{ijk} = g e_{ijk} \; , \quad \epsilon_{ijk}\epsilon^{ijk} = 3! = 6 \; . \tag{1.118}$$

Now we introduce an identity for an arbitrary A_{ij} given in z as follows

$$\epsilon^{ijk} A_{ir} A_{js} A_{kt} = \det(A_{mn})\epsilon_{rst} \; . \tag{1.119}$$

By using the above identity we obtain

$$\begin{aligned} \mathrm{d}v = \epsilon^{ijk}\mathrm{d}z_i^{(1)}\mathrm{d}z_j^{(2)}\mathrm{d}z_k^{(3)} &= \epsilon^{ijk}\frac{\partial z_i}{\partial Z^r}\mathrm{d}Z^{(1)r}\frac{\partial z_j}{\partial Z^s}\mathrm{d}Z^{(2)s}\frac{\partial z_k}{\partial Z^t}\mathrm{d}Z^{(3)t} = \\ = \det\left(\frac{\partial z_i}{\partial Z^j}\right)\epsilon_{rst}\mathrm{d}Z^{(1)r}\mathrm{d}Z^{(2)s}\mathrm{d}Z^{(3)t} &= \det\left(\frac{\partial z_i}{\partial Z^j}\right)\mathrm{d}V = J\mathrm{d}V \; . \end{aligned}$$

The relation $\mathrm{d}v = J\mathrm{d}V$ is the transformation of the volume element from the current frame, $\mathrm{d}v$, to the initial frame, $\mathrm{d}V$. Since the volume element denotes the size, the JACOBIan, J, gives the volume contraction. For the sake of clarity, the deformation gradient in arbitrary coordinates reads

$$F_{ij} = \frac{\partial z_i}{\partial Z^j} \,, \quad F^i{}_j = \frac{\partial z^i}{\partial Z^j} \,. \tag{1.120}$$

The deformation gradient is not an objective tensor of rank two, thus we do not use the metric tensor for lowering or raising the indices. Similar to the volume element we can transform the area element, for example for an area showing in the $z_i^{(1)}$ direction

$$\begin{aligned} n^i da = \mathrm{d}a^i &= \epsilon^{ijk} \mathrm{d}z_j^{(2)} \mathrm{d}z_k^{(3)} = \epsilon^{ijk} \frac{\partial z_j}{\partial Z^s} \mathrm{d}Z^{(2)s} \frac{\partial z_k}{\partial Z^t} \mathrm{d}Z^{(3)t} = \\ &= \left(\frac{\partial z_i}{\partial Z^r} \right)^{-1} \det \left(\frac{\partial z_m}{\partial Z^n} \right) \epsilon_{rst} \mathrm{d}Z^{(2)s} \mathrm{d}Z^{(3)t} = \\ &= (F_{ir})^{-1} J \mathrm{d}A_r = (\boldsymbol{F}^{-1})^{ri} J N_r \mathrm{d}A \,. \end{aligned} \tag{1.121}$$

We have already applied the transformations of volume and area element in Sect. 1.2. In this section we see that these transformations hold for arbitrary coordinate systems, indeed, they hold also for the Cartesian coordinate system. By using these identities we transform the balance equations in the current frame into balance equations in the initial frame. We start with the balance of mass for a *closed* or *material* system, i.e., throughout the simulation the number of particles are preserved. The total mass of the continuum body is constant in time. Of course the mass density changes due to the deformation, however, no particle enters or leaves the continuum body. The total mass of the continuum body is preserved and its rate (change in time) vanishes in case of a material system. The balance of mass in the current frame reads for a material system

$$\left(\int_{\mathcal{B}} \rho \mathrm{d}v \right)^{\bullet} = 0 \,. \tag{1.122}$$

The mass density varies in time, as well as the volume element in the current frame. By transforming it from the current to the initial frame, we obtain

$$\int_{\mathcal{B}_0} (\rho J)^{\bullet} \mathrm{d}V = 0 \,, \tag{1.123}$$

since the volume element in the beginning fails to vary in time, $(\mathrm{d}V)^{\bullet} = 0$. Its local solution is a constant in time:

$$\rho J = \text{const.}|_t = \rho_0 \,, \tag{1.124}$$

where ρ_0 denotes the mass density in the initial frame—we recall that J is the volume contraction. Therefore, the balance of mass in the LAGRANGEan frame is $\rho J = \rho_0$. The balance of mass in the initial frame is an equation, not a differential equation, we have one equation less to solve. Only the balance of momentum needs to be solved. By using $\rho J = \rho_0$ we implicitly satisfy the balance of mass without further ado. In order to transform the balance of momentum in the current frame:

$$\Big(\int_{\mathcal{B}} \rho v_i \mathrm{d}v \Big)^{\bullet} = \int_{\partial\mathcal{B}} \sigma_{ji} \mathrm{d}a^j + \int_{\mathcal{B}} \rho f_i \mathrm{d}v \ , \tag{1.125}$$

into the balance of momentum in the initial frame:

$$\Big(\int_{\mathcal{B}_0} \rho v_i J \mathrm{d}V \Big)^{\bullet} = \int_{\partial\mathcal{B}_0} (\boldsymbol{F}^{-1})^{kj} \sigma_{ji} J N_k \mathrm{d}A + \int_{\mathcal{B}_0} \rho f_i J \mathrm{d}V \ , \tag{1.126}$$

we use Eqs. (1.120), (1.121). After inserting the balance of mass, $\rho J = \rho_0$, and using the fact that neither ρ_0 nor $\mathrm{d}V$ changes in time we acquire

$$\int_{\mathcal{B}_0} \rho_0 \overset{\bullet}{v_i} \mathrm{d}V = \int_{\partial\mathcal{B}_0} (\boldsymbol{F}^{-1})^{kj} \sigma_{ji} J N_k \mathrm{d}A + \int_{\mathcal{B}_0} \rho_0 f_i \mathrm{d}V \ . \tag{1.127}$$

As in Eq. (1.68) we use a shorthand notation and introduce the nominal stress:

$$\begin{aligned} P^k_i &= (\boldsymbol{F}^{-1})^{kj} \sigma_{ji} J \ , \\ \int_{\mathcal{B}_0} \rho_0 \overset{\bullet}{v_i} \mathrm{d}V &= \int_{\partial\mathcal{B}_0} P^k_i N_k \mathrm{d}A + \int_{\mathcal{B}_0} \rho_0 f_i \mathrm{d}V \ . \end{aligned} \tag{1.128}$$

We apply GAUSS's law, however, for an arbitrary coordinate system

$$\int_{\partial\mathcal{B}_0} P^k_i N_k \mathrm{d}A = \int_{\mathcal{B}_0} P^k_{i;k} \mathrm{d}V \ , \tag{1.129}$$

where a semicolon denotes a *covariant* derivative. A Cartesian coordinate system possesses orthonormal base vectors and this triad is constant in whole space. In other words, its metric tensor, δ_{ij}, is constant in space, $\delta_{ij,k} = 0$. For an arbitrary coordinate system the base vectors are oblique and the space is curvilinear such that the triad varies in space. The metric tensor is not a constant in the space, its partial derivative in space fails to vanish, $g_{ij,k} \neq 0$. For example, in a cylindrical coordinate system the lengths of the base vectors remain the same, however, their directions change.

Consider a constant vector field, A_i. In every point of space same length and direction is a constant vector field. We want to express this constant vector field in an arbitrary coordinate system. When the coordinate system is curvilinear then in two different points of space we have different base vectors: metric varies in space. Although A_i is constant, its coordinates in two points in space are different due to the varying base vectors. We obtain $A_{i,j} \neq 0$ for a constant A_i. This fact makes the generalization of the tensor calculation in arbitrary coordinates difficult. It is much more intuitive and easier to analyze if the coordinates of a *constant* vector remains *constant* in space, i.e., derivative of A_i has to vanish. Thus, we want to have a special derivative operation in which the change of the triad is considered implicitly. In other words, since the metric varies in space, the derivative operation shall *co-vary* in the same space, $g_{ij;k} = 0$. This covariant derivative is denoted by a semicolon instead of a comma and it allows us to generalize a space derivative in Cartesian coordinates to arbitrary coordinates. We simply exchange partial derivatives with covariant derivatives. Since the numerical evaluation will be in Cartesian coordinates, we will not make use of it, however, we give the definition for arbitrary coordinate systems for the sake of completeness. For a tensor of rank two with mixed, i.e., co- and contravariant components the covariant derivative reads

$$A^j_{\ i;k} = A^j_{\ i,k} + \Gamma^j_{nk} A^n_{\ i} - \Gamma^m_{ik} A^j_{\ m} , \tag{1.130}$$

where we have introduced the CHRISTOFFEL symbols,[42] giving the curvature of space:

$$\begin{aligned} \Gamma^l_{kn} &= \frac{1}{2} g^{lm} \Big(g_{mk,n} + g_{mn,k} - g_{kn,m} \Big) , \\ \Gamma^l_{kn} &= \Gamma^l_{nk} . \end{aligned} \tag{1.131}$$

For the sake of clarity, a partial derivative is in z in the current frame and in $\boldsymbol{Z}$ in the initial frame. In solid mechanics we work in the initial frame such that the comma notation denotes a partial derivative in $\boldsymbol{Z}$ as follows

$$(\cdot)_{,i} = \frac{\partial(\cdot)}{\partial Z^i} . \tag{1.132}$$

We recall that the chosen arbitrary coordinate system is constant in time. Therefore, we know the coordinate system at the beginning such that we can calculate the metric tensor,

$$g_{ij} = \frac{\partial X_k}{\partial Z^i} \frac{\partial X_k}{\partial Z^j} = X_{k,i} X_{k,j} . \tag{1.133}$$

The notation $\boldsymbol{X}$ denotes to the positions of particles at initial time in the Cartesian coordinate system, $\boldsymbol{Z}$ are the positions of particles at initial time in the curvilinear and oblique coordinate system. The numerical values of metric tensor are the same if calculated by using the current positions, $\boldsymbol{x}$, in the Cartesian coordinate system and in the curvilinear and oblique system, z, as follows

[42] The symbol is named after Elwin Bruno Christoffel.

$$g_{ij} = \frac{\partial x_k}{\partial z^i}\frac{\partial x_k}{\partial z^j} . \tag{1.134}$$

Both coordinate systems are constant in time. Cartesian coordinates are also constant in space. Now by applying GAUSS's law we obtain the balance of linear momentum in LAGRANGEan frame expressed in arbitrary coordinates:

$$\rho_0 v_i^{\cdot} - P^k_{i;k} - \rho_0 f_i = 0 . \tag{1.135}$$

By using Eq. (1.111) we utilize the time discretization

$$\begin{aligned} \rho_0 \frac{v_i - v_i^0}{\Delta t} - P^k_{i;k} - \rho_0 f_i = 0 , \\ \rho_0 \frac{u_i - 2u_i^0 + u_i^{00}}{\Delta t \Delta t} - P^k_{i;k} - \rho_0 f_i = 0 . \end{aligned} \tag{1.136}$$

This formulation is *universal*, in other words, the formulation is valid for any material model. Now we need to use a constitutive equation for rheological materials. In arbitrary coordinates we rewrite the GREEN–LAGRANGE strain tensor from Eq. (1.60) by switching from the partial derivatives to the covariant derivatives

$$E_{ij} = \frac{1}{2} u_{k;i} u^k_{;j} + u_{(i;j)} = \frac{1}{2} g^{kl} u_{k;i} u_{l;j} + u_{(i;j)} . \tag{1.137}$$

The GREEN–LAGRANGE strain tensor is symmetric, $E_{ij} = E_{ji}$, moreover, it is an objective tensor of rank two, thus we can use metric to lower and raise indices. We want to use the second PIOLA–KIRCHHOFF stress tensor from Eq. (1.69):

$$\begin{aligned} S^{kl} = (\boldsymbol{F}^{-1})^{li} P^k_i , \quad P^k_i = F_{il} S^{kl} , \\ F_{il} = \frac{\partial z_i}{\partial Z^l} = \frac{\partial (u_i + Z_i)}{\partial Z^l} = u_{i,l} + (g_{ij} Z^j)_{,l} . \end{aligned} \tag{1.138}$$

As so far we have employed stress as a tensor function of strain, $S_{ij} = S_{ij}(E_{kl})$, so that stress has a dependency solely on the strain. In rheology, more dependency is included such that stress may depend on rates of strain and stress. For simplicity, we include only the first rate of stress and strain. The generic representation of the stress tensor depends on strain, strain rate, and stress rate, $S_{ij} = S_{ij}(E_{kl}, E^{\cdot}_{kl}, S^{\cdot}_{kl})$.

The material response, S_{ij}, which is symmetric, $S_{ij} = S_{ji}$, can be divided into a *volumetric*, $S_{\{ik\}}$, and a *deviatoric*, $S_{|ik|}$, part

$$\begin{aligned} S_{ij} = S_{\{ij\}} + S_{|ij|} , \\ S_{\{ij\}} = \frac{1}{3} S^k_k g_{ij} , \quad S_{|ij|} = S_{ij} - \frac{1}{3} S^k_k g_{ij} . \end{aligned} \tag{1.139}$$

Of course, the same holds true for the symmetric strain tensor of rank two

$$E_{ij} = E_{\{ij\}} + E_{|ij|} \,,$$
$$E_{\{ij\}} = \frac{1}{3} E^k_k g_{ij} \,, \quad E_{|ij|} = E_{ij} - \frac{1}{3} E^k_k g_{ij} \,. \tag{1.140}$$

The volumetric and deviatoric parts can be varied independently. This mathematical fact can be seen by using the stored energy from the last section,

$$w = \frac{1}{2} S_{ij} E^{ij} = \frac{1}{2} \big(S_{\{ij\}} + S_{|ij|} \big) \big(E^{\{ij\}} + E^{|ij|} \big) = \frac{1}{2} \big(S_{\{ij\}} E^{\{ij\}} + S_{|ij|} E^{|ij|} \big) \,, \tag{1.141}$$

since the mixed terms vanish, for example

$$S_{\{ij\}} E^{|ij|} = \frac{1}{3} S^k_k g_{ij} \Big(E^{ij} - \frac{1}{3} E^l_l g^{ij} \Big) = \frac{1}{3} S^k_k E^i_i - \frac{1}{9} S^k_k E^l_l g^i_i \,, \quad g^i_i = \delta^i_i = 3 \,. \tag{1.142}$$

We can easily separate the energy into its volumetric and deviatoric parts

$$w^{\text{vol.}} = \frac{1}{2} S_{\{ij\}} E^{\{ij\}} \,, \quad w^{\text{dev.}} = \frac{1}{2} S_{|ij|} E^{|ij|} \,,$$
$$w = w^{\text{vol.}} + w^{\text{dev.}} \,. \tag{1.143}$$

Energy is supposed to be additive in its independent parts, the well-known example is the separation of energy into kinetic and potential energies in rigid bodies. Since we have found out that the stored energy is additively decomposed into volumetric and deviatoric parts, they must be independent, too. In other words, the volumetric part of the energy can be changed without affecting the deviatoric part. The volumetric part of the stress and the deviatoric part of the stress can be modeled with different material models. Moreover, the latter calculation shows that the volumetric part of the stress depends solely on the volumetric part of the strain and deviatoric part of the stress on the deviatoric part of the strain. The same shall hold for the strain rate, too.

For a better understanding we start with modeling the simplest linear model where stress depends only on strain. First we choose a linear model for the volumetric part of stress depending on the volumetric part of the strain tensor:

$$S_{\{ij\}} = a E_{\{ij\}} = \frac{a}{3} E^k_k g_{ij} \,. \tag{1.144}$$

Secondly, we choose a linear model for the deviatoric part:

$$S_{|ij|} = b E_{|ij|} = b E_{ij} - \frac{b}{3} E^k_k g_{ij} \,. \tag{1.145}$$

Therefore, we obtain the following linear equation:

$$S_{ij} = S_{\{ij\}} + S_{|ij|} = \frac{a+b}{3} E^k_k g_{ij} + b E_{ij} \ . \tag{1.146}$$

The latter becomes the ST. VENANT–KIRCHHOFF material model for isotropic materials by renaming $(a+b)/3 = \lambda$ and $b = \mu$. This method of employing material models for the deviatoric and volumetric parts separately makes a straight-forward generalization of constructing material equations in rheology.

We want to obtain a viscoelastic material model for the simulation in this section. For the volumetric part we choose again a linear elastic model:

$$S_{\{ij\}} = \lambda E^k_k g_{ij} \ . \tag{1.147}$$

This model is often visualized as a linear spring. For the deviatoric part we employ the so-called ZENER model.[43] It is motivated as a parallel connection of a spring with a spring-dashpot in series. The spring is of stiffness E_1 and the spring-dashpot E_2, μ, where the dashpot responds to the rate of displacements, thus, brings in a time lag to response of material—the viscous behavior. The ZENER model is a common linear rheological model. The deviatoric part reads by obtaining a differential equation from the spring parallel to spring-dashpot system

$$S_{|ij|} + \frac{\mu}{E_2} \dot{S}_{|ij|} = E_1 E_{|ij|} + \frac{\mu(E_1+E_2)}{E_2} \dot{E}_{|ij|} \ . \tag{1.148}$$

We have completed the definition of the stress tensor. The material parameters, E_1, E_2, μ, λ, shall be determined upon experiments.

In order to acquire the variational form we use Eq. (1.136)$_2$. For the computation we choose a fixed (in time) Cartesian coordinate system, $g_{ij} = \delta_{ij}$, in the LAGRANGEan frame. Moreover, we simplify the computation by choosing linearized strains,

$$\varepsilon_{ij} = u_{(i,j)} \ . \tag{1.149}$$

After utilizing the implicit time discretization as well as the linearized strains we obtain

$$\begin{aligned}
\frac{\partial S_{|ij|}}{\partial t} &= \frac{S_{|ij|} - S^0_{|ij|}}{\Delta t} \ , \\
S_{|ij|}\Big(1 + \frac{\mu}{E_2 \Delta t}\Big) &= \frac{\mu}{E_2 \Delta t} S^0_{|ij|} + E_1 \varepsilon_{|ij|} + \frac{\mu(E_1+E_2)}{E_2} \frac{\varepsilon_{|ij|} - \varepsilon^0_{|ij|}}{\Delta t} \ , \\
S_{|ij|} &= \frac{\mu}{E_2 \Delta t + \mu} S^0_{|ij|} + \frac{E_1 E_2 \Delta t}{E_2 \Delta t + \mu} \varepsilon_{|ij|} + \frac{\mu(E_1+E_2)}{E_2 \Delta t + \mu} (\varepsilon_{|ij|} - \varepsilon^0_{|ij|}) \ .
\end{aligned} \tag{1.150}$$

[43] It is named after Clarence Melvin Zener.

By including the volumetric part

$$S_{ij} = S_{\{ij\}} + S_{|ij|} \,, \tag{1.151}$$

the constitutive relation reads

$$S_{ij} = \lambda \varepsilon_{kk}\delta_{ij} + \frac{\mu}{E_2 \Delta t + \mu} S^0_{|ij|} + \frac{E_1 E_2 \Delta t}{E_2 \Delta t + \mu}\varepsilon_{|ij|} + \frac{\mu(E_1+E_2)}{E_2\Delta t + \mu}\big(\varepsilon_{|ij|} - \varepsilon^0_{|ij|}\big) \,. \tag{1.152}$$

We implement this constitutive equation. After discretizing in time, we apply an integration by parts, and obtain the weak form in Cartesian coordinates:

$$\begin{aligned} \text{Form} = & \int_{\mathcal{B}_0} \Big(\rho_0 \frac{u_i - 2u_i^0 + u_i^{00}}{\Delta t \Delta t}\,\delta u_i + P_{ki}\,\delta u_{i,k} - \rho_0 f_i\,\delta u_i \Big) \mathrm{d}V - \\ & - \int_{\partial \mathcal{B}_0^{\mathrm{N}}} \hat{t}_i\,\delta u_i \mathrm{d}A \,, \end{aligned} \tag{1.153}$$

where $P_{ki} = F_{ij} S_{kj}$ and $F_{ij} = \delta_{ij} + u_{i,j}$.

We simulate a nanoindentation experiment with fictitious material parameters. The geometry is a small cube which is clamped at its bottom and a tiny needle presses down on the top surface. It is a point loading and we model it as pressing within a small circle. By measuring the force and position of the needle, the material parameters can be determined. Therefore, a nanoindentation device is used for determination of material parameters. Especially for micro-mechanics such an experimental possibility is of interest since for a tensile testing in micrometer length-scale, samples might be difficult to produce.

Consider a small box in 20 µm in each dimensions and a really sharp diamond needle, which is indenting to the sample and deforms a tiny shape in it. The material response is viscoelastic and the output is a force regarding the indentation depth. We apply some refinement techniques to get an accurate solution on point loading, see Figs. 1.5, 1.6, and 1.7. Owing to the high number of elements, the code uses

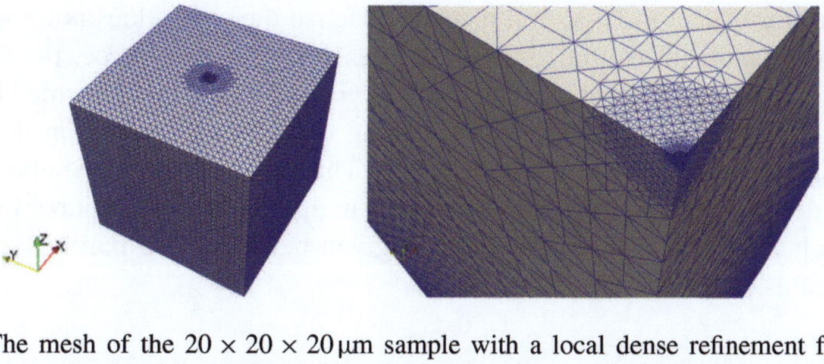

Fig. 1.5 The mesh of the 20 × 20 × 20 µm sample with a local dense refinement for a better loading. The amplitude of the loading evolves in time by means of the indentation height

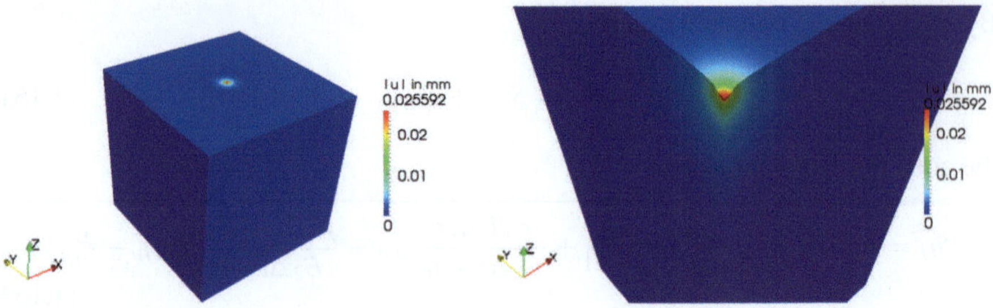

Fig. 1.6 Magnitude of the displacement field is shown for a nanoindentation experiment of $20 \times 20 \times 20\,\mu\text{m}$ cubical viscoelastic material at 0.7 s. Left: Whole sample. Right: Detailed view with deformation scaled 10 times

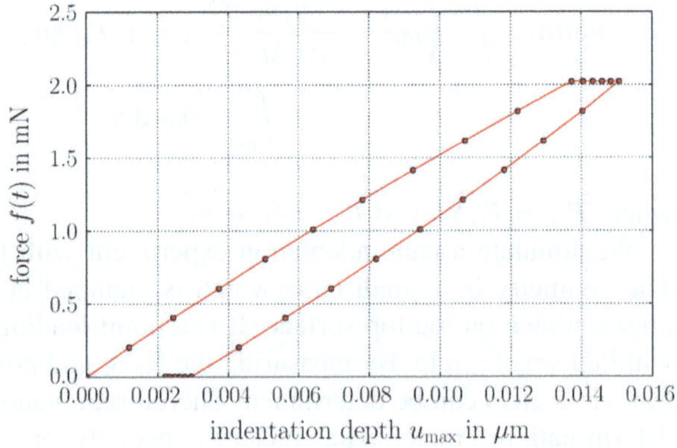

Fig. 1.7 The force to indentation depth from the simulation is often collected as the output of a typical nanoindentation experiment

an iterative solver with a preconditioner. A nanoindentation machine is applying a force until a specific indented depth is arrived and then the machine holds this particular force and measures the deformation under constant stress.[44] The viscous part is responsible for this behavior and upon unloading the material gets back to its initial deformation. This process can be investigated in Fig. 1.7 as the force over indentation depth. In reality the contact area between the indenter and the sample changes, $A = 24.5u^2_{\text{max.}}$, holds for a BERKOVICH indenter. The unloading at the end happens quicker than the material response. Hence the curve does not goes back to the zero displacement (initial) state. After reaching the zero force, the simulation goes on for a couple of time steps in order to visualize how the deformed body tries to attain the initial state. When we wait long enough, depending on the material parameters, then the curve reaches the origin. This is indeed the visco-elasticity. The term elasticity means exactly that phenomenon: upon unloading, sooner or later, the body reaches its initial state. The code is given below in two parts. The first part below creates the geometry and mesh.

[44]This phenomenon is called *creep* in the nanoindentation test, however, creep is a plastic deformation under constant stress such that we omit to use this term in this section.

```
"""Computational reality 04, mesh generation for a nanoindentation simluation."""
__author__ = "B. Emek Abali"
__license__ = "GNU LGPL Version 3.0 or later"
#This code underlies the GNU General Public License, http://www.gnu.org/licenses/gpl-3.0.en.html

from fenics import *
xL=20.0 #micrometer
yL=20.0
zL=20.0
# origin is on the middle of top surface
mesh = BoxMesh(Point(-xL/2, -yL/2, -zL), Point(xL/2, yL/2, 0), 30,30,30)
center=Point(0,0,0)
def refine_interval(radius,mesh):
    markers = MeshFunction("bool", mesh, 3)
    markers.set_all(False)
    # Mark cells for refinement
    for cell in cells(mesh):
        if cell.midpoint().distance(center) < radius:
            markers[cell.index()] = True
    # Refine mesh
    mesh = refine(mesh, markers)
    print 'DOFs: ', mesh.num_vertices()*3
    return mesh

num_refinements = 2
refine_radius1 = 2.0
refine_radius2 = 0.5
refine_radius3 = 0.1
refine_radius4 = 0.05
for i in range(num_refinements):
    mesh=refine_interval(refine_radius1,mesh)
    mesh=refine_interval(refine_radius2,mesh)
    mesh=refine_interval(refine_radius3,mesh)
    mesh=refine_interval(refine_radius4,mesh)
    plot(mesh, title=("Mesh %d" % (i + 1)))

print mesh
pwd='/calcul/CR04/'
meshfile = File(pwd+'mesh.xml')
meshfile << mesh
```

After generating an adequate mesh with necessary refinements around the needle of the indent, the computation is done by using the code below.

```
"""Computational reality 04, simulation of an nanoindentation modeled with a Zener model."""
__author__ = "B. Emek Abali"
__license__ = "GNU GPL Version 3.0 or later"
#This code underlies the GNU General Public License, http://www.gnu.org/licenses/gpl-3.0.en.html

from fenics import *
set_log_level(ERROR)
pwd='/calcul/CR04/'
mesh = Mesh(pwd+'mesh.xml')
mesh.order()
D = mesh.topology().dim()
xL=20.0 #micrometer
yL=20.0
zL=20.0

V = VectorFunctionSpace(mesh, 'P', 1)
T = TensorFunctionSpace(mesh, 'P', 1)

left = CompiledSubDomain('near(x[0],l) && on_boundary',l=-xL/2)
right = CompiledSubDomain('near(x[0],l) && on_boundary',l=xL/2)
back = CompiledSubDomain('near(x[1],l) && on_boundary',l=-yL/2)
front= CompiledSubDomain('near(x[1],l) && on_boundary',l=yL/2)
bottom = CompiledSubDomain('near(x[2],l) && on_boundary',l=-zL)
top = CompiledSubDomain('x[2]==0.0 && x[0]*x[0]+x[1]*x[1] <= r*r',r=0)

cells = CellFunction('size_t', mesh)
facets = FacetFunction('size_t', mesh)
dA = Measure('ds', domain=mesh, subdomain_data=facets)
dV = Measure('dx', domain=mesh, subdomain_data=cells)

facets.set_all(0)

nanoindent = Expression(('0.0','0.0','-Amp'),Amp=1.)

bc = [DirichletBC(V, (0.,0.,0.), bottom)]
f_gr = Constant((0.,0.,0.))

du = TrialFunction(V)
delu = TestFunction(V)
u00 = Function(V)
u0 = Function(V)
u = Function(V)
S=Function(T)
S0=Function(T)

print 'initializing, time in ms'
```

```
t = 0.0
tend = 3000.0
dt = 100.

init = Expression(('0','0','0'))
u.interpolate(init)
u0.assign(u)
u00.assign(u0)

print 'initializing , space'
rho0 = 9000.0E-15 #kg/mikrometer^3
lambada = 90.0 # mN/mikrometer^2 (=GPa)
E1, E2 = 200.0, 200.0 # mN/mikrometer^2 (=GPa)
mu = 2.0E5 #mN ms / mikrometer^2 (=N s/mm^2)

i, j, k, r = indices(4)
delta = Identity(3)

F = as_tensor(delta[i,j]+u[i].dx(j), [i,j])

eps= as_tensor(1.0/2.0*(u[i].dx(k)+u[k].dx(i)), [i,k])
eps0= as_tensor(1.0/2.0*(u0[i].dx(k)+u0[k].dx(i)), [i,k])

eps_dev= as_tensor(eps[i,j]-eps[k,k]*1.0/3.0*delta[i,j], [i,j])
eps0_dev= as_tensor(eps0[i,j]-eps0[k,k]*1.0/3.0*delta[i,j], [i,j])

devS0= as_tensor(S0[i,j]-S0[k,k]*1.0/3.0*delta[i,j], [i,j])

S = as_tensor(lambada*eps[j,j]*delta[i,k] + mu/(E2*dt+mu)*devS0[i,k] \
+ E1*E2*dt/(E2*dt+mu)*eps_dev[i,k] + mu*(E1+E2)/(E2*dt+mu)*(eps_dev[i,k]-eps0_dev[i,k]), [i,k])

N = FacetNormal(mesh)

Form = (rho0/dt/dt*(u-2.*u0+u00)[i]*delu[i] + F[i,k]*S[r,k]*delu[i].dx(r) - \
rho0*f_gr[i]*delu[i])*dV - nanoindent[i]*delu[i]*dA(1)

nz = as_tensor([0.0,0.0,1.0])
forceZ = as_tensor(F[i,k]*S[r,k]*nz[r],[i,])

Gain = derivative(Form, u, du)

file_u = File(pwd+'displacement.pvd')

#file_list = file('list4.xml','w')
u_max = 0.1
A = 24.5*u_max**2
radius = sqrt(A/pi)
top.r =radius
top.mark(facets, 1)
```

```

time_values=[0.0]
u_max_values=[0.0]
forceZ_values=[0.0]
import matplotlib as mpl
mpl.use('Agg')
import matplotlib.pyplot as pylab
pylab.rc('text', usetex=True )
pylab.rc('font', family='serif', serif='cm', size=30 )
pylab.rc('legend', fontsize=30)
pylab.rc(('xtick.major','ytick.major'), pad=15)

#pylab.ion()
fig = pylab.figure(1, figsize=(12,8))
fig.clf()
pylab.subplots_adjust(bottom=0.18)
pylab.subplots_adjust(left=0.16)
pylab.xlabel(r'indentation depth $u_{\mathrm{max}}$ in $\mu$m')
pylab.ylabel(r'force $f(t)$ in mN')
pylab.grid(True)

tic()
while t<tend:
    t += dt
    print 'time: ', t, ' indent: ', u_max, ' in ',toc(),' seconds'
    tic()
    if t<1000.: nanoindent.Amp = 0.002*t/A
    if t>=1000. and t<=1500.: nanoindent.Amp = 2./A
    if t>1500.: nanoindent.Amp = (2.-0.002*(t-1500.))/A
    if t>=2500.:nanoindent.Amp = 0.

    solve(Form==0, u, bc, J=Gain, \
        solver_parameters={"newton_solver":{"linear_solver": "cg", "preconditioner": "hypre_amg", "relative_tolerance": 1E-2, "absolute_tolerance": 1E-5, "maximum_iterations": 30} }, \
        form_compiler_parameters={"cpp_optimize": True, "representation": "quadrature", "quadrature_degree": 2} )

    file_u << (u,t)
    time_values.append(t)
    u_max = abs(u((0,0,0))[2])
    #if u_max<0.02: u_max=0.02
    u_max_values.append(u_max)
    fZ = project(forceZ,V)
    fZvalue = abs(fZ((0,0,0))[2])*A
    forceZ_values.append(fZvalue)
    #print time_values,u_max_values,forceZ_values
    S0.assign(project(S,T))
    u00.assign(u0)
    u0.assign(u)
```

```
        #file_list.write(" time=\"%s\" u(0,0,0)=\"%s\" force
            ↪ (0,0,0)=\"%s\" \n" % (t, u_max, fZvalue))
        pylab.plot(u_max_values, forceZ_values, 'ro-')
        pylab.savefig(pwd+'CR04_nanoindent.pdf')

#file_list.close()
```

To-do

We have included *time* into our computational reality, i.e., from statics we have "upgraded" it to dynamics. For linear rheology there exists many different material models:

- Find couple of different rheological constitutive equations with their corresponding spring/dashpot models.
- Try to draw the ZENER model and obtain Eq. (1.148).
- Rewrite the code for a one-axial tensile testing, plot stress-strain hysteresis plot, and vary the material parameters in order to investigate their effects.
- Try to determine all units, especially for mass density and time. Recall that the numerical computation fails to have a unit system. All units need to be consistent with each other, no matter which system we are using.

1.5 Fractional Rheological Materials

In linear rheology stress depends on strain, strain rate, and stress rate. The first rate of stress is defined

$$\dot{\sigma}_{ij} = \frac{\mathrm{d}^1\sigma_{ij}}{\mathrm{d}t^1} , \tag{1.154}$$

in a fixed Cartesian coordinate system, where the number "1" denoting the first rate is superfluous and normally omitted. The second rate of stress reads

$$\ddot{\sigma}_{ij} = \frac{\mathrm{d}^2\sigma_{ij}}{\mathrm{d}t^2} . \tag{1.155}$$

Integers are used for time derivatives. We have an intuition for derivatives via integers. However, formally, we can define *fractional* time derivatives:

$$\frac{\mathrm{d}^\alpha \sigma_{ij}}{\mathrm{d}t^\alpha} , \tag{1.156}$$

where α is a positive real number. As we have seen in the last section, we can build rheological models by using the volumetric and deviatoric part of stress and strain. For simplicity we assume that the material is *isochoric*, i.e., volume preserving such that $\varepsilon_{kk} = 0$. Then the material equation containing first rates becomes

$$\sigma_{ij} + c_1 \frac{\mathrm{d}\sigma_{ij}}{\mathrm{d}t} = c_2 \varepsilon_{ij} + c_3 \frac{\mathrm{d}\varepsilon_{ij}}{\mathrm{d}t} . \tag{1.157}$$

In this section we ignore material and geometric nonlinearities. Technically, we approximate $F_{ij} \approx \delta_{ij}$, thus $J \approx 1$, hence $\sigma_{ij} \approx S_{ij}$. Furthermore, we use $\varepsilon_{ij} = u_{(i,j)}$ instead of E_{ij}. Since real numbers can be used instead of integers in fractional time derivatives, we can generalize the latter constitutive equation as follows

$$\sigma_{ij} + c_1^\alpha \frac{\mathrm{d}^\alpha \sigma_{ij}}{\mathrm{d}t^\alpha} = c_2 \varepsilon_{ij} + c_3^\gamma \frac{\mathrm{d}^\gamma \varepsilon_{ij}}{\mathrm{d}t^\gamma} , \tag{1.158}$$

where α and γ are positive real numbers. We can rewrite the latter equation:

$$\sigma_{ij} + \tau_0^\alpha \frac{\mathrm{d}^\alpha \sigma_{ij}}{\mathrm{d}t^\alpha} = G_e \Big(\varepsilon_{ij} + \tau_0^\alpha \frac{\mathrm{d}^\alpha \varepsilon_{ij}}{\mathrm{d}t^\alpha} \Big) + G_0 \tau_0^\beta \frac{\mathrm{d}^\beta \varepsilon_{ij}}{\mathrm{d}t^\beta} , \tag{1.159}$$

with α and β denoting the fractional differentiation. The material parameters are τ_0, G_e, G_0. Different well-known material models can be deduced from the latter constitutive equation by choosing specific α and β values. We can choose $\alpha = \beta = 0$ and acquire the linear elastic material equation

$$\sigma_{ij} = \big(G_e + \frac{G_0}{2} \big) \varepsilon_{ij} , \tag{1.160}$$

for an isochoric material. We can choose $\alpha = 0$, $\beta = 1$ and it reduces to the linear viscoelastic material equation

$$\sigma_{ij} = G_e \varepsilon_{ij} + \frac{G_0 \tau_0}{2} \frac{\mathrm{d}\varepsilon_{ij}}{\mathrm{d}t} . \tag{1.161}$$

Another choice is $\alpha = 1$ and $\beta = 0$ and in this case we obtain the linear rheological model called the ZENER model in the last section

$$\sigma_{ij} + \tau_0 \frac{\mathrm{d}\sigma_{ij}}{\mathrm{d}t} = (G_e + G_0)\varepsilon_{ij} + G_e \tau_0 \frac{\mathrm{d}\varepsilon_{ij}}{\mathrm{d}t} . \tag{1.162}$$

Obviously Eq. (1.159) is a general linear material equation. The values of material parameters, viz., G_0, G_e, τ_0, α, β, can be determined by means of experiments.[45]

The process of repeated differentiation and integration is represented by the notation:

$$\frac{\mathrm{d}^n}{\mathrm{d}t^n} \text{ and } \int \ldots \int \mathrm{d}t_1 \ldots . \mathrm{d}t_n . \tag{1.163}$$

For an extended version of these operators we choose n as a positive real number.

Actually, the idea is a very old one to employ real numbers instead of integers for a derivative. In a letter[46] dated September 30, 1695, Guillaume François Antoine de L'Hôpital wrote to Gottfried Wilhelm Leibniz asking him about a particular notation he had used in his publications for the nth-derivative of the linear function $f(x) = x$ and $\mathrm{d}^n f/\mathrm{d}x^n$. L'Hôpital posed the question to Leibniz, what the result would be if $n = 0.5$. Leibniz's response: "An apparent paradox, from which one day useful consequences will be drawn." In these words fractional calculus was born.

There are different definitions for a fractional derivative. Mathematicians prefer to use the RIEMANN-LOUVILLE definition[47] of the fractional derivative, however, we will use the GRUNWALD-LETNIKOV definition[48] as it is easier to implement in numerical calculations. The two definitions are equivalent.[49] The GRUNWALD-LETNIKOV definition of the fractional derivative reads

$$\frac{\mathrm{d}^\alpha f}{\mathrm{d}t^\alpha} = \lim_{h \to 0} \frac{1}{h^\alpha} \sum_{m=0}^{t/h} (-1)^m \frac{\Gamma|_{(\alpha+1)}}{m! \, \Gamma|_{(\alpha-m+1)}} \, f|_{(t-mh)} , \tag{1.164}$$

where h is the time step and the so-called *gamma* function is

$$\Gamma\Big|_x = \Gamma(x) = \int_0^\infty \exp(-t) t^{(x-1)} \mathrm{d}t . \tag{1.165}$$

---

45 See [9] for such experiments.

46 This historical note is taken from [11].

47 It is named after Georg Friedrich Bernhard Riemann and Joseph Liouville.

48 It is named for Anton Karl Grünwald and Alexey Vasilievich Letnikov.

49 See [18] for these definitions and their equivalence.

The fractional derivative utilizes a time step, h, which we can exchange with Δt for the time discretization in our computation, as exactly we have done it for the first and second derivatives in the last section. For the left hand side of the constitutive Eq. (1.159) we apply the fractional derivative

$$\begin{aligned}
&\sigma_{ij} + \tau_0^\alpha \sum_{m=0}^{t/\Delta t} \frac{1}{\Delta t^\alpha} (-1)^m \frac{\Gamma|_{(\alpha+1)}}{m!\,\Gamma|_{(\alpha-m+1)}} \sigma_{ij}\big|_{(t-m\Delta t)} = \\
&= \sigma_{ij} + \frac{\tau_0^\alpha \sigma_{ij}}{\Delta t^\alpha} + \tau_0^\alpha \sum_{m=1}^{t/\Delta t} \frac{1}{\Delta t^\alpha} (-1)^m \frac{\Gamma|_{(\alpha+1)}}{m!\,\Gamma|_{(\alpha-m+1)}} \sigma_{ij}\big|_{(t-m\Delta t)} = \\
&= \sigma_{ij}\Big(1 + \frac{\tau_0^\alpha}{\Delta t^\alpha}\Big) + \tau_0^\alpha \sum_{m=1}^{t/\Delta t} \frac{1}{\Delta t^\alpha} (-1)^m \frac{\Gamma|_{(\alpha+1)}}{m!\,\Gamma|_{(\alpha-m+1)}} \sigma_{ij}\big|_{(t-m\Delta t)} \,.
\end{aligned} \tag{1.166}$$

By using the latter we express the stress tensor:

$$\begin{aligned}
\sigma_{ij} = \frac{1}{1 + \tau_0^\alpha/\Delta t^\alpha} \Big(& - \tau_0^\alpha \sum_{m=1}^{t/\Delta t} \frac{1}{\Delta t^\alpha} (-1)^m \frac{\Gamma|_{(\alpha+1)}}{m!\,\Gamma|_{(\alpha-m+1)}} \sigma_{ij}\big|_{(t-m\Delta t)} + \\
& + G_e \Big(\varepsilon_{ij} + \tau_0^\alpha \frac{\mathrm{d}^\alpha \varepsilon_{ij}}{\mathrm{d}t^\alpha} \Big) + G_0 \tau_0^\beta \frac{\mathrm{d}^\beta \varepsilon_{ij}}{\mathrm{d}t^\beta} \Big) \,,
\end{aligned} \tag{1.167}$$

where

$$\frac{\mathrm{d}^a \varepsilon_{ij}}{\mathrm{d}t^a} = \frac{1}{\Delta t^a} \sum_{m=0}^{t/\Delta t} (-1)^m \frac{\Gamma|_{(a+1)}}{m!\,\Gamma|_{(a-m+1)}} \varepsilon_{ij}\big|_{(t-m\Delta t)} \,, \quad a = \alpha, \beta \,. \tag{1.168}$$

The weak form remains the same, we recall that we have simplified the problem by neglecting the geometric nonlinearities (large deformations), $F_{ij} \approx \delta_{ij}$, $J \approx 1$, $\sigma_{ij} \approx S_{ij}$, $\varepsilon_{ij} = u_{(i,j)}$ instead of E_{ij}. In this constellation the form reads

$$\begin{aligned}
\text{Form} = \int_{\mathcal{B}_0} \Big(\rho_0 \frac{u_i - 2u_i^0 + u_i^{00}}{\Delta t \Delta t} \delta u_i + \sigma_{ji}\, \delta u_{i,j} - \rho_0 f_i\, \delta u_i \Big) \mathrm{d}V - \\
- \int_{\partial \mathcal{B}_0^{\mathrm{N}}} \hat{t}_i\, \delta u_i \mathrm{d}A \,.
\end{aligned} \tag{1.169}$$

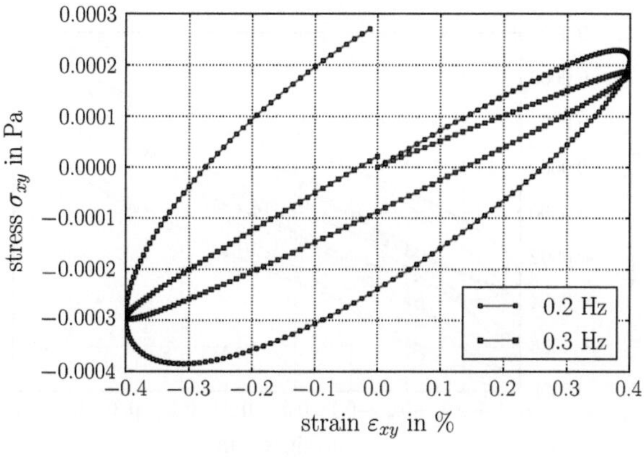

Fig. 1.8 Hysteresis plot of polycarbonate under shear loading for different loading rates

Consider a polycarbonate[50] typically used for CDs, DVDs, BluRays, and also in Nokia N9's case, IPhone 5c's case, or Samsung Galaxy III's battery cover. It is a transparent, hard, and impact resistant thermoplastic polymer with the following materials data:

$$\begin{aligned} \alpha &= 0.844\ , \quad \beta = 0.844 - 0.364\ , \quad \tau_0 = 214\,700\,\mathrm{s}\ , \\ G_0 &= 2.152 \cdot 10^2\,\mathrm{Pa}\ , \quad G_e = 1.185 \cdot 10^4\,\mathrm{Pa}\ , \quad \rho_0 = 1200\,\mathrm{kg/m^3}\ . \end{aligned} \tag{1.170}$$

We shear a simple 2D geometry out of Makrolon M2200 sinusoidally and plot the hysteresis plot for two different loading rates. The rate-dependency of the material can be observed in Fig. 1.8. This behavior is due to the viscous character since the model includes a strain rate and stress rate dependency. The rate dependency is applied by fractional time rates. The hysteresis curve in Fig. 1.8 is typical for a viscoelastic response subject to a cyclic loading. The enclosed area denotes the dissipated energy due to the viscous behavior. Under a slower loading rate, the material possesses more time to respond viscously and dissipates more energy. If the loading rate is so high that the response time is larger than a complete cycle's period, then the material shows an elastic behavior.

In order to comprehend the effect of the fractional time rate, we apply a harmonic loading for many cycles, see Fig. 1.9. In the case of integer rates we obtain a steady-state in the hysteresis plot after couple of cycles. The integer rate dependency

[50] Makrolon M2200 is a commercial polycarbonate manufactured by Bayer, for the material parameters with 5 % multi-wall carbon nanotubes from Baytubes, see [10].

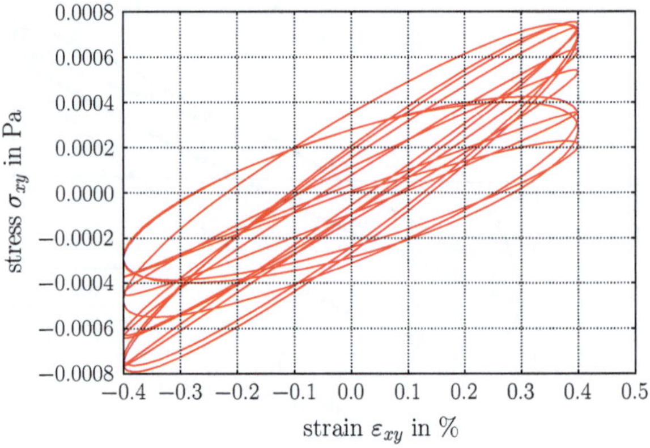

Fig. 1.9 Hysteresis plot of polycarbonate under shear loading for the loading rate 0.1 Hz and 10 cycles

represents a limited *memory* of the material. In other words, the deformation in history affects the stress and this history is implemented by using a rate dependency.[51] For a viscoelastic model with first rates of stress and strain, we have a short memory in the material such that after one or two cycles we expect the steady-state. For a second rate model, more history is involved such that we may expect after four or five cycles. These are all rough estimations in other to explain the effect of a fractional time rate. In the definition of the fractional time rate we incorporate the *whole* history. Owing to the gamma function the effects diminish exponentially, concretely the parameter τ_0 is controlling how many seconds of history is involved in. The used material has a huge τ_0 parameter with respect to the total simulation time. Therefore, the steady-state does not occur even in the first 10 cycles. Therefore, the fractional time rate facilitates a history dependency by using a material parameter. In a model with integer time rates, the materials dependence on history is limited by the highest number of the time rate. For computation we use an efficient implementation of the gamma function from SciPy packages. A simple two-dimensional rectangle is used as the geometry, where right and left boundaries are handled as periodic boundaries. This boundary condition has the physical interpretation of a material much wider than the geometry used in the computation. The following code has been used for simulations.

[51] Mathematically, we should expand in time by using a TAYLOR expansion up to nth integer rate. Then the history is involved in for a limited interval of time given by nth derivative.

```
"""Computational reality 05, linear rheology with fractional time rates"""
__author__ = "B. Emek Abali"
__license__ = "GNU GPL Version 3.0 or later"
#This code underlies the GNU General Public License, http://www.gnu.org/licenses/gpl-3.0.en.html

from fenics import *
import scipy
from scipy.special import gamma
from scipy.misc import factorial
import numpy as np
set_log_level(ERROR)
pwd='/calcul/CR05/'

xlength =0.100 #[m]
ylength =0.025 #[m]
mesh = RectangleMesh(Point(0,0),Point(xlength,ylength),20,5)

class PeriodicBoundary(SubDomain):
    def inside(self, x, on_boundary):
        return near(x[0],0) and on_boundary
    def map(self, y, x):
        #this maps right side x[0]=xlength to the left
        x[0] = y[0]-xlength
        x[1] = y[1]

PeriodicBC = PeriodicBoundary()

Space = VectorFunctionSpace(mesh,'P',1, constrained_domain=PeriodicBC)
TensorSpace = TensorFunctionSpace(mesh,'P',1, constrained_domain=PeriodicBC)

delu = TestFunction(Space)
du = TrialFunction(Space)
u = Function(Space)
u0 = Function(Space)
u00 = Function(Space)

print 'initializing,time'
t_start=0.0
t_end = 10.0

# a class in dolfin saves the data (binary) in time
stress_history = TimeSeries(mesh.mpi_comm(), pwd+'hist/stressHist')
strain_history = TimeSeries(mesh.mpi_comm(), pwd+'hist/strainHist')
stress_history.parameters["clear_on_write"] = False
strain_history.parameters["clear_on_write"] = False

#setting initial conditions
initialcond = Expression(('0','0'))
```

```
u0.interpolate(initialcond)
u00.assign(u0)

#Defining boundary conditions
left = CompiledSubDomain('near(x[0],0) && on_boundary')
right = CompiledSubDomain('near(x[0],l) && on_boundary',l=xlength)
bottom = CompiledSubDomain('near(x[1],0) && on_boundary')
top = CompiledSubDomain('near(x[1],l) && on_boundary',l=ylength)

cells = CellFunction('size_t', mesh)
facets = FacetFunction('size_t', mesh)
dA = Measure('ds', domain=mesh, subdomain_data=facets)
dV = Measure('dx', domain=mesh, subdomain_data=cells)

facets.set_all(0)
# a cyclic shear on top
shear = Expression(('0.0001*sin(2.*pi*f*time)','0.0'),f=0,time=0)
bc1 = DirichletBC(Space,shear,top)
bc2 = DirichletBC(Space,(0.0,0.0),bottom)
bc = [bc1,bc2]

def fractional(index,arg,t,h,power):
    m=index+1
    temp=as_tensor([[0.,0.],[0.,0.]])
    while m <= t/h:
        hist=int(t/h-m)
        tensor=Function(TensorSpace)
        arg.retrieve(tensor.vector(),hist)
        temp += as_tensor((-1.)**m*gamma(power+1.)\
        /gamma(power-m+1.)/factorial(m)*tensor[i,j]\
        /(h**power),[i,j])
        m += 1
    return temp

file_u = File (pwd+'displacements.pvd')

#Plotting stress v/s strain curves
import matplotlib as mpl
mpl.use('Agg')
import matplotlib.pyplot as pylab
pylab.rc('text', usetex=True )
pylab.rc('font', family='serif', serif='cm', size=30 )
pylab.rc('legend', fontsize=30)
pylab.rc(('xtick.major','ytick.major'), pad=15)

def compute(freq,loops):
    t=t_start
    dt = 0.005/freq
    #Setting up the parameters in kg, s, N, m, Pa
    alpha = 0.844
    beta = 0.844-0.364
```

```
    tau0 = 214700.
    G0 = 2.152E2
    Ge = 1.185E4
    rho0 = 1200. #[kg/m3]
    # index notation
    i,j,k,l = indices(4)
    delta= Identity(2)
    epsilon= as_tensor(1.0/2.0*(u[i].dx(j)+u[j].dx(i)),(i,j))
    epsilon0= as_tensor(1.0/2.0*(u0[i].dx(j)+u0[j].dx(i)),(i,j))

    sigma = as_tensor( 1.0/(1.0+ tau0**alpha/dt**alpha) * ( \
        - tau0**alpha*fractional(1,stress_history,t,dt,alpha)[i,j] \
        + Ge*( epsilon[i,j] \
        + tau0**alpha*fractional(0,strain_history,t,dt,alpha)[i,j] ) \
        + G0*tau0**beta*fractional(0,strain_history,t,dt,beta)[i,j]) \
        , (i,j))
    f= Constant((0.0,0.0))

    Form = (rho0*(u[i]-2.*u0[i]+u00[i])/(dt*dt)*delu[i] \
        + sigma[j,i]*delu[i].dx(j) \
        - rho0*f[i]*delu[i] )*dx
    Gain = derivative(Form,u,du)

    stresses = []
    strains = []
    temp_array= []
    time= []
    stress_history.clear()
    strain_history.clear()
    u.interpolate(initialcond)
    u0.assign(u)
    u00.assign(u0)
    shear.f = freq
    tend=loops/freq

    while t<=tend:
        shear.time = t
        #print 'time: ', t
        solve(Form==0, u, bc, J=Gain, \
            solver_parameters={"newton_solver":{"linear_solver": "lu", "relative_tolerance": 1e-4} }, \
            form_compiler_parameters={"cpp_optimize": True, "representation": "quadrature", "quadrature_degree": 2} )
        if t==0.25/freq : file_u <<(u,t)
        if t==0.75/freq : file_u <<(u,t)
        u00.assign(u0)
        u0.assign(u)
        actualStress = project(sigma, TensorSpace)
```

```
            stress_history.store(actualStress.vector(),t)
            actualStrain = project(epsilon,TensorSpace)
            strain_history.store(actualStrain.vector(),t)
            P=(xlength/2.,ylength)
            sigma12 = actualStress(P)[1]
            epsilon12 = u(P)[0]/ylength
            stresses.append(sigma12) #in Pa
            strains.append(epsilon12*100.) #in %
            t = t + dt

    return stresses, strains


fig = pylab.figure(1, figsize=(12,8))
pylab.subplots_adjust(bottom=0.15)
pylab.subplots_adjust(left=0.20)
pylab.xlabel(r'strain $\varepsilon_{xy}$ in $\%$')
pylab.ylabel(r'stress $\sigma_{xy}$ in Pa')
pylab.grid(True)

tic()
freq = 0.2
stresses, strains = compute(freq,1)
#np.save('01Hz_stress.npy', stresses)
#np.save('01Hz_strain.npy', strains)
#stresses=np.load('01Hz_stress.npy')
#strains=np.load('01Hz_strain.npy')
pylab.plot(strains,stresses,color='red',marker='o',markersize
    ↪ =5,label='%(0).1f Hz'%{'0':freq})
pylab.savefig(pwd+'CompReal05_hysteresis1.pdf')
print 'it took ',toc(),' seconds'
tic()
freq=0.3
stresses, strains = compute(freq,1)
pylab.plot(strains,stresses,color='blue',marker='s',
    ↪ markersize=5,label='%(0).1f Hz'%{'0':freq})
pylab.legend(loc='best')
pylab.savefig(pwd+'CompReal05_hysteresis1.pdf')
print 'it took ',toc(),' seconds'

pylab.cla()
pylab.clf()
fig = pylab.figure(2, figsize=(12,8))
pylab.subplots_adjust(bottom=0.15)
pylab.subplots_adjust(left=0.20)
pylab.xlabel(r'strain $\varepsilon_{xy}$ in $\%$')
pylab.ylabel(r'stress $\sigma_{xy}$ in Pa')
pylab.grid(True)
tic()
freq = 0.2
stresses, strains = compute(freq,10)
pylab.plot(strains,stresses,'r-')
pylab.savefig(pwd+'CompReal05_hysteresis2.pdf')
print 'it took ',toc(),' seconds'
```

To-do

In order to obtain a better understanding, vary the parameters α, β and compare the stress-strain hysteresis:

- Try first to guess the stress-strain plot for $\alpha = \beta = 0$ and then apply.
- What is the proper name of the rheological model in case of setting $\alpha = 0, \beta = 1$? In this model, would the strain affect the hysteresis curve?
- Try to explain the case of $\alpha = 0, \beta = 2$.
- What is the effect of setting $\alpha \neq 0$?

In order to gain a perception of the implemented periodic boundary condition, try to redo the simulation by removing this condition and using a free surface condition on the boundary. Then double or triple the length in x-direction for eliminating the boundary effects and compare the solution to the one with the periodic boundary condition.

1.6 Associated Plasticity

Particles in their initial positions, X_i, move and displace as a consequence of a mechanical loading. This displacement, u_i, for every particle, X_i, at the current time, t, is a function in space and time, $u_i = u_i(X_j, t)$. We compute the displacement with the balance of linear momentum augmented by the constitutive equation. The balance of linear momentum possesses stress. The constitutive equation relates stress to displacements over strains. This connection is a mathematical equation, every stress value is related to a unique strain value. For example zero stress is related to zero strain. If we compute a loading and unloading scenario—stress increases and then decreases—the particles move under loading and move back to their initial positions after unloading. Before loading, at zero stress, no deformation occurs (zero strain). After loading and unloading, at zero stress, zero strain has to occur again. In other words, the process is reversible and the displacements are recoverable.

In the so-called *elastic* behavior, the displacement vanishes after unloading. The process is reversible and for several cases it is admissible. For engineering materials like steel, copper, magnesium, and aluminum, the admissible strains are less than $0.2\,\% = 0.002$. Above this threshold a *plastic* deformation starts occurring such that after unloading some of the displacements remain in the continuum body. The process is not reversible; some of displacements are recovered, not all. During this plastic deformation the material behavior changes, too. We need different material models for elasticity and plasticity. Hence, we need to distinguish between elastic and plastic *regimes*.

We start with elasticity. As discussed in Sect. 1.4, stress tensor consists of volumetric and deviatoric parts. We can motivate this decomposition for small strains by using another argumentation. Consider a cubic body expressed in Cartesian coordinates with its origin in one corner. Lengths of its sizes are simply identical to unit vectors of the coordinate system. The volume reads

$$V = X_1 X_2 X_3 = 1 \,. \tag{1.171}$$

Suppose that its length changes due to a mechanical loading. The displacement, u_1, u_2, u_3 along X_1, X_2, X_3, respectively, can be used to calculate the volumetric change:

$$\begin{aligned} V + \Delta V = (X_1 + u_1)(X_2 + u_2)(X_3 + u_3) = \\ = X_1X_2X_3 + X_1X_2u_3 + X_1u_2X_3 + X_1u_2u_3 + u_1X_2X_3 + u_1u_2X_3 + u_1u_2u_3 \,. \end{aligned} \tag{1.172}$$

By multiplying the latter by $V/X_1X_2X_3 = 1$ and then neglecting the nonlinear terms, viz.,

$$\begin{aligned} \frac{u_2u_3}{X_2X_3} = \varepsilon_{22}\varepsilon_{33} \le 0.002^2 \approx 0 \,, \quad \frac{u_1u_2}{X_1X_2} = \varepsilon_{11}\varepsilon_{22} \le 0.002^2 \approx 0 \,, \\ \frac{u_1u_2u_3}{X_1X_2X_3} = \varepsilon_{11}\varepsilon_{22}\varepsilon_{33} \le 0.002^3 \approx 0 \,, \end{aligned} \tag{1.173}$$

since small strains (smaller than 0.002) occur in the elastic regime, we obtain

$$\begin{aligned} V + \Delta V = V + \frac{u_3}{X_3}V + \frac{u_2}{X_2}V + \frac{u_1}{X_1}V \\ \frac{\Delta V}{V} = \frac{u_3}{X_3} + \frac{u_2}{X_2} + \frac{u_1}{X_1} = \varepsilon_{kk} \,. \end{aligned} \tag{1.174}$$

We can use a simplified notation:

$$\varepsilon_{kk} = 3e \,, \tag{1.175}$$

where the parameter e is simply the measure of the volumetric change (dilation) in Cartesian coordinates. The deviatoric strains are responsible for a distortion without dilation

$$\varepsilon_{|ij|} = \varepsilon_{ij} - e\delta_{ij} \,. \tag{1.176}$$

Employing HOOKE's law for isotropic materials, we obtain the linear relation between the symmetric CAUCHY stress and symmetric strain

$$\sigma_{ij} = c_1 e\delta_{ij} + c_2\varepsilon_{|ij|} = \lambda\delta_{ij}\varepsilon_{kk} + 2\mu\varepsilon_{ij} \,. \tag{1.177}$$

This tensor equation of rank two has to hold in its lower ranks, too. We can reduce the rank by contracting indices, for a Cartesian coordinate system we multiply by the KRONECKER delta[52]

$$\delta_{ij}\sigma_{ij} = \sigma_{ii} = \lambda\delta_{ii}3e + 2\mu\varepsilon_{ii} = \lambda 9e + 2\mu 3e \,, \tag{1.178}$$

and introduce a simplified notation:

$$\sigma_{ii} = 3s \Rightarrow s = (3\lambda + 2\mu)e \,. \tag{1.179}$$

Another bulk quantity, s, for stress has been used, the deviatoric part reads

$$\sigma_{|ij|} = \sigma_{ij} - s\delta_{ij} \,. \tag{1.180}$$

Finally, we observe a simple relation

$$\begin{aligned} \sigma_{ij} &= \lambda\delta_{ij}\varepsilon_{kk} + 2\mu\varepsilon_{ij} \,, \\ \sigma_{|ij|} + s\delta_{ij} &= \lambda\delta^{ij}3e + 2\mu\big(\varepsilon_{|ij|} + e\delta_{ij}\big) \,, \\ \sigma_{|ij|} + (3\lambda + 2\mu)e\delta_{ij} &= \lambda\delta_{ij}3e + 2\mu e\delta_{ij} + 2\mu\varepsilon_{|ij|} \,, \\ \sigma_{|ij|} &= 2\mu\varepsilon_{|ij|} \,. \end{aligned} \tag{1.181}$$

In other words, under the assumption of small strains, deviatoric and volumetric parts can be (additively) decomposed and related to each other separately. Indeed, we have seen this decomposition already by using the energy concept, however, herein we present the same result without using the notion of energy. The assumption of small strains in the elastic regime is adequate for engineering materials like steel, aluminum, magnesium, and copper.

By excessing the yield stress, σ_Y, body starts to flow with the velocity, $v_i = v_i(X_j, t)$, of particles X_i. Since velocity is the rate of displacement, we obtain

$$\frac{\partial v_{(i}}{\partial X^{j)}} = \frac{\partial^2 u_{(i}}{\partial X^{j)}\partial t} = \frac{\partial^2 u_{(i}}{\partial t \partial X^{j)}} = \frac{\partial \varepsilon_{ij}}{\partial t} = \varepsilon^{\bullet}_{ij} \,, \tag{1.182}$$

where $\varepsilon^{\bullet}_{ij}$ is the strain rate. The strain rate or equally the symmetric part of velocity gradient causes a viscous flow. If a *yield condition* is fulfilled such velocities occur. Hence this type of deformation is elasto-plastic and the flow of continuum body can be expressed by strain rate. In a simple tensile test, a loading above the yield stress causes a plastic deformation, which remains in the body after unloading. Thus, we can simply measure the elastic and plastic elongations. By dividing the elastic and plastic

[52]We lower the rank by contracting indices. In order to contract two indices we multiply by the metric tensor. KRONECKER delta is also the metric tensor in Cartesian coordinates.

elongations by the length of the beam, we obtain elastic and plastic strains.[53] In order to model the plastic behavior we need the strain rate or symmetric velocity gradients. According to the VON MISES yield criterion, the second invariant[54] of deviatoric part of the stress tensor should be greater than an experimentally determined quantity. We know from a tensile test that the material starts deforming plastically above the yield stress, σ_Y. This scalar value is representing the threshold of the plasticity. In a tensile test the stress tensor attains the yield stress in one component as the material starts yielding

$$\sigma_{ij} = \begin{pmatrix} \sigma_Y & 0 & 0 \\ 0 & 0 & 0 \\ 0 & 0 & 0 \end{pmatrix} . \tag{1.183}$$

Since we want to use the VON MISES yield criterion, we calculate the second invariant

$$\sigma_{|ij|} = \sigma_{ij} - s\delta_{ij} = \begin{pmatrix} \frac{2}{3}\sigma_Y & 0 & 0 \\ 0 & -\frac{1}{3}\sigma_Y & 0 \\ 0 & 0 & -\frac{1}{3}\sigma_Y \end{pmatrix} , \tag{1.184}$$

and relate it to the yield stress

$$\begin{aligned} \sigma_{|ij|}\sigma_{|ij|} &= \frac{6}{9}\sigma_Y^2 , \\ \sigma_Y &= \sqrt{\frac{3}{2}\sigma_{|ij|}\sigma_{|ij|}} , \end{aligned} \tag{1.185}$$

which is the yield criterion. The value of σ_Y is specific to the material. For any deformation we can calculate

$$\sigma_{eq} = \sqrt{\frac{3}{2}\sigma_{|ij|}\sigma_{|ij|}} , \tag{1.186}$$

and compare it to the yield stress obtained from the tensile test. When the plasticity starts, $\sigma_{eq} = \sigma_Y$, the yield criterion is fulfilled.

As a consequence of mechanical loading, a deformation occurs. Upon unloading, plastic part of the deformation remains in the body whereas the elastic part is recovered. Since we express the deformation by using the strain tensor, a simple approach of modeling such a behavior reads

$$\varepsilon_{ij} = {}^{e}\varepsilon_{ij} + {}^{p}\varepsilon_{ij} , \tag{1.187}$$

---

[53] This consideration has been used in [20, 22], therefore, the associated plasticity is also called PRANDTL–REUSS plasticity, see [15, Chap. 11]. The PRANDTL–REUSS plasticity is named for Ludwig Prandtl and András Reuß (Endre Reuss).

[54] There are three invariants in three-dimensional space of the stress tensor. The first invariant of stress is the bulk quantity $s = \sigma_{ii}$, the second invariant is $\sigma_{ij}\sigma_{ij}$ and the third invariant is, $\sigma_{ij}\sigma_{jk}\sigma_{ki}$.

where ${}^{\mathrm{e}}\varepsilon_{ij}$ denotes the elastic part of the strain tensor and ${}^{\mathrm{p}}\varepsilon_{ij}$ the plastic part. This additive decomposition of strain tensor is a phenomenological fact. For many engineering materials with small deformations and small strains the approach gives accurate results. For the elastic part of strains we have applied HOOKE's law:

$$\sigma_{ij} = C_{ijkl}\,{}^{\mathrm{e}}\varepsilon_{ij} = C_{ijkl}\big(\varepsilon_{kl} - {}^{\mathrm{p}}\varepsilon_{kl}\big) \,. \tag{1.188}$$

The plasticity is given by strain rate. Hence, the latter equation is rewritten

$$\sigma^{\bullet}_{ij} = C_{ijkl}\big(\varepsilon^{\bullet}_{kl} - {}^{\mathrm{p}}\varepsilon^{\bullet}_{kl}\big) \,, \tag{1.189}$$

since the stiffness tensor is constant in time. We need a constitutive relation for ${}^{\mathrm{p}}\varepsilon^{\bullet}_{ij}$. In a former section we have utilized a scalar function, stored energy, in order to define the stress as in Eq. (1.91). Stored energy has a first integral, i.e., it is a potential. The same concept is used for plasticity and we assume that a flow potential f exists, leading to

$${}^{\mathrm{p}}\varepsilon^{\bullet}_{ij} = \Lambda^{\bullet}\frac{\partial f}{\partial \sigma_{ij}} \,, \tag{1.190}$$

where we need a (positive) multiplier $\Lambda^{\bullet}$ since the plastic strain cannot be expressed with a first integral. In other words, the evolution of plastic strain is important, we cannot use the start and end states for calculating plastic strain. Therefore, a flow potential, f, fails to define the plastic strain and we need a multiplier. Both will be defined in the following by using the yield criterion. This approach is called *associated plasticity* in the literature.

1.6.1 Isotropic Hardening

We start by defining the flow potential, f. For many engineering materials, the VON MISES yield criterion is used in order to generate a function resulting in 0 in the case of plasticity

$$\begin{aligned} f &= \frac{1}{3}\sigma_{\mathrm{eq}}^2 - \frac{1}{3}k^2 \,, \\ f &= \frac{1}{2}\sigma_{|ij|}\sigma_{|ij|} - \frac{1}{3}k^2 \,. \end{aligned} \tag{1.191}$$

The value of k changes with respect to the plastic deformation. Consider a tensile test, the value of $k = \sigma_{\mathrm{Y}}$ in the elastic regime. Obviously, the flow potential is negative, $f < 0$. The axial force increases such that the equivalent stress approaches the yield stress and f goes to zero. At the moment, when the yield criterion is fulfilled, f vanishes and plasticity starts. If the force increases further, we would have a positive

f if the value of k remains as $k = \sigma_Y$. However, the value of k increases as the plasticity is occurring such that $f = 0$ as long as a plastic deformation is performed. The flow potential is zero in the plastic regime and negative in the elastic regime:

$$f \leq 0 \,, \quad f \leq \begin{cases} = 0 & \text{plastic regime} \\ < 0 & \text{elastic regime} \end{cases} . \tag{1.192}$$

Of course, we need to model k depending on the plastic deformation. For simplicity, consider a linear function in the plastic strain:

$$k = \sigma_Y + h\, {}^{\mathrm{p}}\varepsilon_{\mathrm{eq}} \,, \tag{1.193}$$

where ${}^{\mathrm{p}}\varepsilon_{\mathrm{eq}}$ denotes the equivalent plastic strain. This approach[55] is obviously the simplest case. Many engineering materials show such a simple *hardening* behavior. An explanation of this behavior is based on arising dislocations in case of plastic yielding, where the high density of dislocations slows down the plastic flow. From a phenomenological point of view, we observe in a tensile experiment a behavior as in Eq. (1.193) and model it by determining the material constants σ_Y and h without considering a microscopic reasoning. Since we determine the parameters from a tensile test we need to use the VON MISES equivalent stress and strain.

For example an AISI steel 1010 has the *initial* yield stress $\sigma_Y = 305\,\mathrm{MPa}$. This value remains constant. For the elastic regime the flow potential is below zero, $f < 0$, since the equivalent stress is smaller than $k = \sigma_Y$. When the loading causes an equivalent stress higher than σ_Y, the value of k increases such that $f = 0$ during the plasticity.[56] As we have seen in Eq. (1.193), k depends on the plastic strain and is independent on the stress. The flow potential depends on k and stress,

$$f = f(k, \sigma_{ij}) \,. \tag{1.194}$$

During plastic deformation $f = 0$, moreover, f remains zero:

$$\begin{aligned} f^{\bullet} &= 0 \,, \\ f^{\bullet} &= \frac{\partial f}{\partial \sigma_{ij}} \sigma_{ij}^{\bullet} + \frac{\partial f}{\partial k} k^{\bullet} = 0 \,. \end{aligned} \tag{1.195}$$

---

[55] See [16].

[56] This approach gives the so-called KARUSH–KUHN–TUCKER conditions:

$$\Lambda^{\bullet} \geq 0 \,, \quad f \leq 0 \,, \quad \Lambda^{\bullet} f = 0 \,,$$

since in the elastic regime $f < 0$ and $\Lambda^{\bullet} = 0$ whereas in the plastic regime $f = 0$ and $\Lambda^{\bullet} > 0$. We will not make much use of these relations, they are mostly used in conditional optimization.

By using Eq. (1.191) we obtain

$$\frac{\partial f}{\partial \sigma_{ij}} = \frac{\partial f}{\partial \sigma_{|kl|}} \frac{\partial \sigma_{|kl|}}{\partial \sigma_{ij}} = \sigma_{|kl|}\big(\delta_{ki}\delta_{lj} - \frac{1}{3}\delta_{kl}\delta_{ni}\delta_{nj}\big) = \sigma_{|ij|} - \frac{1}{3}\delta_{ij}\sigma_{|kk|} = \sigma_{|ij|} , \tag{1.196}$$

as well as

$$\frac{\partial f}{\partial k} = -\frac{2}{3}k , \tag{1.197}$$

thus the condition in Eq. (1.195) results in

$$f^{\bullet} = 0 = \sigma_{|ij|}\sigma^{\bullet}_{ij} - \frac{2}{3}kk^{\bullet} . \tag{1.198}$$

The latter can be rewritten in terms of the rate of k as follows

$$k^{\bullet} = \frac{3\sigma_{|ij|}\sigma^{\bullet}_{ij}}{2k} . \tag{1.199}$$

By combining the latter with the rate of k obtained from the linear isotropic hardening model in Eq. (1.193),

$$k^{\bullet} = h\,{}^{\mathrm{p}}\varepsilon^{\bullet}_{\mathrm{eq}} , \tag{1.200}$$

we acquire the so-called *evolution equation* for plastic equivalent strain:

$$\begin{aligned} k^{\bullet} = h\,{}^{\mathrm{p}}\varepsilon^{\bullet}_{\mathrm{eq}} &= \frac{3\sigma_{|ij|}\sigma^{\bullet}_{ij}}{2k} , \\ {}^{\mathrm{p}}\varepsilon^{\bullet}_{\mathrm{eq}} &= \frac{3\sigma_{|ij|}\sigma^{\bullet}_{ij}}{2kh} . \end{aligned} \tag{1.201}$$

The evolution equation describes the change of plastic equivalent strain. The plastic strain accumulates in the continuum body due to the evolution equation

$${}^{\mathrm{p}}\varepsilon_{\mathrm{eq}} = \int {}^{\mathrm{p}}\varepsilon^{\bullet}_{\mathrm{eq}}\mathrm{d}t . \tag{1.202}$$

The plastic strain lacks a first integral, we cannot write $\int \mathrm{d}\,{}^{\mathrm{p}}\varepsilon_{\mathrm{eq}}$. Therefore, we need an evolution equation also for the three-dimensional case. In order to obtain ${}^{\mathrm{p}}\varepsilon^{\bullet}_{ij}$ we start by inserting Eq. (1.189)

$${}^{\mathrm{p}}\varepsilon^{\bullet}_{\mathrm{eq}} = \frac{3\sigma_{|ij|}C_{ijkl}\big(\varepsilon^{\bullet}_{kl} - {}^{\mathrm{p}}\varepsilon^{\bullet}_{kl}\big)}{2kh} , \tag{1.203}$$

and continue by utilizing Eq. (1.190) with Eq. (1.196)

$$\begin{aligned} {}^{\mathrm{p}}\varepsilon^{\bullet}_{ij} &= \Lambda^{\bullet}\frac{\partial f}{\partial \sigma_{ij}} = \Lambda^{\bullet}\sigma_{|ij|} , \\ {}^{\mathrm{p}}\varepsilon^{\bullet}_{\mathrm{eq}} &= \frac{3\sigma_{|ij|}C_{ijkl}\big(\varepsilon^{\bullet}_{kl} - \Lambda^{\bullet}\sigma_{|kl|}\big)}{2kh} . \end{aligned} \tag{1.204}$$

During plasticity, $f = 0$, we postulate that the power calculated by the equivalent stress and strain equals to the power calculated by the three-dimensional stress and strain states:

$$\sigma_{\mathrm{eq}}\, {}^{\mathrm{p}}\varepsilon^{\bullet}_{\mathrm{eq}} = \sigma_{ij}\, {}^{\mathrm{p}}\varepsilon^{\bullet}_{ij} . \tag{1.205}$$

The plastic flow of a solid is equal to a viscous flow of a fluid. In a viscous flow the shear deformation results in an incompressible flow without volumetric change. This phenomenon occurs in the plastic flow, too

$${}^{\mathrm{p}}\varepsilon_{kk} = 0 \Rightarrow {}^{\mathrm{p}}\varepsilon^{\bullet}_{kk} = 0 , \tag{1.206}$$

thus, there is only a deviatoric (traceless) plastic strain rate, ${}^{\mathrm{p}}\varepsilon^{\bullet}_{|ij|} = {}^{\mathrm{p}}\varepsilon^{\bullet}_{ij}$. The aforementioned postulate leads to

$$\begin{aligned} \sigma_{\mathrm{eq}}\, {}^{\mathrm{p}}\varepsilon^{\bullet}_{\mathrm{eq}} &= \sigma_{|ij|}\, {}^{\mathrm{p}}\varepsilon^{\bullet}_{|ij|} , \\ \sigma_{\mathrm{eq}} = \sqrt{\frac{3}{2}\sigma_{|ij|}\sigma_{|ij|}} &\Rightarrow {}^{\mathrm{p}}\varepsilon^{\bullet}_{\mathrm{eq}} = \sqrt{\frac{2}{3}\, {}^{\mathrm{p}}\varepsilon^{\bullet}_{|ij|}\, {}^{\mathrm{p}}\varepsilon^{\bullet}_{|ij|}} . \end{aligned} \tag{1.207}$$

Now, the postulate can be rewritten

$${}^{\mathrm{p}}\varepsilon^{\bullet}_{\mathrm{eq}} = \frac{1}{\sigma_{\mathrm{eq}}}\sigma_{|ij|}\, {}^{\mathrm{p}}\varepsilon^{\bullet}_{|ij|} . \tag{1.208}$$

Moreover, during plasticity, from Eq. (1.191) we acquire

$$f = 0 \Rightarrow k = \sqrt{\frac{2}{3}\sigma_{|ij|}\sigma_{|ij|}} = \sigma_{\mathrm{eq}} . \tag{1.209}$$

By using the latter and Eq. $(1.204)_1$ we obtain

$${}^{\mathrm{p}}\varepsilon^{\bullet}_{\mathrm{eq}} = \frac{1}{k}\sigma_{|ij|}\, {}^{\mathrm{p}}\varepsilon^{\bullet}_{|ij|} = \frac{1}{k}\sigma_{|ij|}\Lambda^{\bullet}\sigma_{|ij|} = \frac{1}{k}\Lambda^{\bullet}\frac{2}{3}k^2 = \frac{2}{3}\Lambda^{\bullet}k . \tag{1.210}$$

Finally, we can explicitly define $\Lambda^{\bullet}$ by combining Eq. $(1.204)_2$ with the latter

$$\begin{aligned} \Lambda^{\bullet} &= \frac{9\sigma_{|ij|}C_{ijkl}\big(\varepsilon^{\bullet}_{kl} - \Lambda^{\bullet}\sigma_{|kl|}\big)}{4k^2h} \,, \\ \Lambda^{\bullet}\Big(1 + \frac{9}{4k^2h}\sigma_{|ij|}C_{ijkl}\sigma_{|kl|}\Big) &= \frac{9}{4k^2h}\sigma_{|ij|}C_{ijkl}\varepsilon^{\bullet}_{kl} \,, \\ \Lambda^{\bullet} &= \frac{\sigma_{|ij|}C_{ijkl}\varepsilon^{\bullet}_{kl}}{\frac{4}{9}k^2h + \sigma_{|ij|}C_{ijkl}\sigma_{|kl|}} \,. \end{aligned} \tag{1.211}$$

We have reached the material model of the plastic multiplier in case of the linear hardening. We recall Eq. (1.204)$_1$:

$$^{\mathrm{p}}\varepsilon^{\bullet}_{ij} = \Lambda^{\bullet}\sigma_{|ij|} \,, \tag{1.212}$$

such that the evolution of the plastic strain can be computed by means of the additional parameter, h, from the used linear hardening model. This parameter shall be obtained by a tensile testing. We aim at defining a constitutive equation for stress, which reads

$$\sigma^{\bullet}_{ij} = C_{ijkl}\big(\varepsilon^{\bullet}_{kl} - {}^{\mathrm{p}}\varepsilon^{\bullet}_{kl}\big) = \Big(C_{ijmn} - C_{ijkl}\sigma_{|kl|}\frac{\sigma_{|op|}C_{opmn}}{\frac{4}{9}k^2h + \sigma_{|ij|}C_{ijkl}\sigma_{|kl|}}\Big)\varepsilon^{\bullet}_{mn} \,. \tag{1.213}$$

In order to allow plasticity to occur above the yield stress, we can define a conditional parameter:

$$\langle\gamma\rangle = \begin{cases} 1 & \text{if } k \geq \sigma_{\mathrm{Y}} \\ 0 & \text{otherwise} \end{cases} \,, \tag{1.214}$$

where the conditional parameter is written with MACAULAY brackets[57] $\langle\cdot\rangle$. This parameter is used as follows

$$\sigma^{\bullet}_{ij} = \Big(C_{ijmn} - \langle\gamma\rangle\frac{C_{ijkl}\sigma_{|kl|}\sigma_{|op|}C_{opmn}}{\frac{4}{9}k^2h + \sigma_{|ij|}C_{ijkl}\sigma_{|kl|}}\Big)\varepsilon^{\bullet}_{mn} \,. \tag{1.215}$$

Formally, the parameter γ can be computed after having computed the displacement field (and thus stress). This approach is computationally costly, therefore, we will employ it regarding the displacement field from the last time step. By choosing appropriately small time steps, the computation will be accurate.

From the point of algorithmic ease the MACAULAY brackets can be implemented in an unusual way:

$$He(a) = \frac{1}{|a|}\Big(\frac{a + |a|}{2}\Big) = \begin{cases} 1 & \text{if } a > 0 \\ 0 & \text{otherwise} \end{cases} \,. \tag{1.216}$$

[57] They are named for William Herrick Macaulay.

This step-function is known as the HEAVISIDE function.[58] Especially in signal processing and system control and dynamics, by using the time instead of a, HEAVISIDE function is used frequently. We can use the same idea for the MACAULAY brackets and implement it in this way:

$$\langle\gamma\rangle = He(k - \sigma_\mathrm{Y}) = \frac{1}{|k - \sigma_\mathrm{Y}|}\Big(\frac{k - \sigma_\mathrm{Y} + |k - \sigma_\mathrm{Y}|}{2}\Big) . \tag{1.217}$$

In the code below we use a boolean query for obtaining $\langle\gamma\rangle$, it is quicker than the HEAVISIDE function.

Since we apply the displacement field from the last time step for obtaining $\langle\gamma\rangle$, the behavior of plastic flow is characterized with a time lag, which converges to reality by choosing small time steps. During plasticity the deformation is partly elastic and partly plastic. The plastic deformation is a viscous flow and we can imagine this phenomenon as a consequence of the velocity gradient. Elastic deformation is modeled by using the displacement gradient (strain). Displacement and velocity are coupled; but they are independent.[59] Therefore, elasticity and plasticity are coupled and independent phenomena.

1.6.2 *Kinematic Hardening*

In the associated plasticity an isotropic hardening rule has some limitations. The yield stress, k, increases in all directions (isotropic), hence, it can only be used for a monotonic loading. For the case of a cyclic loading the BAUSCHINGER effect[60] can be modeled by amending the hardening rule. The so-called *kinematic* hardening uses a *back stress*, β_{ij}, which results in a dependence on loading in the hardening model. Instead of the aforementioned flow potential with linear isotropic hardening:

$$f = \frac{1}{2}\sigma_{|ij|}\sigma_{|ij|} - \frac{1}{3}(\sigma_\mathrm{Y} + h\,{}^\mathrm{p}\varepsilon_\mathrm{eq})^2 , \tag{1.218}$$

[58]It is named after Oliver Heaviside.

[59]Formally, a sinusoidal displacement, $u = a\sin(bt)$, and thus the velocity, $v = a\,b\,\cos(bt)$, are independent, $\int u\,v\mathrm{d}t = 0$, since sinus and cosinus are orthogonal. This independence means that a variation in one does not change the other. A velocity in the current time causes a displacement in a *future* time, therefore, they are affecting each other in the subsequent times, however, independent at the current time.

[60]This effect has been discussed in [4] for the first time and therefore it is named after Johann Bauschinger.

we use a flow potential[61] in this form:

$$f = \frac{1}{2}(\sigma_{|ij|} - \beta_{ij})(\sigma_{|ij|} - \beta_{ij}) - \frac{1}{3}\sigma_Y^2 . \tag{1.219}$$

We aim at modeling the back stress, β_{ij}, in an adequate way. By starting with zero back stress, we can model the back stress by using an evolution equation for its rate, $\beta^{\bullet}_{ij}$. We search for a model of the rate of back stress. The simplest model[62] is a linear relation:

$$\beta^{\bullet}_{ij} = c\,{}^{\mathrm{p}}\varepsilon^{\bullet}_{ij} , \tag{1.220}$$

where the material parameter c has to be determined instead of h in isotropic hardening. Starting from the latter relation for back stress, a theoretical treatise[63] results in

$$\beta^{\bullet}_{ij} = (\sigma_{ij} - \beta_{ij})\Gamma^{\bullet} , \quad \Gamma^{\bullet} \geq 0 , \tag{1.221}$$

where we have introduced $\Gamma^{\bullet}$, which has to be obtained in a way that Eq. (1.220) holds. We redo the same steps as in isotropic hardening. The flow potential in Eq. (1.219) is a function of stress and back stress, $f = f(\sigma_{ij}, \beta_{ij})$. While a plastic yielding is occurring,

$$f^{\bullet} = 0 = \frac{\partial f}{\partial \sigma_{ij}}\sigma^{\bullet}_{ij} + \frac{\partial f}{\partial \beta_{ij}}\beta^{\bullet}_{ij} = \big(\sigma_{|ij|} - \beta_{ij}\big)\sigma^{\bullet}_{|ij|} - \big(\sigma_{|ij|} - \beta_{ij}\big)\beta^{\bullet}_{ij} , \tag{1.222}$$

we obtain the following relation by using Eq. (1.221):

$$\begin{aligned} \big(\sigma_{|ij|} - \beta_{ij}\big)\sigma^{\bullet}_{|ij|} &= \big(\sigma_{|ij|} - \beta_{ij}\big)\beta^{\bullet}_{ij} , \\ \big(\sigma_{|ij|} - \beta_{ij}\big)\sigma^{\bullet}_{|ij|} &= \big(\sigma_{|ij|} - \beta_{ij}\big)\big(\sigma_{ij} - \beta_{ij}\big)\Gamma^{\bullet} , \\ \Gamma^{\bullet} &= \frac{\big(\sigma_{|ij|} - \beta_{ij}\big)\sigma^{\bullet}_{|ij|}}{\big(\sigma_{|ij|} - \beta_{ij}\big)\big(\sigma_{ij} - \beta_{ij}\big)} . \end{aligned} \tag{1.223}$$

Equation (1.220) states a traceless (deviatoric) back stress, $\beta_{ij} = \beta_{|ij|}$, since ${}^{\mathrm{p}}\varepsilon^{\bullet}_{ii} = 0$. Moreover, during plasticity, $f = 0$, we insert Eq. (1.219) into the latter and find

$$\Gamma^{\bullet} = \frac{\big(\sigma_{|ij|} - \beta_{ij}\big)\sigma^{\bullet}_{ij}}{\big(\sigma_{|ij|} - \beta_{ij}\big)\big(\sigma_{|ij|} - \beta_{ij}\big)} = \frac{\big(\sigma_{|ij|} - \beta_{ij}\big)\sigma^{\bullet}_{ij}}{\frac{2}{3}\sigma_Y^2} . \tag{1.224}$$

[61] See [13].

[62] See [19].

[63] See [23].

The starting assumption with the flow potential for kinematic hardening,

$$ {}^{\mathrm{p}}\varepsilon^{\bullet}_{ij} = \Lambda^{\bullet} \frac{\partial f}{\partial \sigma_{ij}} = \Lambda^{\bullet}\big(\sigma_{|ij|} - \beta_{ij}\big) , \tag{1.225} $$

allows us to obtain from Eq. (1.220) with Eq. (1.221) the evolution of back stress:

$$ \beta^{\bullet}_{ij} = c\Lambda^{\bullet}\big(\sigma_{|ij|} - \beta_{ij}\big) = \big(\sigma_{|ij|} - \beta_{ij}\big)\Gamma^{\bullet} , \tag{1.226} $$

for arbitrary σ_{ij} and β_{ij} values. Hence, it reads

$$ c\Lambda^{\bullet} = \Gamma^{\bullet} , \quad \Lambda^{\bullet} = \frac{\big(\sigma_{|ij|} - \beta_{ij}\big)\sigma^{\bullet}_{ij}}{\frac{2}{3}c\sigma_{\mathrm{Y}}^2} . \tag{1.227} $$

In order to eliminate the rate of stress, we use again HOOKE's law,

$$ \Lambda^{\bullet} = \frac{\big(\sigma_{|ij|} - \beta_{ij}\big)C_{ijkl}\big(\varepsilon^{\bullet}_{kl} - {}^{\mathrm{p}}\varepsilon^{\bullet}_{kl}\big)}{\frac{2}{3}c\sigma_{\mathrm{Y}}^2} , \quad {}^{\mathrm{p}}\varepsilon^{\bullet}_{kl} = \Lambda^{\bullet}\big(\sigma_{|kl|} - \beta_{kl}\big) , \tag{1.228} $$

and acquire

$$ \begin{aligned} &\Lambda^{\bullet}\left(1 + \frac{\big(\sigma_{|ij|} - \beta_{ij}\big)C_{ijkl}\big(\sigma_{|kl|} - \beta_{kl}\big)}{\frac{2}{3}c\sigma_{\mathrm{Y}}^2}\right) = \frac{\big(\sigma_{|ij|} - \beta_{ij}\big)C_{ijkl}\varepsilon^{\bullet}_{kl}}{\frac{2}{3}c\sigma_{\mathrm{Y}}^2} , \\ &\Lambda^{\bullet} = \frac{\big(\sigma_{|ij|} - \beta_{ij}\big)C_{ijkl}\varepsilon^{\bullet}_{kl}}{\frac{2}{3}c\sigma_{\mathrm{Y}}^2 + \big(\sigma_{|ij|} - \beta_{ij}\big)C_{ijkl}\big(\sigma_{|kl|} - \beta_{kl}\big)} . \end{aligned} \tag{1.229} $$

Often $c = 2/3h$ is chosen for a better correspondence to the isotropic hardening. In this case we obtain the constitutive equation for kinematic hardening:

$$ \sigma^{\bullet}_{ij} = C_{ijmn}\big(\varepsilon^{\bullet}_{mn} - {}^{\mathrm{p}}\varepsilon^{\bullet}_{mn}\big) = C_{ijmn}\big(\varepsilon^{\bullet}_{mn} - \Lambda^{\bullet}(\sigma_{|mn|} - \beta_{mn})\big) , \tag{1.230} $$

by using the conditional parameter $\langle\gamma\rangle$ from Eq. (1.217) the constitutive equation can be rewritten

$$ \sigma^{\bullet}_{ij} = \left(C_{ijmn} - \langle\gamma\rangle \frac{C_{ijkl}\big(\sigma_{|kl|} - \beta_{kl}\big)\big(\sigma_{|op|} - \beta_{op}\big)C_{opmn}}{\frac{4}{9}\sigma_{\mathrm{Y}}^2 h + \big(\sigma_{|ij|} - \beta_{ij}\big)C_{ijkl}\big(\sigma_{|kl|} - \beta_{kl}\big)}\right)\varepsilon^{\bullet}_{mn} . \tag{1.231} $$

The latter equation is the counterpart of Eq. (1.215) with the isotropic hardening. For the case of the isotropic hardening the yield stress, k, evolves with the plastic strain; whereas for the case of the kinematic hardening, the back stress, β_{ij}, evolves with the plastic strain.

In order to implement the constitutive equation, time discretization is used in the LAGRANGEan frame

$$\sigma^{\bullet}_{ij} = \frac{\partial \sigma_{ij}}{\partial t} = \frac{\sigma_{ij} - \sigma^0_{ij}}{\Delta t} , \\ \sigma_{ij} = \sigma^0_{ij} + \Delta t \sigma^{\bullet}_{ij} . \tag{1.232}$$

By considering linearized strains for small deformation the variational form becomes

$$\text{Form} = \int_{\mathcal{B}_0} \Big(\rho_0 \frac{u_i - 2u_i^0 + u_i^{00}}{\Delta t \Delta t} \delta u_i + \sigma_{ji}\, \delta u_{i,j} - \rho_0 f_i\, \delta u_i \Big) \mathrm{d}V \\ - \int_{\partial \mathcal{B}_0^{\mathrm{N}}} \hat{t}_i\, \delta u_i \mathrm{d}A , \tag{1.233}$$

where stress in Eq. (1.232) is complemented with Eq. (1.215) or Eq. (1.231). Unfortunately, we need to know σ_{ij} in Eq. (1.215) or Eq. (1.231) for computing the current stress, σ_{ij}. The correct way of programming relies on an iterative schema,[64] which is computationally costly. Therefore, we use the value of stress from the last time step and approximate the rate of stress, for example, for kinematic hardening we acquire

$$\sigma^{\bullet}_{ij} = \Bigg(C_{ijmn} - \langle \gamma \rangle \frac{C_{ijkl}\big(\sigma^0_{|kl|} - \beta^0_{kl}\big)\big(\sigma^0_{|op|} - \beta^0_{op}\big) C_{opmn}}{\frac{4}{9}\sigma^2_{\mathrm{Y}} h + \big(\sigma^0_{|ij|} - \beta^0_{ij}\big) C_{ijkl} \big(\sigma^0_{|kl|} - \beta^0_{kl}\big)} \Bigg) \varepsilon^{\bullet}_{mn} , \\ \beta_{ij} = \beta^0_{ij} + \Delta t \beta^{\bullet}_{ij} , \quad \beta^{\bullet}_{ij} = \frac{(\sigma^0_{|kl|} - \beta^0_{kl}) \sigma^{\bullet}_{kl}}{\frac{2}{3}\sigma^2_{\mathrm{Y}}} \big(\sigma^0_{|ij|} - \beta^0_{ij}\big) . \tag{1.234}$$

For small time increments the numerical solution is accurate and the computational time is reasonable.

Consider a one-axial tensile testing where a quadratic beam is under a mechanical loading. The machine is controlled by displacement. Suppose that a cyclic loading is set. We give below the code for the kinematic hardening and the hysteresis plot can be seen in Fig. 1.10. The computation is in three-dimensions as seen in Fig. 1.11. HOOKE's law incorporates the transverse strain, although the loading is only axial. The plasticity model affects the material behavior only during the plasticity, which is decided by using two conditions: $f = 0$ and $f^{\bullet} = 0$. Both of them are computed with a boolean query in each time step. The second condition enables an elastic response as a consequence of unloading. The code is given below.

---

[64] See [24, Chap. 3].

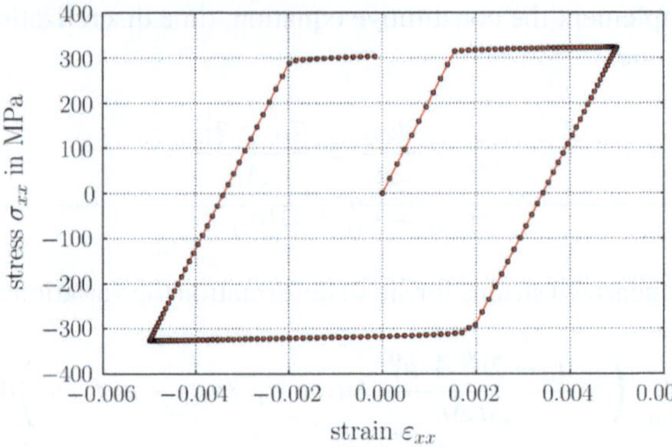

Fig. 1.10 Hysteresis plot for a tensile test simulation of plasticity with kinematic hardening

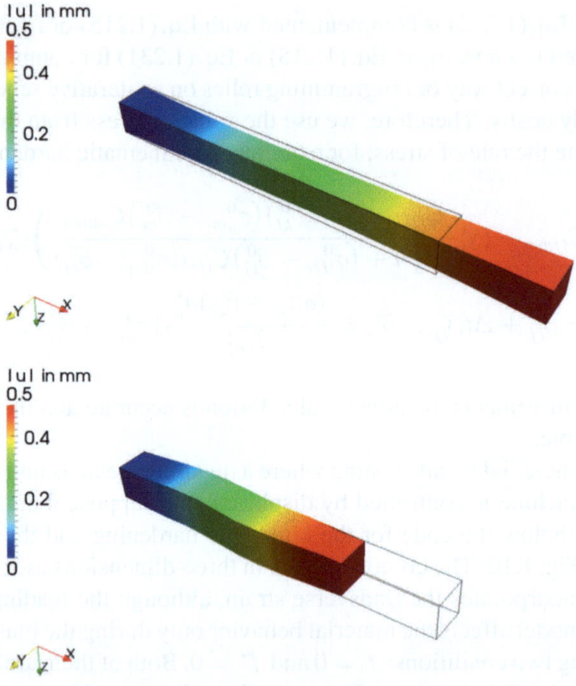

Fig. 1.11 Deformation is presented by using a scale factor of 50. *Top* After 1/4 cycle. *Bottom* After 3/4 cycle. The initial geometry is outlined

```
"""Computational reality 06, plasticity"""
__author__ = "B. Emek Abali"
__license__ = "GNU GPL Version 3.0 or later"
#This code underlies the GNU General Public License,
    ↪ http://www.gnu.org/licenses/gpl-3.0.en.html

from fenics import *
import numpy
set_log_level(ERROR)
xlength =100. #[mm]
ylength =10. #[mm]
zlength =10. #[mm]
mesh = BoxMesh(Point(0, 0, 0), Point(xlength, ylength,
    ↪ zlength), 10, 3, 3)
Coeff = FunctionSpace(mesh,'P',1)
Space = VectorFunctionSpace(mesh,'P',1)
Tensor = TensorFunctionSpace(mesh,'P',1)
delu = TestFunction(Space)
du = TrialFunction(Space)
u = Function(Space)
u0 = Function(Space)
u00 = Function(Space)

cells = CellFunction('size_t', mesh)
facets = FacetFunction('size_t', mesh)
dA = Measure('ds', domain=mesh, subdomain_data=facets)
dV = Measure('dx', domain=mesh, subdomain_data=cells)

print 'initializing,time'
t=0.0
t_end = 10.0
dt = 0.05

#Defining boundary conditions
left = CompiledSubDomain('near(x[0],0) && on_boundary')
right = CompiledSubDomain('near(x[0],l) && on_boundary',l=
    ↪ xlength)

boundaries = FacetFunction('uint',mesh)
boundaries.set_all(0)
# a cyclic displacement on the right end
displ = Expression(('0.5*sin(2.*pi*f*time)','0.0','0.0'),f
    ↪ =0.1,time=0)
bc1 = DirichletBC(Space,displ,right)
bc2 = DirichletBC(Space,(0.0,0.0,0.0),left)
bc = [bc1,bc2]

#Setting up the material parameters in tonne, seconds, Newton
    ↪ , milimeter
rho0 = 8.3E-9 #tonne/mm^3
nu, E = 0.3, 200000.0 # [-], [MPa]
h=0.01*E
lambada=E*nu/(1.0+nu)/(1.0-2.0*nu)
mu=E/(2.0+2.0*nu)
```

```
#Y=Function(Coeff)
sigmaY=Constant(300.0) # [MPa]
# index notation
i,j,k,l,m,n,o,p = indices(8)
delta= Identity(3)
eps= as_tensor(1.0/2.0*(u[i].dx(j)+u[j].dx(i)),(i,j))
eps0= as_tensor(1.0/2.0*(u0[i].dx(j)+u0[j].dx(i)),(i,j))
epsDot=as_tensor((eps[i,j]-eps0[i,j])/dt,(i,j))
gamma=Function(Coeff)

C = as_tensor(lambada*delta[i,j]*delta[k,l]\
+mu*delta[i,k]*delta[j,l]+mu*delta[i,l]*delta[j,k],(i,j,k,l))

sigma0 = Function(Tensor)
dev_sigma0 = as_tensor(sigma0[i,j]-1./3.*sigma0[k,k]*delta[i,
    ↪ j],(i,j))
beta0=Function(Tensor)

sigmaDot = as_tensor( (C[i,j,m,n]-gamma*C[i,j,k,l]*(
    ↪ dev_sigma0[k,l]\
-beta0[k,l])*(dev_sigma0[o,p]-beta0[o,p])*C[o,p,m,n]/(4./9.*
    ↪ sigmaY**2*h\
+(dev_sigma0[i,j]-beta0[i,j])*C[i,j,k,l]*(dev_sigma0[k,l]\
-beta0[k,l]) ) )*epsDot[m,n], (i,j) )

sigma = as_tensor(sigma0[i,j]+dt*sigmaDot[i,j],(i,j))
dev_sigma = as_tensor(sigma[i,j]-1./3.*sigma[k,k]*delta[i,j
    ↪ ],(i,j))

betaDot = as_tensor( gamma*(dev_sigma0[k,l]-beta0[k,l])*
    ↪ sigmaDot[k,l]\
/(2.0/3.0*sigmaY**2)*(dev_sigma0[i,j]-beta0[i,j]),(i,j))

beta=as_tensor(beta0[i,j]+dt*betaDot[i,j],(i,j))

f= Constant((0.,0.,0.))

Form = (rho0*(u[i]-2.*u0[i]+u00[i])/(dt*dt)*delu[i] \
    + sigma[j,i]*delu[i].dx(j) \
    - rho0*f[i]*delu[i] )*dV
Gain = derivative(Form,u,du)

#Plotting stress vs. strain curves
import matplotlib as mpl
mpl.use('Agg')
import matplotlib.pyplot as pylab
pylab.rc('text', usetex=True )
pylab.rc('font', family='serif', serif='cm', size=30 )
pylab.rc('legend', fontsize=30)
pylab.rc(('xtick.major','ytick.major'), pad=15)

#pylab.ion()
fig = pylab.figure(1, figsize=(12,8))
fig.clf()
```

```
pylab.subplots_adjust(bottom=0.18)
pylab.subplots_adjust(left=0.16)
pylab.xlabel(r'strain $\varepsilon_{xx}$')
pylab.ylabel(r'stress $\sigma_{xx}$ in MPa')
pylab.grid(True)

stress_plot = []
strain_plot = []
temp_array= []
time= []
uinit = Expression(('0.0','0.0','0.0'))
u0.interpolate(uinit)
u00.assign(u0)
file_u = File ('/calcul/CR06/displacements.pvd')

while t <= t_end :
    displ.time = t
    print 'time: ', t
    solve(Form==0, u, bc, J=Gain, \
        solver_parameters={"newton_solver":{"linear_solver":"mumps", "relative_tolerance": 1e-3} }, \
        form_compiler_parameters={"cpp_optimize": True, "representation": "quadrature", "quadrature_degree": 2} )

    sigma_ = project(sigma,Tensor, solver_type="mumps",\
        form_compiler_parameters={"cpp_optimize": True, "representation": "quadrature", "quadrature_degree": 2} )
    strain_plot.append(u(xlength,ylength/2.,zlength/2.)[0]/xlength)
    stress_plot.append(sigma_(xlength/2.,ylength/2.,zlength/2.)[0])
    file_u << (u,t)
    sigma0.assign(sigma_)
    beta_ = project(beta,Tensor, solver_type="mumps",\
        form_compiler_parameters={"cpp_optimize": True, "representation": "quadrature", "quadrature_degree": 2} )
    beta0.assign(beta_)
    flow_ = project(1./2.*(dev_sigma0[i,j]-beta0[i,j])*(dev_sigma0[i,j]-beta0[i,j]) -1./3.*sigmaY**2, Coeff)
    flow_bool = flow_.vector().array() >= 0.
    direction_ = project((dev_sigma0[i,j]-beta0[i,j])*epsDot[i,j], Coeff, solver_type="mumps",\
        form_compiler_parameters={"cpp_optimize": True, "representation": "quadrature", "quadrature_degree": 2} )
    direction_bool=1./2.*(numpy.sign(direction_.vector().array())+1.)
    gamma.vector()[:] = numpy.array(flow_bool*direction_bool, dtype=int)

```

```
        u00.assign(u0)
        u0.assign(u)
        t = t + dt

        pylab.plot(strain_plot,stress_plot,color='red',marker='o'
            ,markersize=5)
        #pylab.draw()

        pylab.savefig('/calcul/CR06/CompReal06_plast_tensile.pdf'
            )
```

To-do

We have implemented the kinematic hardening law such that the stress-strain hysteresis curve is enclosed in a cyclic loading.

- Try to implement the isotropic hardening and compare the hysteresis curves.
- What is the so-called BAUSCHINGER effect? Which hardening law is more realistic?
- Depends the plasticity modeling on the loading rate? How is the response of the material subject to a quicker loading?

1.7 Linear Viscous Fluids

Under a mechanical loading, material deforms. A plastic solid and a viscous fluid perform a similar behavior under mechanical loading. Categorizing a material as solid or fluid is challenging. A solid is understood as a body keeping its shape under gravitational forces; the molecular bounds are strong. Therefore, the LAGRANGEan frame works well, where particles are associated by their reference positions. In plasticity, above the yield stress the material starts flowing if the applied force exceeds the intramolecular force. In case of a fluid, the molecular bound is so weak that it takes the shape of the container under gravitational forces. Basically, it flows under *any* mechanical loading.[65] We can visualize a channel, through which the fluid flows. In a LAGRANGEan frame we may identify the particles by their positions at a reference time to be seen as $\boldsymbol{X}$ in Fig. 1.12. The velocity of a fluid particle can be measured at a position in space expressed in $\boldsymbol{x}$. At two different instants of time, at the same *spatial* position, $\boldsymbol{x}$, we measure the velocity of two different fluid particles, see Fig. 1.12. The spatial position, $\boldsymbol{x}$, is occupied by a material particle. At a time instant we can

[65] We simply ignore the surface stress, which holds a drop together.

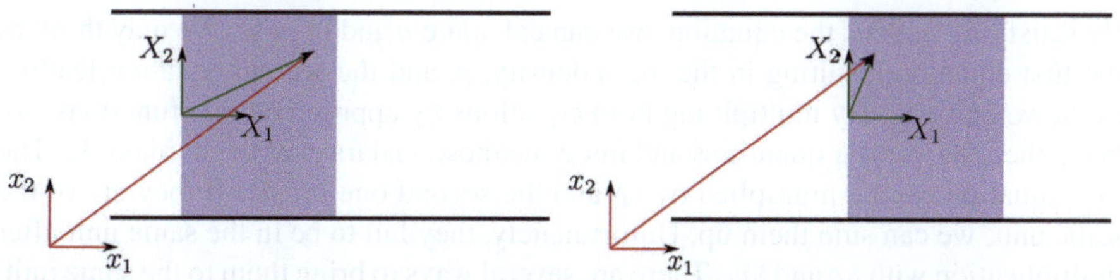

Fig. 1.12 Two snapshots at different time instants are drawn for a fluid flowing in a channel. In both drawings the same spatial position is indicated in the *red*, $\boldsymbol{x}$, and *green*, $\boldsymbol{X}$, coordinate systems. The coordinates in the EULERian frame, $\boldsymbol{x}$, are identical, since it is the same spatial position. The coordinates in the LAGRANGEan frame, $\boldsymbol{X}$, varies denoting that different particles occupy the same spatial position at different instants of time

identify a material particle for each spatial position; however, the position denotes a point in space, not a particle. A EULERian frame is the collection of spatial positions indicating points in the physical (ordinary) space independent on the underlying material. This frame is used for fluid flow problems.

In (solid or fluid) mechanics we solve the balance equations of mass and (linear) momentum:

$$\begin{aligned} &\frac{\partial \rho}{\partial t} + \frac{\partial}{\partial x_i}(\rho v_i) = 0 \,, \\ &\frac{\partial \rho v_i}{\partial t} + \frac{\partial}{\partial x_j}(\rho v_i v_j - \sigma_{ji}) - \rho f_i = 0 \,, \end{aligned} \tag{1.235}$$

in a EULERian frame for every material particle occupying a spatial position x_i. Their derivation will be discussed in the next section. These balance equations will be solved at the current (present) time at every x_i in Ω called a *domain* or a *control volume*. Since we will solve them simultaneously, we may subtract the first from the second and obtain

$$\begin{aligned} &\frac{\partial \rho}{\partial t} + \frac{\partial}{\partial x_i}(\rho v_i) = 0 \,, \\ &\rho \frac{\partial v_i}{\partial t} + \rho v_j \frac{\partial v_i}{\partial x_j} - \frac{\partial \sigma_{ji}}{\partial x_j} - \rho f_i = 0 \,. \end{aligned} \tag{1.236}$$

Now, we employ the time rate with finite difference method:

$$\begin{aligned} &\frac{1}{\Delta t}(\rho - \rho^0) + v_i \frac{\partial \rho}{\partial x_i} + \rho \frac{\partial v_i}{\partial x_i} = 0 \,, \\ &\frac{\rho}{\Delta t}(v_i - v_i^0) + \rho v_j \frac{\partial v_i}{\partial x_j} - \frac{\partial \sigma_{ji}}{\partial x_j} - \rho f_i = 0 \,. \end{aligned} \tag{1.237}$$

By satisfying both of the equations we can calculate ρ and v_i in Ω. We may think as the first equation resulting in the mass density, ρ, and the second equation leading to the velocity, v_i. By multiplying both equations by appropriate test functions, we bring them into scalar quantities and integrate these scalars over the domain, Ω. The first equation can be multiplied by $\delta\rho$ and the second one by δv_i. If they are of the same unit, we can sum them up. Unfortunately, they fail to be in the same unit after multiplication with $\delta\rho$ and δv_i. There are several ways to bring them to the same unit. The prominent way is to choose a dimensionless form by introducing a *characteristic* length, ℓ, a *characteristic* time, τ, and a *characteristic* velocity, $\omega = \ell/\tau$. By using these characteristic quantities we can obtain dimensionless space, time, and velocity as follows

$$\bar{x}_i = \frac{x_i}{\ell}\ ,\quad \bar{t} = \frac{t}{\tau}\ ,\quad \bar{v}_i = \frac{v_i}{\omega}\ . \tag{1.238}$$

Hence the balance equations can be rewritten as

$$\begin{aligned}
&\frac{1}{\tau\Delta\bar{t}}(\rho-\rho^0) + \omega\bar{v}_i\frac{1}{\ell}\frac{\partial\rho}{\partial\bar{x}_i} + \rho\frac{\omega}{\ell}\frac{\partial\bar{v}_i}{\partial\bar{x}_i} = 0\ ,\\
&\frac{\rho}{\tau\Delta\bar{t}}\omega(\bar{v}_i-\bar{v}_i^0) + \rho\frac{\omega^2}{\ell}\bar{v}_j\frac{\partial\bar{v}_i}{\partial\bar{x}_j} - \frac{1}{\ell}\frac{\partial\sigma_{ji}}{\partial\bar{x}_j} - \rho f_i = 0\ .
\end{aligned} \tag{1.239}$$

By multiplying the balance of mass by τ/ρ and the balance of linear momentum by $\tau/(\rho\omega)$, we obtain two dimensionless balance equations:

$$\begin{aligned}
&\frac{1}{\Delta\bar{t}}\Big(1-\frac{\rho^0}{\rho}\Big) + \frac{1}{\rho}\bar{v}_i\frac{\partial\rho}{\partial\bar{x}_i} + \frac{\partial\bar{v}_i}{\partial\bar{x}_i} = 0\ ,\\
&\frac{1}{\Delta\bar{t}}(\bar{v}_i-\bar{v}_i^0) + \bar{v}_j\frac{\partial\bar{v}_i}{\partial\bar{x}_j} - \frac{\tau}{\omega\ell\rho}\frac{\partial\sigma_{ji}}{\partial\bar{x}_j} - \frac{\tau}{\omega}f_i = 0\ .
\end{aligned} \tag{1.240}$$

which can be *weighted* by dimensionless test functions and summed up within the domain

$$\begin{aligned}
&\int_\Omega \Bigg(\Big(\frac{1}{\Delta\bar{t}}\Big(1-\frac{\rho^0}{\rho}\Big) + \frac{1}{\rho}\bar{v}_i\frac{\partial\rho}{\partial\bar{x}_i} + \frac{\partial\bar{v}_i}{\partial\bar{x}_i}\Big)\delta\bar{\rho} +\\
&+\Big(\frac{1}{\Delta\bar{t}}(\bar{v}_i-\bar{v}_i^0) + \bar{v}_j\frac{\partial\bar{v}_i}{\partial\bar{x}_j} - \frac{\tau}{\omega\ell\rho}\frac{\partial\sigma_{ji}}{\partial\bar{x}_j} - \frac{\tau}{\omega}f_i\Big)\delta\bar{v}_i\Bigg)\mathrm{d}v = 0\ .
\end{aligned} \tag{1.241}$$

We search for the mass density, $\rho = \rho(x_i, t)$, and the velocity, $v_i = v_i(x_j, t)$, in space, x_i, and time, t. Indeed, we first need to define a constitutive equation for the CAUCHY stress, σ_{ij}, in order to close the integral form. Thereafter, we can solve the integral form numerically.

The simplest constitutive equation for a viscous fluid is called NAVIER–STOKES's equation:[66]

$$\sigma_{ji} = \Big(- p + \lambda \frac{\partial v_k}{\partial x_k} \Big) \delta_{ji} + 2\mu \frac{\partial v_{(j}}{\partial x_{i)}} , \qquad (1.242)$$

where p denotes pressure, λ and μ are the material parameters (constants). For materials like water this linear constitutive equation provides accurate results. It is a linear model since it models a linear relation between the symmetric part of the velocity gradient, $\partial v_{(i}/\partial x_{j)}$, and the CAUCHY stress tensor, σ_{ij}. By defining stress we have introduced a quantity called hydrostatic pressure, p. This pressure p shall be given by a constitutive equation, too. For a viscous fluid it is difficult to find an adequate equation of p. There is a way of circumventing this problem, at least for incompressible flows, which is often the case in water. The incompressibility is easy to measure. We can fill a bicycle pump with water and try to change the volume of the water by holding the outlet closed and compressing the pump. By increasing the pressure we should be compressing the water in the pump and decreasing the volume. Suppose we have increased the pressure to a specific value and the volume contraction is negligible. Then up to this specific pressure, the fluid flow is *incompressible*. For an incompressible fluid flow the mass density remains unchanged, $\rho = \rho^0$. Initially, if the material is homogeneous, $\rho_{,i} = 0$, i.e., same mass density within the domain, then we have a constant value of the mass density throughout the simulation. For an incompressible fluid flow, the balance of mass reads

$$\frac{\partial v_i}{\partial x_i} = 0 . \qquad (1.243)$$

The mechanical pressure:[67]

$$-\frac{1}{3}\sigma_{ii} = p - \Big(\lambda + \frac{2}{3}\mu\Big) \frac{\partial v_i}{\partial x_i} , \qquad (1.244)$$

applied on the fluid equals to the hydrostatic pressure, p, for an incompressible fluid flow.

In the case of an incompressible fluid flow, the mass density is known. Therefore, we can use the balance of mass for calculating the pressure, p, which is unknown. In order to emphasize that the pressure is now a primitive variable instead of mass density, we rewrite the integral form with $\delta \bar{p}$ instead of $\delta \bar{\rho}$ (both are dimensionless)

---

[66] The constitutive equation is named for Claude Louis Marie Henri Navier and George Gabriel Stokes.

[67] Positive stress attains a lengthening, whereas pressure shortens the structure. Hence, the mechanical pressure is a *minus* trace of stress tensor.

$$\int_\Omega \Big(\bar v_{i,i}\,\delta\bar p + \Big(\frac{(\bar v_i - \bar v_i^0)}{\Delta\bar t} + \bar v_j \bar v_{i,j} + \frac{\tau}{\omega\ell\rho} p_{,j}\delta_{ji} - \lambda\frac{1}{\omega\ell\rho}\bar v_{k,kj}\delta_{ji} - \\ -\frac{2\mu}{\omega\ell\rho}\bar v_{(j,i)j} - \frac{\tau}{\omega} f_i \Big)\,\delta\bar v_i \Big)\mathrm{d}v = 0\,, \tag{1.245}$$

where we have introduced, for the sake of simplicity, the following notation:

$$\frac{\partial(\cdot)}{\partial \bar x_i} = (\cdot)_{,i}\,. \tag{1.246}$$

We can now introduce dimensionless pressure, stress, and body forces:

$$\bar p = \frac{1}{\omega^2\rho} p = \frac{\tau}{\omega\ell\rho} p\,,\ \ \bar\sigma_{ij} = \frac{\tau}{\omega\ell\rho}\sigma_{ij}\,,\ \ \bar f_i = \frac{\tau}{\omega} f_i\,, \tag{1.247}$$

as well as the so-called REYNOLDS number:[68]

$$Re = \frac{\omega\ell\rho}{2\mu}\,. \tag{1.248}$$

The list of unknowns, $\{\bar p, \bar v_1, \bar v_2, \bar v_3\}$, are the primitive variables in this section. We use GALERKIN type finite elements, in other words, the primitive and their test functions are of the same class.[69] A term like $\bar v_{(j,i)k}\,\delta\bar v_l$ necessitates at least C^2 because of the second derivative in the primitive variable. We can shift one derivative to the test function and lower the necessary differentiability, so the continuity requirement of the integral form is weaken such that we acquire a *weak form*. By integrating by parts and using GAUSS's law on the aforementioned term

$$-\int_\Omega \bar v_{(j,i)k}\,\delta\bar v_l \mathrm{d}v = \int_\Omega \bar v_{(j,i)}\,\delta\bar v_{l,k}\mathrm{d}v - \int_{\partial\Omega} \bar v_{(j,i)}\,\delta\bar v_l n_k \mathrm{d}a\,, \tag{1.249}$$

we attain the weak form for the linear viscous fluid flow problem:

$$\text{Form} = \int_\Omega \Big(\bar v_{i,i}\,\delta\bar p + \frac{(\bar v_i - \bar v_i^0)}{\Delta\bar t}\delta\bar v_i + \bar v_j\bar v_{i,j}\,\delta\bar v_i + \bar p_{,i}\,\delta\bar v_i + \frac{\lambda}{2\mu Re}\bar v_{k,k}\,\delta\bar v_{i,i} + \\ + \frac{1}{Re}\bar v_{(j,i)}\,\delta\bar v_{i,j} - \bar f_i\,\delta\bar v_i\Big)\mathrm{d}v - \int_{\partial\Omega}\Big(\frac{\lambda}{2\mu}\frac{1}{Re}\bar v_{k,k}\,\delta\bar v_i n_i + \frac{1}{Re}\bar v_{(j,i)}\,\delta\bar v_i n_j\Big)\mathrm{d}a\,. \tag{1.250}$$

After inserting the constitutive equation and introducing a dimensionless number, $a = 2\mu/\lambda$, the weak form becomes

[68] It is named after Osborne Reynolds.

[69] A function of the class C^k has a continuous kth derivative. Since we use linear finite elements (polynomial order one) the first derivative exists, only C^1 can be represented in one finite element.

Fig. 1.13 Drawing of a two-dimensional channel filled with a viscous fluid in the EULERian frame

$$\text{Form} = \int_\Omega \bigg(\bar{v}_{i,i}\,\delta\bar{p} + \frac{(\bar{v}_i - \bar{v}_i^0)}{\Delta\bar{t}}\,\delta\bar{v}_i + \bar{v}_j\bar{v}_{i,j}\,\delta\bar{v}_i + \bar{p}_{,i}\,\delta\bar{v}_i + \frac{1}{aRe}\bar{v}_{k,k}\,\delta\bar{v}_{i,i} + \\ + \frac{1}{Re}\bar{v}_{(j,i)}\,\delta\bar{v}_{i,j} - \bar{f}_i\,\delta\bar{v}_i \bigg)\mathrm{d}v - \int_{\partial\Omega} \Big(\bar{\sigma}_{ji} + \bar{p}\delta_{ji}\Big)\,\delta\bar{v}_i n_j \mathrm{d}a\,. \tag{1.251}$$

Consider a channel filled with water (linear viscous fluid) in the EULERian frame expressed in Cartesian coordinates as drawn in Fig. 1.13. We simulate in two dimensions, where on left and right ends, water may cross the boundaries of Ω. On top and bottom, walls omit a leakage. Moreover, water adheres to the walls such that particles possess the same velocity as the walls. We simulate for fixed walls:

$$v_i = \bar{v}_i = 0 \text{ on } \partial\Omega_{\text{top}} \cup \partial\Omega_{\text{down}}\,. \tag{1.252}$$

This condition is a DIRICHLET condition for velocity, satisfied for all times. Hence, the boundary integral in the weak form vanishes at the top and bottom boundaries. At the beginning, water rests under the normal atmospheric pressure:

$$\begin{aligned} p(\boldsymbol{x}, t = 0) &= 1\,\text{bar} \mathrel{\hat{=}} 10^5\,\text{Pa}\,, \quad \bar{p}(\bar{\boldsymbol{x}}, t = 0) = 100\,, \\ v_i(\boldsymbol{x}, t = 0) &= 0\,\text{m/s}\,, \quad \bar{v}_i(\bar{\boldsymbol{x}}, t = 0) = 0\,. \end{aligned} \tag{1.253}$$

It shall be recalled that we have not defined the characteristic length and time, however, we will choose them such that $\omega^2\rho = 10^3$ Pa. Moreover, the ratio of volume to shear viscosity, $a = 2\mu/\lambda$, is a dimensionless number. Since water is nearly incompressible, it is difficult to measure the volume viscosity λ. For numerical reasons we need a value of a; we choose it to be $a = 0.1$. The mechanical pressure is varied on the right and left ends such that p_{in} on left becomes greater than p_{out} on right. Then water flows from left to right. Thus, the mechanical pressure:

$$\begin{aligned} -\frac{1}{3}\bar{\sigma}_{kk} &= p_{\text{in}} \text{ on } \partial\Omega_{\text{left}}\,, \\ -\frac{1}{3}\bar{\sigma}_{kk} &= p_{\text{out}} \text{ on } \partial\Omega_{\text{right}}\,, \end{aligned} \tag{1.254}$$

is directed inward the domain on the left and right boundaries

$$
\begin{aligned}
\hat{t}_i &= -n_i p_{\text{in}} \text{ on } \partial\Omega_{\text{left}} \, , \\
\hat{t}_i &= -n_i p_{\text{out}} \text{ on } \partial\Omega_{\text{right}} \, .
\end{aligned}
\tag{1.255}
$$

We apply NEUMANN conditions for velocity by defining $\hat{t}_i = n_j \bar{\sigma}_{ji}$ on right and left boundaries. As a special case for incompressible fluid flows, we obtain the hydrostatic pressure on the boundaries

$$
\begin{aligned}
\bar{p} &= -\frac{1}{3}\bar{\sigma}_{kk} = p_{\text{in}} \text{ on } \partial\Omega_{\text{left}} \, , \\
\bar{p} &= -\frac{1}{3}\bar{\sigma}_{kk} = p_{\text{out}} \text{ on } \partial\Omega_{\text{right}} \, .
\end{aligned}
\tag{1.256}
$$

Hence, for the pressure, $\bar{p}$, we apply DIRICHLET conditions on the left and right boundaries. After applying the boundary conditions, boundary terms in the weak form read

$$
-\int_{\partial\Omega^*} \Big(\hat{t}_i + \bar{p} n_i\Big) \, \delta\bar{v}_i \mathrm{d}a \, , \tag{1.257}
$$

with $\partial\Omega^* = \partial\Omega_{\text{left}} \cup \partial\Omega_{\text{right}}$. From Eqs. (1.255), (1.256) it is obvious that the boundary terms vanish. Therefore, for an incompressible fluid flow we implement the mechanical pressure via hydrostatic pressure and the boundary integrals vanish. We have attained the weak form to be implemented:

$$
\begin{aligned}
\text{Form} = \int_\Omega \bigg(& \bar{v}_{i,i} \, \delta\bar{p} + \frac{(\bar{v}_i - \bar{v}_i^0)}{\Delta\bar{t}} \, \delta\bar{v}_i + \bar{v}_j \bar{v}_{i,j} \, \delta\bar{v}_i + \bar{p}_{,i} \, \delta\bar{v}_i + \frac{1}{aRe} \bar{v}_{k,k} \, \delta\bar{v}_{i,i} + \\
& + \frac{1}{Re} \bar{v}_{(j,i)} \, \delta\bar{v}_{i,j} - \bar{f}_i \, \delta\bar{v}_i \bigg) \mathrm{d}v \, ,
\end{aligned}
\tag{1.258}
$$

which is nonlinear in primitive variables. Thus, we use again the NEWTON–RAPHSON linearization by a symbolic differentiation. The pressure difference induces a flow within the channel. We start from an equilibrium (no pressure difference) and apply a pressure difference linear in time up to the time $\bar{t} = 0.5$. Then the pressure difference is held constantly. After a while under the constant pressure difference, a steady-state of the flow is reached. In Fig. 1.14 the solutions at two different time instants for the REYNOLDS number $Re = 100$ can be seen.

The channel is of dimensions 5×1 in $\bar{x}$. Its real length depends on the characteristic length ℓ, which is the thickness of the channel. The chosen REYNOLDS number, $Re = 100$, for the flow of water[70] having $\mu = 1001.6 \cdot 10^{-6}\,\text{Ns/m}^2$ and $\rho = 998.21\,\text{kg/m}^3$ results in

[70] See [29, p. 32] at 20 °C.

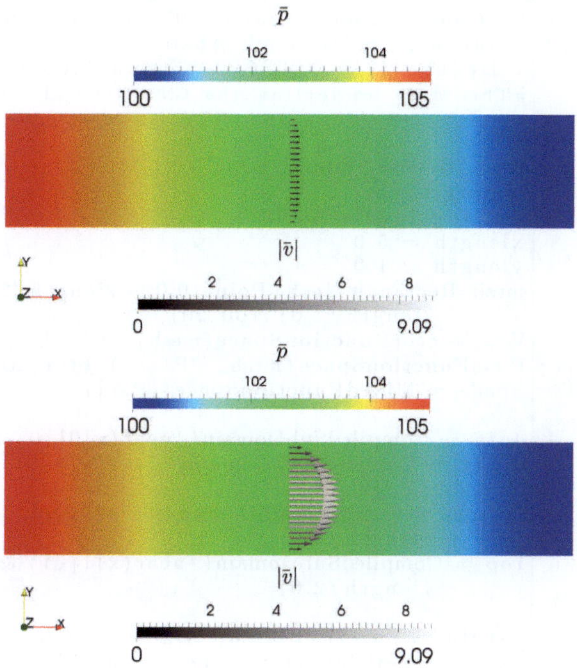

Fig. 1.14 Normalized pressure distribution as colors and velocity distribution as *arrows* at $\bar{t} = 2.0$ on the *upper* and at $\bar{t} = 10.0$ on the *lower* visualization. At $\bar{t} = 1.0$ the pressure 105 on *left* is achieved and held, where $\bar{p} = 100$ corresponds to $p = 1\,\text{bar}$

$$Re = \frac{\omega \ell \rho}{2\mu} = \frac{\ell^2 \rho}{\tau 2\mu}\,, \quad \frac{\ell^2}{\tau} = 2 \cdot 10^{-4}\,\text{m}^2\,. \tag{1.259}$$

Since we have chosen $\bar{p}$ such that

$$\omega^2 \rho = \frac{\ell^2}{\tau^2}\rho = 10^3\,\text{Pa}\,, \quad \frac{\ell^2}{\tau} = \frac{\tau}{\rho}10^3\,, \tag{1.260}$$

we obtain

$$\begin{aligned} &\frac{\tau}{\rho}10^3 = 2 \cdot 10^{-4}\,, \quad \tau = 2\rho 10^{-7} = 2 \cdot 10^{-4}\,\text{s}\,, \\ &\ell = 2 \cdot 10^{-4}\,\text{m}\,, \quad \omega = \frac{\ell}{\tau} = 1\,\text{m/s}\,. \end{aligned} \tag{1.261}$$

Since the characteristic length is chosen as the width of the channel, we have simulated a channel of $1 \times 0.2\,\text{mm}$ for a total time of $2\,\text{ms}$. For a longer simulation in a bigger channel, we need to change the REYNOLDS number. The code is given below. The primitive variables are solved monolithically by using a mixed function space.

```
"""Computational reality 07, linear fluid flow"""
__author__ = "B. Emek Abali"
__license__ = "GNU GPL Version 3.0 or later"
#This code underlies the GNU General Public License, http://www.gnu.org/licenses/gpl-3.0.en.html

from fenics import *
import numpy
set_log_level(ERROR)
xlength = 5.0
ylength = 1.0
mesh=RectangleMesh(Point(0.0,-ylength/2.0), Point(xlength, ylength/2.0),100,20)
V = VectorFunctionSpace(mesh, 'P', 1) #for velocity
P = FunctionSpace(mesh, 'P', 1) #for pressure
Space = MixedFunctionSpace([P,V])

left = CompiledSubDomain('near(x[0],0) && on_boundary')
right = CompiledSubDomain('near(x[0],l) && on_boundary',l=xlength)
bottom = CompiledSubDomain('near(x[1],l) && on_boundary',l=-ylength/2.0)
top = CompiledSubDomain('near(x[1],l) && on_boundary',l=+ylength/2.0)

facets = FacetFunction('size_t',mesh)
cells = CellFunction('size_t',mesh)
da = Measure('ds', domain=mesh, subdomain_data=facets)
dv = Measure('dx', domain=mesh, subdomain_data=cells)
n = FacetNormal(mesh)

facets.set_all(0)
left.mark(facets, 1)
right.mark(facets, 2)

#setting the Dirichlet conditions
v_noslip = Constant(('0.0', '0.0'))
p_in = Expression('100.0+10.0*t',t=0)
p_out = Constant('100.0')
tL = -p_in*n
tR = -p_out*n
p_bc1=DirichletBC(Space.sub(0), p_in, left)
p_bc2=DirichletBC(Space.sub(0), p_out, right)
v_bc1=DirichletBC(Space.sub(1), v_noslip, bottom)
v_bc2=DirichletBC(Space.sub(1), v_noslip, top)
v_bc3=DirichletBC(Space.sub(1).sub(1), 0.0, left)
v_bc4=DirichletBC(Space.sub(1).sub(1), 0.0, right)

bc=[p_bc1,p_bc2,v_bc1,v_bc2,v_bc3,v_bc4]
f = Expression(('0.0', '0.0'))
# material constants
a = 0.1
Re = 100.
```

```
i, j, k, l = indices(4)
t = 0.0
t_bc = 0.5
t_end = 10.0
dt = 0.05

test = TestFunction(Space)
du = TrialFunction(Space)
u0 = Function(Space) # solution from the last time slice
u = Function(Space) # current solution
u_init = Expression(('p0','0.0','0.0'), p0=100.0)
u0.interpolate(u_init)
u.interpolate(u_init)
# split() sets pointer to the memory address,
# so it is a direct access, no deep copy
p0, v0= split(u0)
p, v = split(u)
del_p, del_v = split(test)
Form = (v[i].dx(i)*del_p + (v-v0)[i]/dt*del_v[i] \
    + v[j]*v[i].dx(j)*del_v[i] + p.dx(i)*del_v[i] \
    + 1.0/(a*Re)*v[k].dx(k)*del_v[i].dx(i) \
    + 1.0/Re*sym(grad(v))[j,i]*del_v[i].dx(j) \
    - f[i]*del_v[i] )*dv

Gain = derivative(Form, u, du)
file_p = File('/calcul/CR07/pres.pvd')
file_v = File('/calcul/CR07/velo.pvd')
# time loop
tic()
while t<= t_end:
    t = t + dt
    if t<t_bc: p_in.t = t
    else: p_in.t = t_bc
    print 'time: ' ,t
    solve(Form==0, u, bc, J=Gain, \
        solver_parameters={"newton_solver":{"linear_solver":
        ↪ "lu", "relative_tolerance": 1e-3} }, \
        form_compiler_parameters={"cpp_optimize": True, "
        ↪ "representation": "quadrature", "
        ↪ quadrature_degree": 2} )

    file_p << (u.split()[0],t)
    file_v << (u.split()[1],t)
    u0.assign(u)

print 'the total simulation time in seconds: ', toc()
```

To-do

A linear viscous fluid flowing in a channel is implemented.

- Show the nonlinear terms in Eq. (1.258).
- Inspect the code above and change the boundary conditions such that the fluid may enter and leave with a vertical velocity. Explain the changes in the pressure and velocity distributions.

- Vary the value of a and report if there is a significant change in the solution.
- Try to change the REYNOLDS number for solving a fluid flow in a bigger channel of $1 \times 0.2\,\text{m}$. Is the code working for this Re?
- Do the same calculations without normalizing and obtain the weak form in the unit of power:

$$\text{Form} = \int_\Omega \Big(v_{i,i}\, \delta p + \frac{\rho}{\Delta t}(v_i - v_i^0)\, \delta v_i + \rho v_j v_{i,j}\, \delta v_i + p_{,i}\, \delta v_i + \\ + \lambda v_{k,k}\, \delta v_{i,i} + 2\mu v_{(j,i)}\, \delta v_{i,j} - \rho f_i\, \delta v_i \Big) \mathrm{d}v \,.$$

  Try to implement the latter form into the code and observe a simulation.
- Explain the type of flow for $Re \to \infty$. See [8, 14], and [21, Sect. 4.8] for various methods trying to prevent the numerical problems occurring in simulations with a high REYNOLDS number.

1.8 Nonlinear Viscous Fluids

For suspensions and polymer melts, a linear constitutive equation fails to be adequate. Especially polymer melts are often called BINGHAM–ILYUSHIN fluids[71] showing a nonlinear viscous behavior. In this section we will introduce a nonlinear constitutive equation modeling such a behavior. Moreover, we discuss the linearization in the partial differentiation's level as utilized in the former sections and handle the linearization manually.

Consider a two-dimensional domain, Ω, where $x_i = (x_1, x_2)$ label the spatial positions on Ω, which is expressed in the Cartesian coordinate system. The goal is to compute the velocity and pressure fields for every coordinates, $\boldsymbol{x}$, in Ω. The domain is fixed, i.e., the coordinates x_i are constant in time. The fixed domain can be a channel with walls on top and bottom; with open boundaries on right and left. Material enters and leaves the domain—it is an open system. The mass is a conserved quantity,[72] hence the change of the mass within the domain can be tracked by measuring the change across the (fixed) boundaries of the domain

$$\Big(\int_\Omega \rho \mathrm{d}v \Big)^{\boldsymbol{\cdot}} = - \int_{\partial\Omega} \rho n_i v_i \mathrm{d}a \,. \tag{1.262}$$

[71] They are named after Eugene Cook Bingham and Alexey Antonovich Il'yushin.

[72] A conserved quantity lacks a production term, it cannot vanish or be produced out of nothing.

The right hand side corresponds to the *convective* term modeling the particles entering the domain. Since the plane normal, n_i, points outward the domain, a minus sign in front of the convective term denotes an increase of the mass. Analogously, the linear momentum is a conserved quantity; its balance equation lacks a production term. For an open system with the fixed domain, the balance of linear momentum reads

$$\Big(\int_\Omega \rho v_i \mathrm{d}v\Big)^{\cdot} = -\int_{\partial\Omega} \rho v_i n_j v_j \mathrm{d}a + \int_{\partial\Omega} t_i \mathrm{d}a + \int_\Omega \rho f_i \mathrm{d}v \ . \tag{1.263}$$

Besides the convective term, $-\rho v_i n_j v_j$, a *non-convective* term, t_i, may also increase the linear momentum. Without any particles entering or leaving the system, the non-convective term affects the linear momentum. For example, the mechanical pressure on the boundary may increase the linear momentum within the domain. The CAUCHY tetrahedron argument:[73]

$$t_i = n_j \sigma_{ji} \ , \tag{1.264}$$

leads to the CAUCHY stress tensor, σ_{ji}. For the formulations in this section, we ignore the body forces, $f_i = 0$. We obtain two residuals by subtracting the right hand side from the left hand side, after applying GAUSS's law, they read

$$\begin{aligned} \mathrm{R}_\rho &= \int_\Omega \Big(\frac{\partial \rho}{\partial t} + x_i^{\cdot} \frac{\partial \rho}{\partial x_i}\Big)\mathrm{d}v + \int_\Omega \rho (\mathrm{d}v)^{\cdot} + \int_\Omega \frac{\partial}{\partial x_i}(\rho v_i)\mathrm{d}v = 0 \ , \\ \mathrm{R}_v &= \int_\Omega \Big(\frac{\partial \rho v_i}{\partial t} + x_j^{\cdot} \frac{\partial \rho v_i}{\partial x_j}\Big)\mathrm{d}v + \int_\Omega \rho v_i (\mathrm{d}v)^{\cdot} + \int_\Omega \frac{\partial}{\partial x_j}(\rho v_i v_j - \sigma_{ji})\mathrm{d}v = 0 \ . \end{aligned} \tag{1.265}$$

Since the domain is fixed, these relations hold

$$x_i^{\cdot} = 0 \ , \quad (\mathrm{d}v)^{\cdot} = 0 \ . \tag{1.266}$$

The spatial position, x_i, has no dependence on the underlying material. The domain itself might have a velocity, too. In this section we consider a fixed domain leading to the latter relations. Then the residuals read

$$\begin{aligned} \mathrm{R}_\rho &= \int_\Omega \Big(\frac{\partial \rho}{\partial t} + \frac{\partial}{\partial x_i}(\rho v_i)\Big)\mathrm{d}v = 0 \ , \\ \mathrm{R}_v &= \int_\Omega \Big(\frac{\partial \rho v_i}{\partial t} + \frac{\partial}{\partial x_j}(\rho v_i v_j - \sigma_{ji})\Big)\mathrm{d}v = 0 \ . \end{aligned} \tag{1.267}$$

By using the first residual in the second, we obtain

[73] It is formulated by Augustin-Louis Cauchy. For a detailed explanation see [28, Sect. 203].

$$
\begin{aligned}
\mathrm{R}_\rho &= \int_\Omega \Big(\frac{\partial \rho}{\partial t} + \frac{\partial \rho v_i}{\partial x_i} \Big) \mathrm{d}v = 0 \,, \\
\mathrm{R}_v &= \int_\Omega \Big(\rho \frac{\partial v_i}{\partial t} + \rho v_j \frac{\partial v_i}{\partial x_j} - \frac{\partial \sigma_{ji}}{\partial x_j} \Big) \mathrm{d}v = 0 \,.
\end{aligned}
\tag{1.268}
$$

Moreover, we apply the incompressibility assumption:

$$
\frac{\partial \rho}{\partial t} = 0 \,, \quad \frac{\partial \rho}{\partial x_i} = 0 \,, \tag{1.269}
$$

such that

$$
\begin{aligned}
\mathrm{R}_\rho &= \int_\Omega \frac{\partial v_i}{\partial x_i} \mathrm{d}v \,, \\
\mathrm{R}_v &= \int_\Omega \Big(\rho \frac{\partial v_i}{\partial t} + \rho v_j \frac{\partial v_i}{\partial x_j} - \frac{\partial \sigma_{ji}}{\partial x_j} \Big) \mathrm{d}v \,.
\end{aligned}
\tag{1.270}
$$

The residuals vanish for the correct choice of primitive variables, $\{p, v_i\}$, therefore, we can multiply or *weight* them by arbitrary test functions, $\{\delta p, \delta v_i\}$, respectively. After multiplication by the test functions, we check the units and find out that both are in the unit of power such that we can sum them up and get the variational form:

$$
\text{Form} = \int_\Omega \frac{\partial v_i}{\partial x_i} \, \delta p \mathrm{d}v + \int_\Omega \Big(\rho \frac{\partial v_i}{\partial t} + \rho v_j \frac{\partial v_i}{\partial x_j} - \frac{\partial \sigma_{ji}}{\partial x_j} \Big) \, \delta v_i \mathrm{d}v \,. \tag{1.271}
$$

The flux of linear momentum is given by the CAUCHY stress tensor, σ_{ji}. This tensor is of rank two and it is symmetric for non-polar materials, $\sigma_{ji} = \sigma_{ij}$. For viscous fluids, the CAUCHY stress is a function of the symmetric velocity gradient:

$$
\sigma_{ij} = \sigma_{ij}(d_{ij}) \,, \quad d_{ij} = \frac{\partial v_{(i}}{\partial x_{j)}} = \frac{1}{2} \Big(\frac{\partial v_i}{\partial x_j} + \frac{\partial v_j}{\partial x_i} \Big) \,. \tag{1.272}
$$

This functional dependency can be analyzed mathematically in order to attain the generic functional form of a tensor depending on another tensor. Representing theorems of tensors deal with such functional relations. For an isotropic material, the generic representation of a tensor of rank two, σ_{ij}, depending on one (symmetric) variable, d_{ij}, reads[74]

$$
\sigma_{ji} = a\delta_{ji} + b d_{ji} + c d_{jk} d_{ki} \,, \tag{1.273}
$$

where a, b, c are scalar functions of the so-called *invariants* of d_{ij} and are given as follows

[74]For a detailed derivation see [25].

$$a = a(I, II, III)\ ,\quad b = b(I, II, III)\ ,\quad c = c(I, II, III)\ ,$$
$$I = d_{ii}\ ,\quad II = \frac{1}{2} d_{ij} d_{ij}\ ,\quad III = \frac{1}{3} d_{ij} d_{jk} d_{ki}\ . \tag{1.274}$$

The rank two of the tensor limits the number of invariants to three, I, II, III, in three-dimensional space according to the HAMILTON–CAYLEY theorem.[75] The stress, σ_{ij}, is quadratic in its argument, d_{ij}. If we assume $c = 0$ then the stress is linear in its argument. By choosing

$$a = -p + \lambda I\ ,\quad b = 2\mu\ ,\quad c = 0\ , \tag{1.275}$$

where p denotes pressure, λ is the volume viscosity, and μ indicates the shear viscosity; the stress definition results in NAVIER–STOKES's equation. In this definition a, b, c are constants in the invariants (p depends on space and time).

Analogously we can employ a more complicated material model. For example, we adopt a *sigmoid* model:[76]

$$a = -p + \lambda I\ ,\quad b = 2\mu + \frac{2k}{\pi\sqrt{II}} \arctan\Big(\frac{\sqrt{II}}{B}\Big)\ ,\quad c = 0\ . \tag{1.276}$$

The new term with arctan() brings in an effect that we all experience at breakfast—the fight against honey! Honey is a viscous material and it flows down the knife. If we move the knife quickly, the same material behaves differently and "sticks" to the knife. By a higher velocity gradient, which we achieve intuitively by rotating the knife, the viscosity of honey is increased. In other words, the viscosity of the material, given by the parameter b, depends on the velocity gradient, $b(II)$. Such materials are called BINGHAM–ILYUSHIN fluids. In rheology the *prominent* constitutive equations modeling nonlinear viscous fluids are the CARREAU model,[77] the power-law model, and the BINGHAM model.[78]

The relation between stress and velocity gradient is component-wise. In order to get a better understanding of the constitutive equation:

$$\sigma_{ji} = (-p + \lambda I)\delta_{ji} + \Big(2\mu + \frac{2k}{\pi\sqrt{II}} \arctan\Big(\frac{\sqrt{II}}{B}\Big)\Big) d_{ji}\ , \tag{1.277}$$

we consider a simple example. In a shear test on x_1x_2-plane, the following symmetric velocity gradient arises with the invariants:

---

75 The theorem is named after Arthur Cayley and William Rowan Hamilton.

76 For this model see [30, 31] where the volume viscosity has been ignored due to the incompressibility, however, we include the volume viscosity for a better numerical stability.

77 It is named after Pierre Carreau.

78 See [21, Sect. 6.2.3] for equations of these models.

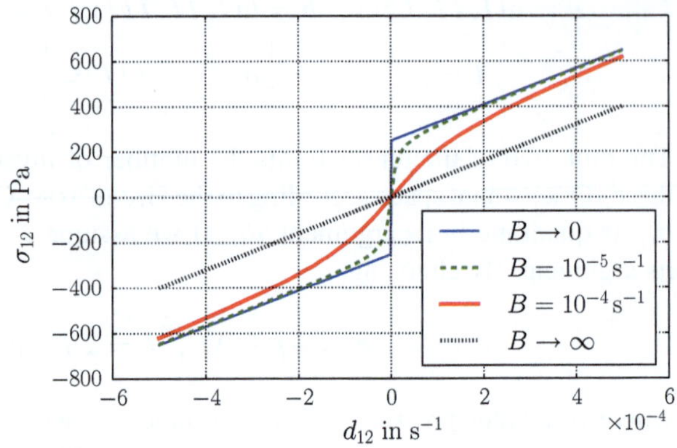

Fig. 1.15 Shear stress versus shear velocity gradient. By increasing the value of B the sigmoid model transforms into NAVIER–STOKES's equation. If the value of B converges zero, there occurs a limit stress k and a visually discontinuous change above the limit stress. Actually, the function is continuous and differentiable due to arctan()

$$
d_{ij} = \begin{pmatrix} 0 & d_{12} & 0 \\ d_{12} & 0 & 0 \\ 0 & 0 & 0 \end{pmatrix} , \quad I = 0 , \quad II = d_{12}^2 , \quad III = 0 . \tag{1.278}
$$

Hence, in the shear test, the shear component of stress reads

$$
\sigma_{12} = \left(2\mu + \frac{2k}{\pi \left| d_{12} \right|} \arctan \left(\frac{\left| d_{12} \right|}{B} \right) \right) d_{12} . \tag{1.279}
$$

For presenting the capabilities of the sigmoid model, we plot this relation for realistic values and vary the parameter B in Fig. 1.15. Owing to the shape of the function in Fig. 1.15 we have called it a *sigmoid* model. By setting the parameter B small enough we obtain a material model with a limit stress. Caused by arctan() the function is differentiable in the whole domain. Therefore, the sigmoid material model is beneficial, however, it makes the set of equations highly nonlinear. In order to investigate the nonlinearity further, we generate the weak form. After integrating by parts, using the finite difference method for time discretization, and observing $(\cdot)_{,i} = \partial(\cdot)/\partial x_i$, the weak form reads

$$
\begin{aligned}
\text{Form} = & \int_\Omega \Big(v_{i,i}\, \delta p + \frac{\rho}{\Delta t}(v_i - v_i^0)\, \delta v_i + \rho v_j v_{i,j}\, \delta v_i + p_{,i}\, \delta v_i + \lambda I\, \delta v_{i,i} + \\
& + b d_{ji}\, \delta v_{i,j} \Big) \mathrm{d}v - \int_{\partial\Omega} (\lambda I\, \delta v_i n_i + b d_{ji}\, \delta v_i n_j) \mathrm{d}a = \\
= & \int_\Omega \Big(v_{i,i}\, \delta p + \frac{\rho}{\Delta t}(v_i - v_i^0)\, \delta v_i + \rho v_j v_{i,j}\, \delta v_i + p_{,i}\, \delta v_i + \lambda I\, \delta v_{i,i} + \\
& + b d_{ji}\, \delta v_{i,j} \Big) \mathrm{d}v - \int_{\partial\Omega} (\hat{\sigma}_{ji} + p\delta_{ji}) n_j\, \delta v_i \mathrm{d}a .
\end{aligned} \tag{1.280}
$$

On boundaries with a given velocity, we apply a DIRICHLET condition of velocity such that the test function of velocity vanishes. On other boundaries we apply a

DIRICHLET condition of pressure. Since for incompressible flows the mechanical and hydrostatic pressure are equal, the boundary term in the integral form vanishes. Thus, the weak form to be implemented becomes

$$\text{Form} = \int_\Omega \Big(v_{i,i}\,\delta p + \frac{\rho}{\Delta t}(v_i - v_i^0)\,\delta v_i + \rho v_j v_{i,j}\,\delta v_i + p_{,i}\,\delta v_i + \lambda I\,\delta v_{i,i} + \\ + \Big(2\mu + \frac{2k}{\pi\sqrt{II}}\arctan\Big(\frac{\sqrt{II}}{B}\Big)\Big) d_{ji}\,\delta v_{i,j}\Big)\mathrm{d}v\,. \tag{1.281}$$

This weak form is obviously nonlinear, we have to linearize it. We basically perform an abstract linearization using NEWTON's method at the partial differential level.[79] The latter form is a functional of primitive variables $\boldsymbol{P} = \{p, v_i\}$ and their variations (test functions) $\delta\boldsymbol{P} = \{\delta p, \delta v_i\}$; we write it as Form= $F(\boldsymbol{P}, \delta\boldsymbol{P})$. The form is fulfilled initially since we know the correct values of $\boldsymbol{P}$ at t_0. We can search for the next time step, $t = \Delta t + t_0$, by considering the known $\boldsymbol{P}$. For the subsequent time steps the same approach holds, since the values from the last time steps are known. We can describe the algorithm in the following way:

$$\begin{aligned} \text{given: } & \boldsymbol{P}(t) \text{ for } \boldsymbol{x}\,, \\ \text{find: } & \boldsymbol{P}(t+\Delta t) \text{ at } \boldsymbol{x}\,, \\ \text{satisfying: } & F(\boldsymbol{P}(t+\Delta t), \delta\boldsymbol{P}) = 0\,. \end{aligned} \tag{1.282}$$

We can rewrite the unknowns $\boldsymbol{P}(t + \Delta t)$ in terms of the known values

$$\boldsymbol{P}(t+\Delta t) = \boldsymbol{P}(t) + \Delta\boldsymbol{P}(t)\,, \tag{1.283}$$

and search for $\Delta\boldsymbol{P}(t)$ instead of $\boldsymbol{P}(t + \Delta t)$. If Δt is sufficiently small, then the solution is near to the known solution, i.e., $\Delta\boldsymbol{P}(t)$ is small. For a small $\Delta\boldsymbol{P}(t)$ we can utilize a TAYLOR expansion[80] around the known values, $\boldsymbol{P}(t)$, up to the order one

$$F(\boldsymbol{P} + \Delta\boldsymbol{P}, \delta\boldsymbol{P}) = F(\boldsymbol{P}, \delta\boldsymbol{P}) + \nabla_{\boldsymbol{P}} F(\boldsymbol{P}, \delta\boldsymbol{P}) \cdot \Delta\boldsymbol{P}\,, \tag{1.284}$$

where we have suppressed the time argument for the sake of a simplified notation. We neglect the quadratic terms in the TAYLOR expansion owing to the small $\Delta\boldsymbol{P}$; so we have to use small time steps in the simulation, otherwise the solution will not converge. We have applied a condition that the formulation is of order one in $\Delta\boldsymbol{P}$. The same condition has to be satisfied for the differentiation operator, $\nabla_{\boldsymbol{P}}(\cdot)$, too. Therefore, we apply the so-called *directed* or GATEAUX derivative:[81]

[79] We follow the ideas in [6, Part I, Sect. 2.2.3].

[80] It is named for Brook Taylor.

[81] It is named after René Eugène Gâteaux.

$$\nabla_{\boldsymbol{P}} F(\boldsymbol{P}, \delta \boldsymbol{P}) \cdot \Delta \boldsymbol{P} = \lim_{\epsilon \to 0} \frac{\mathrm{d}}{\mathrm{d}\epsilon} F(\boldsymbol{P} + \epsilon \Delta \boldsymbol{P}, \delta \boldsymbol{P}) \,, \tag{1.285}$$

where ϵ is an arbitrary parameter. This directed derivative is the derivative in $\boldsymbol{P}$ projected in the direction $\Delta \boldsymbol{P}$. Technically, we first differentiate in ϵ and then set the parameter ϵ equal to zero. Hence, only terms of order one in $\Delta \boldsymbol{P}$ remain in the solution. The higher order terms vanish. By introducing the so-called JACOBIan:[82]

$$\boldsymbol{J}(\boldsymbol{P}, \delta \boldsymbol{P}) = \nabla_{\boldsymbol{P}} F(\boldsymbol{P}, \delta \boldsymbol{P}) \,, \tag{1.286}$$

we can reformulate the algorithm:

$$\begin{aligned} \text{given: } & \boldsymbol{P} \text{ for } \boldsymbol{x} \,, \\ \text{find: } & \Delta \boldsymbol{P} \text{ at } \boldsymbol{x} \,, \\ \text{satisfying: } & F(\boldsymbol{P}, \delta \boldsymbol{P}) + \boldsymbol{J}(\boldsymbol{P}, \delta \boldsymbol{P}) \cdot \Delta \boldsymbol{P} = 0 \,. \end{aligned} \tag{1.287}$$

This linear function in $\Delta \boldsymbol{P}$ can be solved and applied to update the solution:

$$\boldsymbol{P} := \boldsymbol{P} + \Delta \boldsymbol{P} \,, \tag{1.288}$$

where ":=" is an assign operator in computational algebra.[83] Finally, we obtain the following algorithm:

$$\begin{aligned} & \text{while } |\Delta \boldsymbol{P}| > \text{TOL.} \\ & \qquad \text{solve } \Delta \boldsymbol{P}, \text{ where } F(\boldsymbol{P}, \delta \boldsymbol{P}) + \boldsymbol{J}(\boldsymbol{P}, \delta \boldsymbol{P}) \cdot \Delta \boldsymbol{P} = 0 \\ & \qquad \boldsymbol{P} := \boldsymbol{P} + \Delta \boldsymbol{P} \end{aligned} \tag{1.289}$$

The term $\boldsymbol{J} \cdot \Delta \boldsymbol{P}$ can be computed automatically.[84] We have used the names "Form" for $F(\boldsymbol{P}, \delta \boldsymbol{P})$ and "Gain" for $\boldsymbol{J}(\boldsymbol{P}, \delta \boldsymbol{P}) \cdot \Delta \boldsymbol{P}$ in the codes, where "Gain" has been computed automatically. In this section we will obtain it manually by calculating the directed derivative.

For the form in Eq. (1.281) we obtain $F(p + \epsilon \Delta p, v_i + \epsilon \Delta v_i, \delta p, \delta v_i)$ as follows

---

[82] It is named after Carl Gustav Jakob Jacobi.

[83] An assign operator is not a mathematical equality. The value of the object $\boldsymbol{P}$ is updated as $\boldsymbol{P} + \Delta \boldsymbol{P}$. The assign operator in computational algebra is also denoted by $\leftarrow$ so we may write $\boldsymbol{P} \leftarrow \boldsymbol{P} + \Delta \boldsymbol{P}$. In mathematics, the assign operator := is used in the meaning of a definition, i.e., an equality introduced for the first time. We use = for every mathematical equation and := for a computational assignment of values.

[84] The symbolic computation of the JACOBIan is realized by SyFi in FEniCS project, see [2, 3].

$$
\begin{aligned}
p &\to p+\epsilon\Delta p \\
v_i &\to v_i+\epsilon\Delta v_i\ , \\
d_{ij}=v_{(i,j)} &\to \frac{1}{2}(v_{i,j}+\epsilon\Delta v_{i,j}+v_{j,i}+\epsilon\Delta v_{j,i})=\frac{1}{2}(v_{i,j}+v_{j,i})+ \\
&+\frac{1}{2}\epsilon(\Delta v_{i,j}+\Delta v_{j,i})=v_{(i,j)}+\epsilon\Delta v_{(i,j)}\ , \\
II=\frac{1}{2}v_{(k,l)}v_{(k,l)} &\to \frac{1}{2}S\ ,\quad S=(v_{(k,l)}v_{(k,l)}+2\epsilon v_{(k,l)}\Delta v_{(k,l)}+\epsilon^2\Delta v_{(k,l)}\Delta v_{(k,l)})
\end{aligned}
\tag{1.290}
$$

and

$$
\begin{aligned}
F(p,v_i,\delta p,\delta v_i)\to F(p+\epsilon\Delta p,v_i+\epsilon\Delta v_i,\delta p,\delta v_i)=\int_\Omega\Big((v_i+\epsilon\Delta v_i)_{,i}\,\delta p+ \\
+\rho\frac{(v_i+\epsilon\Delta v_i-v_i^0)}{\Delta t}\,\delta v_i+(v_j+\epsilon\Delta v_j)\rho(v_i+\epsilon\Delta v_i)_{,j}\,\delta v_i+(p+\epsilon\Delta p)_{,i}\,\delta v_i+ \\
+\lambda(v_{k,k}+\epsilon\Delta v_{k,k})\,\delta v_{i,i}+2\mu(v_{(k,j)}+\epsilon\Delta v_{(k,j)})\,\delta v_{j,k}+ \\
+\frac{2k\sqrt{2}}{\pi\sqrt{S}}\arctan\left(\frac{\sqrt{S}}{\sqrt{2}B}\right)\big(v_{(k,j)}+\epsilon\Delta v_{(k,j)}\big)\,\delta v_{j,k}\Big)\mathrm{d}v\ .
\end{aligned}
\tag{1.291}
$$

We can now calculate "Gain" by using the directed derivative[85]

$$
\begin{aligned}
\text{Gain}=\frac{\mathrm{d}}{\mathrm{d}\epsilon}F(v_i+\epsilon\Delta v_i,\,p+\epsilon\Delta p,\,\delta p,\,\delta v_i)\Big|_{\epsilon=0}=\int_\Omega\Big(\Delta v_{i,i}\,\delta p+ \\
+\frac{\rho}{\Delta t}\Delta v_i\,\delta v_i+\Delta v_j\rho(v_{i,j}+\epsilon\Delta v_{i,j})\,\delta v_i+(v_j+\epsilon\Delta v_j)\rho\Delta v_{i,j}\,\delta v_i+ \\
+\Delta p_{,i}\,\delta v_i+\lambda\Delta v_{k,k}\,\delta v_{i,i}+2\mu\Delta v_{(k,j)}\,\delta v_{j,k}-\frac{k\sqrt{2}}{\pi\,S^{3/2}}\big(2v_{(i,l)}\Delta v_{(i,l)}+ \\
+2\epsilon\Delta v_{(i,l)}\Delta v_{(i,l)}\big)\arctan\left(\frac{\sqrt{S}}{\sqrt{2}B}\right)\big(v_{(k,j)}+\epsilon\Delta v_{(k,j)}\big)\,\delta v_{j,k}+\frac{2k\sqrt{2}}{\pi\,S^{1/2}}\frac{2B^2}{2B^2+S}\times \\
\times\frac{2v_{(m,n)}\Delta v_{(m,n)}+2\epsilon\Delta v_{(m,n)}\Delta v_{(m,n)}}{2\sqrt{2}BS^{1/2}}(v_{(k,j)}+\epsilon\Delta v_{(k,j)})\,\delta v_{j,k}+ \\
+\frac{2k\sqrt{2}}{\pi\,S^{1/2}}\arctan\left(\frac{S^{1/2}}{\sqrt{2}B}\right)\Delta v_{(k,j)}\,\delta v_{j,k}\Big)\mathrm{d}v\Big|_{\epsilon=0}\ ,
\end{aligned}
\tag{1.292}
$$

---

[85] The derivative of arctan() reads

$$
\frac{\partial}{\partial x}\Big(\arctan(f)\Big)=\frac{1}{1+f^2}\frac{\partial f}{\partial x}\ ,
\tag{1.291}
$$

according to [5, Sect. 21.5.4.3, nr. 473].

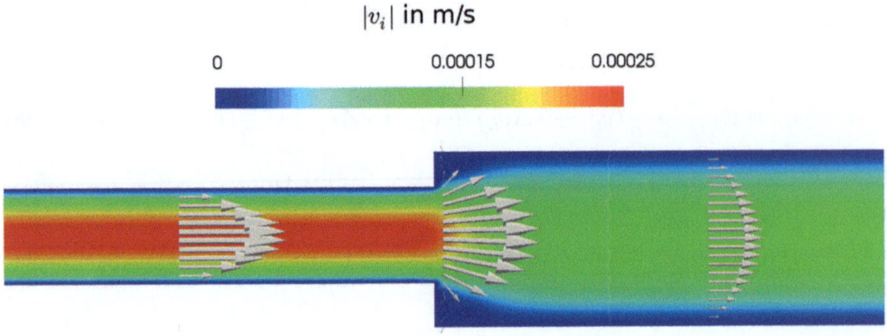

Fig. 1.16 Velocity distribution at $t = 10\,\text{s}$ for the flow of the HDPE. The color denotes the magnitude and the arrows their direction

such that we obtain

$$\begin{aligned}\text{Gain} = \int_\Omega \bigg(& \Delta v_{i,i}\, \delta p + \frac{\rho}{\Delta t} \Delta v_i\, \delta v_i + \Delta v_j \rho v_{i,j}\, \delta v_i + v_j \rho \Delta v_{i,j}\, \delta v_i + \Delta p_{,i}\, \delta v_i + \\ & + \lambda \Delta v_{k,k}\, \delta v_{i,i} + 2\mu \Delta v_{(k,j)}\, \delta v_{j,k} - \frac{2k\sqrt{2}}{\pi \big(v_{(m,n)} v_{(m,n)}\big)^{3/2}} v_{(i,l)} \Delta v_{(i,l)} \times \\ & \times \arctan\bigg(\frac{\sqrt{v_{(m,n)} v_{(m,n)}}}{\sqrt{2}B}\bigg) v_{(k,j)}\, \delta v_{j,k} + \frac{4Bk v_{(m,n)} \Delta v_{(m,n)}}{\pi\, v_{(i,l)} v_{(i,l)} \big(2B^2 + v_{(p,r)} v_{(p,r)}\big)} \times \\ & \times v_{(k,j)}\, \delta v_{j,k} + \frac{2\sqrt{2}k}{\pi \sqrt{v_{(i,l)} v_{(i,l)}}} \arctan\bigg(\frac{\sqrt{v_{(m,n)} v_{(m,n)}}}{\sqrt{2}B}\bigg) \Delta v_{(k,j)}\, \delta v_{j,k} \bigg) \mathrm{d}v \,. \end{aligned} \tag{1.293}$$

We want to simulate a thick polymer melt. A high-density polyethylene (HDPE) is a thermoplastic made from petroleum and its material parameters read

$$\begin{gathered} \rho = 1450\,\text{kg/m}^3 \,, \quad \mu = 400\,000\,\text{Pa s} \,, \quad \lambda = 100\,000\,\text{Pa s} \,, \\ k = 250\,\text{Pa} \,, \quad B = 0.00001\,\text{s}^{-1} \,. \end{gathered} \tag{1.294}$$

We combine two channel flows such that the polymer melt is pressed down to a bigger channel. The velocity distribution at $t = 10\,\text{s}$ is visualized in Fig. 1.16. The pressure distribution in Fig. 1.17 shows some perturbations like a *chess* board. These perturbations indicate that the converged solution encounters some *numerical* problems. The chess board solution of the pressure field is wrong. Especially for incompressible fluid flows, there are various strategies for rectifying the solution. The most prominent amendment of the solution is to choose different shape functions for the primitive variables. We have chosen the same polynomial degree for elements building the functional space for velocity and pressure. However, the so-called LADYZHEN-

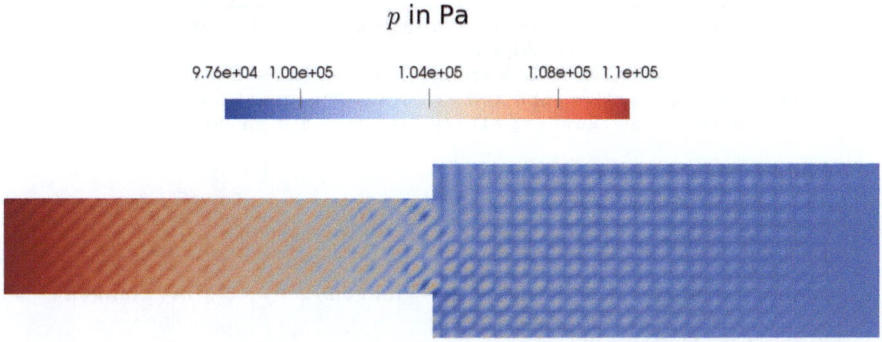

Fig. 1.17 Pressure distribution at $t = 10\,$s for the flow of the HDPE (high-density polyethylene)

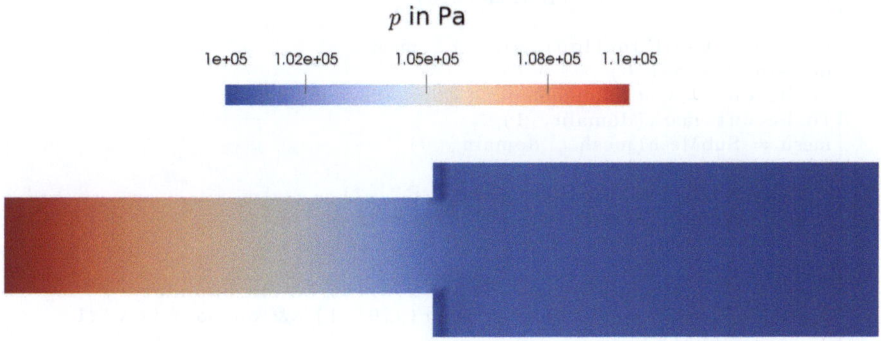

Fig. 1.18 Pressure distribution at $t = 10\,$s for the flow of the HDPE (high-density polyethylene) calculated by using Taylor–Hood elements

skaya–Babuska–Brezzi (LBB)[86] or inf-sup conditions suggest to use a higher order element for velocity than pressure. This type of elements belong to Taylor–Hood family[87] as introduced in [27]. For example, we can use quadratic elements for velocity (P2) and linear elements for pressure (P1). With the implementation of Taylor–Hood elements, the pressure distribution in Fig. 1.18 is achieved. The code below shows the implementation of the LBB-conditions. Their generalization to the case of many primitive variables (more than two) seems to be difficult.[88]

---

[86] LBB conditions are named after Olga Aleksandrovna Ladyzhenskaya, Ivo Babuška, and Franco Brezzi.

[87] It is named for Cedric Taylor and Paul Hood.

[88] There are also other approaches to obtain a stable solution method for fluids in finite element method, see for example [1].

```python
"""Computational reality 08, nonlinear fluid flow"""
__author__ = "B. Emek Abali"
__license__ = "GNU GPL Version 3.0 or later"
#This code underlies the GNU General Public License, http://www.gnu.org/licenses/gpl-3.0.en.html

from fenics import *
set_log_level(ERROR)

xlength = 5.0 # m
ylength = 1.0 # m
mesh_=RectangleMesh(Point(0.0,-ylength/2.0), Point(xlength, ylength/2.0), 100,20)
class Cut(SubDomain):
    def inside(self, x, on_boundary):
        return x[0]<xlength/2.0 and (x[1]>+ylength/4.0 or x[1]<-ylength/5.0)

domain = CellFunction('size_t',mesh_)
domain.set_all(0)
to_be_cut = Cut()
to_be_cut.mark(domain, 1)
mesh = SubMesh(mesh_, domain, 0)

V = VectorFunctionSpace(mesh, 'P', 2)
P = FunctionSpace(mesh, 'P', 1)
Space = MixedFunctionSpace([P,V])

left = CompiledSubDomain('near(x[0],0) && on_boundary')
right = CompiledSubDomain('near(x[0],l) && on_boundary',l=xlength)
bottom1 = CompiledSubDomain('x[0]>xl/2.0 && near(x[1],-yl/2.0)',xl=xlength,yl=ylength)
bottom2 = CompiledSubDomain('x[0]<xl/2.0 && x[1]<-yl/5.0',xl=xlength,yl=ylength)
top1 = CompiledSubDomain('x[0]>xl/2.0 && near(x[1],yl/2.0)',xl=xlength,yl=ylength)
top2 = CompiledSubDomain('x[0]<xl/2.0 && x[1]>yl/4.0',xl=xlength,yl=ylength)
opening1 = CompiledSubDomain('near(x[0],xl/2.0) && x[1]<-yl/5.0',xl=xlength,yl=ylength)
opening2 = CompiledSubDomain('near(x[0],xl/2.0) && x[1]>yl/4.0',xl=xlength,yl=ylength)

facets = FacetFunction('size_t',mesh)
cells = CellFunction('size_t',mesh)
da_ = Measure('ds', domain=mesh, subdomain_data=facets)
dv_ = Measure('dx', domain=mesh, subdomain_data=cells)

v_noslip = Constant((0.0, 0.0))
pL = Expression('100000.0+1000.0*t',t=0)
pR = Constant(100000.0)
p_bc1=DirichletBC(Space.sub(0), pL, left)
p_bc2=DirichletBC(Space.sub(0), pR, right)
```

```
p_bc3=DirichletBC(Space.sub(0), pR, opening1)
p_bc4=DirichletBC(Space.sub(0), pR, opening2)
v_bc1=DirichletBC(Space.sub(1), v_noslip, bottom1)
v_bc2=DirichletBC(Space.sub(1), v_noslip, bottom2)
v_bc3=DirichletBC(Space.sub(1), v_noslip, top1)
v_bc4=DirichletBC(Space.sub(1), v_noslip, top2)
v_bc5=DirichletBC(Space.sub(1).sub(1), 0.0, left)
v_bc6=DirichletBC(Space.sub(1).sub(1), 0.0, right)

bc=[p_bc1,p_bc2,p_bc3,p_bc4,v_bc1,v_bc2,v_bc3,v_bc4,v_bc5,v_bc6]
u_init = Expression(('p0','0.0','0.0'), p0=100000.0)

i, j, k, l, m, n =indices(6)
t = 0.0
t_end = 10.0
dt = 0.1

test = TestFunction(Space)
du = TrialFunction(Space)
u0 = Function(Space)
u = Function(Space)
u = interpolate(u_init,Space)
u0.assign(u)
p0, v0 = split(u0)
p, v = split(u)
delp, delv = split(test)
dp, dv = split(du)
delta = Identity(2)

# approximate values of HDPE melt, rho [kg/m^3], mu [Pa s], k [Pa]
rho, mu, lambada, k_fluid, B_fluid = 1450.0, 400000.0, 1E6, 250.0, 0.00001

# if II=0 (which is physically possible) Form will be singular,
# to avoid that, add a small enough number to approximate an
# accurate solution
II = as_tensor(1.0/2.0*sym(grad(v))[m,n]*sym(grad(v))[m,n]+0.000001, ())
I = as_tensor(sym(grad(v))[k,k], ())
#scalar is a tensor of rank zero


Form = ( v[i].dx(i)*delp \
    + rho/dt*(v-v0)[i]*delv[i] \
    + rho*v[j]*v[i].dx(j)*delv[i] \
    + p.dx(i)*delv[i] \
    + lambada*I*delv[i].dx(i) \
    + (2.0*mu + 2.0*k_fluid/pi/sqrt(II) \
    *atan(sqrt(II)/B_fluid))*sym(grad(v))[j,i]\
    *delv[i].dx(j) \
    )*dv_
```

```

S=2.0*II
Gain = (  dv[i].dx(i)*delp \
    + rho/dt*dv[i]*delv[i] \
    + dv[j]*rho*v[i].dx(j)*delv[i] \
    + v[j]*rho*dv[i].dx(j)*delv[i] \
    + dp.dx(i)*delv[i] \
    + lambada*dv[k].dx(k)*delv[i].dx(i) \
    + 2.0*mu*sym(grad(dv))[k,j]*delv[j].dx(k) \
    - 2.0*k_fluid*2.0**0.5/(pi*S**(3.0/2.0))*sym(grad(v))[i,l] \
    *sym(grad(dv))[i,l]*atan(S**0.5/(2.0**0.5*B_fluid))\
    *sym(grad(v))[k,j]*delv[j].dx(k) \
    + 4.0*B_fluid*k_fluid*sym(grad(v))[m,n]*sym(grad(dv))[m,n] \
    /(pi*S*(2.0*B_fluid**2+S))*sym(grad(v))[k,j]*delv[j].dx(k) \
    + 2.0*k_fluid*2.0**0.5/(pi*S**0.5)*atan(S**0.5/(2.0**0.5*B_fluid)) \
    *sym(grad(dv))[k,j]*delv[j].dx(k) )*dv_

#Gain = derivative(Form, u, du)

pwd='/calcul/CR08/'
file_p = File(pwd+'pressure.pvd')
file_v = File(pwd+'velocity.pvd')

# ------------------------------------------time loop
tic()
while t<t_end:
    t += dt
    print 'time:' ,t
    pL.t = t
    solve(Form==0, u, bc, J=Gain, \
        solver_parameters={"newton_solver":{"linear_solver":"mumps", "relative_tolerance": 1e-3} }, \
        form_compiler_parameters={"cpp_optimize": True, "representation": "quadrature", "quadrature_degree": 2}  )
    file_p << (u.split()[0], t)
    file_v << (u.split()[1], t)
    II_ = project( II, FunctionSpace(mesh,'P',1), solver_type="mumps",\
        form_compiler_parameters={"cpp_optimize": True, "representation": "quadrature", "quadrature_degree": 2})
    sigma = as_tensor(-p*delta[j,i]+(2.0*mu+2.0*k_fluid/pi/sqrt(II_) \
    *atan(sqrt(II_)/B_fluid))*sym(grad(v))[j,i],(j,i))
    sigma_=project(sigma,TensorFunctionSpace(mesh,'P',1), solver_type="mumps",\
        form_compiler_parameters={"cpp_optimize": True, "representation": "quadrature", "quadrature_degree": 2})
```

```
        print 'sigma12 :',sigma_(xlength/2.0,ylength/4.0)[1],' Pa'
        u0.assign(u)
    print 'it took ',toc(),' seconds'
```

To-do

A nonlinear viscous fluid flowing in two subsequent channels with different heights is implemented.

- Inspect the code and determine applied boundary conditions.
- Try to change the geometry to an L-shape channel.
- Find a shear viscosity for ketchup or mayonnaise and report if there is a significant change in the flow profile.
- For a linear viscous fluid like water, we have to wait for the steady-state. If the simulation is held under the pressure at 10 s, then the steady-state will be achieved just in one time step. Which term is responsible for this behavior?

1.9 Fluid Structure Interaction

In solid mechanics we have used a LAGRANGEan frame. For fluid mechanics a EULERian frame is more beneficial. If we want to model a deformable solid embedded in a viscous fluid, we need to use the so-called *arbitrary* EULER–LANGRANGEan frame.[89] In this section we will set the preliminaries and solve an elastic beam embedded in a viscous flow, see Fig. 1.19 on p. 100. The beam itself will be modeled in the LAGRANGEan frame whereas the fluid in the EULERian frame.

The motion of a particle is computed by using the balance of mass and the balance of linear momentum. Consider a material body $\mathcal{B}_0$ at time $t = 0$. Each of its particles is associated with its corresponding position X_i at any reference frame. The initial position is known. Therefore, we use X_i as the initial position to identify particles. In other words, we select the initial frame as the reference frame, where X_i denote particles. Suppose that we track the motion of many particles in $\mathcal{B}_0$ simultaneously. We can visualize this notion as points marked on the body. A grid connecting these points leads to a mesh used in the computation. All points change their spatial positions in case of a deformation. The deviation from the initial position is the displacement, ${}^{s}u_i$, of a particle, X_i, at time t,

$$ {}^{s}u_i = {}^{s}u_i(\boldsymbol{X}, t) \; . \tag{1.295} $$

This primitive variable will be computed for each particle of the deformable solid body. We employ Cartesian coordinates for the LAGRANGEan frame and write out the balance of linear momentum:

---

[89] In many textbooks the word *configuration* is used instead of frame.

Fig. 1.19 Beam in *red* and the surrounding water in *blue* in two dimensional continuum

$$\int_{\mathcal{B}_0} {}^{s}\!\rho_0\, {}^{s}\ddot{u}_i \,\mathrm{d}V = \int_{\partial\mathcal{B}_0} {}^{s}\!P_{ki} N_k \mathrm{d}A + \int_{\mathcal{B}_0} {}^{s}\!\rho_0 f_i \mathrm{d}V \,, \tag{1.296}$$

for calculating the displacement, where the balance of mass in the initial frame, $\rho_0 = J\rho$, has already been implemented into the balance of linear momentum.[90] We have written the balance equation in the LAGRANGEan frame, where $\mathrm{d}V$ and $\mathrm{d}A$ refer to the initial volume and surface elements. The direction of the surface element is given by N_i of a unit length (length of one). The nominal stress, ${}^{s}\!P_{ki}$, is obtained by the transformation of the surface element from the current to the initial frame[91]

$${}^{s}\!P_{ki} = {}^{s}\!J({}^{s}\!\boldsymbol{F}^{-1})_{kj}\, {}^{s}\!\sigma_{ji} \,, \tag{1.297}$$

where ${}^{s}\!\sigma_{ij}$ is the CAUCHY stress.[92] The deformation gradient, ${}^{s}\!F_{ki}$, and its determinant, ${}^{s}\!J$, have been defined as follows

$${}^{s}\!F_{ki} = \frac{\partial\, {}^{s}u_k}{\partial X_i} + \delta_{ki} \,, \quad {}^{s}\!J = \det({}^{s}\!\boldsymbol{F}) \,. \tag{1.298}$$

We need a constitutive relation for ${}^{s}\!P_{ki}$ in order to obtain a differential equation in displacements, i.e., the *field* equation for displacement. Before discussing the constitutive relation, we bring the global balance Eq. (1.296) into the *local* form by using GAUSS's theorem

$${}^{s}\!\rho_0\, {}^{s}\ddot{u}_i - \frac{\partial\, {}^{s}\!P_{ki}}{\partial X_k} - {}^{s}\!\rho_0 f_i = 0 \,. \tag{1.299}$$

---

[90] The balance of mass in the initial frame is not a differential equation, see Sect. 1.4 for its derivation.

[91] See Sect. 1.2 for the derivation this transformation.

[92] Especially in materials science, the nominal stress is called the engineering stress and CAUCHY's stress is called the true stress.

In order to acquire the field equation for the displacement, ${}^{s}u_i$, we employ the constitutive equation for linear elastic isotropic bodies:

$$
\begin{aligned}
{}^{s}P_{ki} &= F_{ij} S_{kj} \ , \quad {}^{s}S_{ij} = {}^{s}\lambda\, {}^{s}E_{kk}\delta_{ij} + 2\, {}^{s}\mu\, {}^{s}E_{ij} \ , \\
{}^{s}E_{ij} &= \frac{1}{2}({}^{s}C_{ij} - \delta_{ij}) \ , \quad {}^{s}C_{ij} = {}^{s}F_{ki}\, {}^{s}F_{kj} \ ,
\end{aligned}
\tag{1.300}
$$

where the deformation gradient, ${}^{s}F_{ij}$, is used for obtaining the right CAUCHY–GREEN deformation tensor, ${}^{s}C_{ij}$, from which the GREEN–LAGRANGE strain tensor, ${}^{s}E_{ij}$, has been derived. Now, Eq. (1.299) can be brought to the weak form after discretizing in time and employing integration by parts

$$
\begin{aligned}
\overset{s}{\mathrm{F}} = \int_{\mathcal{B}_0} \Big({}^{s}\rho_0 \frac{{}^{s}u_i - 2\, {}^{s}u_i^0 + {}^{s}u_i^{00}}{\Delta t^2} \delta u_i + {}^{s}P_{ki} \frac{\partial\, \delta u_i}{\partial X_k} - {}^{s}\rho_0 f_i\, \delta u_i \Big) \mathrm{d}V - \\
- \int_{\partial \mathcal{B}_0^{\mathrm{N}}} \hat{t}_i\, \delta u_i \mathrm{d}A \ .
\end{aligned}
\tag{1.301}
$$

The traction vector, $\hat{t}_i = {}^{s}P_{ki} N_k = {}^{s}J({}^{s}\boldsymbol{F}^{-1})_{kj}\, {}^{s}\sigma_{ji} N_k$, on the NEUMANN boundary, $\partial \mathcal{B}_0^{\mathrm{N}}$, is given as the boundary condition. This boundary is the hull of the solid body. Since $\mathcal{B}_0$ is embedded in a fluid, the traction vector is due to the fluid particles being on the boundary. This *interaction* will be modeled by using the true stress. No matter what the underlying material is, the true stress applied from the fluid to the structure and its reaction from structure to the fluid has to be the same on the interaction. Therefore, we have $n_j\, {}^{s}\sigma_{ji} = n_j\, {}^{f}\sigma_{ji}$ on boundary. This choice is of utmost importance and the modeling of interaction is chosen differently in the scientific literature. Herein, we set the traction vector on the boundary of solid equal to the traction vector on the boundary of fluid and write $\hat{t}_i = n_j\, {}^{f}\sigma_{ji} = {}^{s}J({}^{s}\boldsymbol{F}^{-1})_{kj}\, {}^{f}\sigma_{ji} N_k$.[93]

Defining particles via their initial positions works well for a solid body, as long as the body is defined as a material system. A material system means that no massive particles enter or leave the domain, $\mathcal{B}$. The particles may move, however, within the computational domain the same particles exist throughout the simulation. If the solid deforms, the domain shall co-deform to prevent a convection across the boundaries.

For fluids the material system fails to be an adequate choice. We measure the velocity of a fluid particle in a spatial position, x_i, and we model fluids as open systems where fluid enters and leaves a domain, Ω. This domain of interest is called a *control volume*. Within the control volume, Ω, at every spatial position, x_i, the velocity, ${}^{f}v_i$, and the mass density, ${}^{f}\rho$, are the primitive variables to be computed for fluid particles occupying x_i as follows

$$
{}^{f}v_i = {}^{f}v_i(\boldsymbol{x}, t) \ , \quad {}^{f}\rho = {}^{f}\rho(\boldsymbol{x}, t) \ . \tag{1.302}
$$

[93] We skip a lengthy discussion for this fact. Actually, the balance of linear momentum on singular surfaces suggests this fact. In case of neglecting the surface tension, the traction on the interface—it is a singular surface without mass—is continuous. In order words, the traction vectors experiencing solid and fluid are identical.

In reality mass density and velocity are measured simultaneously in every position within the control volume. We can visualize a mesh in Ω and suppose that in each node we are measuring the primitive variables of particles currently occupying nodes at $\boldsymbol{x}$. These measurements have no dependence on the underlying particles. In other words, we may measure constantly at the same position or we can even move the control volume such that we measure at different spatial positions at different time instants. The control volume is the chosen computational domain, Ω, with a known (given) velocity, $w_i = x_i^{\bullet}$. The domain velocity, w_i, has no connection to the underlying fluid. It is essential to distinguish between the domain velocity, w_i, and the fluid velocity, v_i. The balances of mass and linear momentum for an open and moving domain read

$$\begin{aligned}\Big(\int_\Omega {}^{\mathrm{f}}\!\rho \,\mathrm{d}v\Big)^{\bullet} &= -\int_{\partial\Omega} {}^{\mathrm{f}}\!\rho({}^{\mathrm{f}}v_i - w_i)n_i \mathrm{d}a \;,\\ \Big(\int_\Omega {}^{\mathrm{f}}\!\rho\, {}^{\mathrm{f}}v_i \mathrm{d}v\Big)^{\bullet} &= -\int_{\partial\Omega} {}^{\mathrm{f}}\!\rho v_i({}^{\mathrm{f}}v_j - w_j)n_j \mathrm{d}a + \int_{\partial\Omega} {}^{\mathrm{f}}\sigma_{ji} n_j \mathrm{d}a + \int_\Omega {}^{\mathrm{f}}\!\rho f_i \mathrm{d}v \;.\end{aligned} \tag{1.303}$$

These balance equations in the moving EULERian frame are given at the *present* or *current* time. Therefore, we use the current volume element, $\mathrm{d}v$, and current surface element, $\mathrm{d}a$, with its direction, n_i. In order to acquire the local forms of the balance equations we need to "shift" the time rate into the integral. By using the following identity for the rate of the volume element in Cartesian coordinates:

$$\begin{aligned}\mathrm{d}v^{\bullet} &= \big(\mathrm{d}x_1\mathrm{d}x_2\mathrm{d}x_3\big)^{\bullet} = \big(\mathrm{d}x_1\mathrm{d}x_2\mathrm{d}x_3\big)^{\bullet}\frac{\mathrm{d}v}{\mathrm{d}x_1\mathrm{d}x_2\mathrm{d}x_3} = \\ &= \Big(\frac{\mathrm{d}x_1^{\bullet}}{\mathrm{d}x_1} + \frac{\mathrm{d}x_1^{\bullet}}{\mathrm{d}x_1} + \frac{\mathrm{d}x_1^{\bullet}}{\mathrm{d}x_1}\Big)\mathrm{d}v = \frac{\partial w_k}{\partial x_k}\mathrm{d}v \;,\end{aligned} \tag{1.304}$$

and by employing the product rule we attain

$$\begin{aligned}\int_\Omega \Big(\frac{\partial\, {}^{\mathrm{f}}\!\rho}{\partial t} + \frac{\partial\, {}^{\mathrm{f}}\!\rho}{\partial x_i} w_i + {}^{\mathrm{f}}\!\rho \frac{\partial w_k}{\partial x_k}\Big)\mathrm{d}v &= -\int_{\partial\Omega} {}^{\mathrm{f}}\!\rho({}^{\mathrm{f}}v_i - w_i)n_i \mathrm{d}a \;,\\ \int_\Omega \Big(\frac{\partial\, {}^{\mathrm{f}}\!\rho\, {}^{\mathrm{f}}v_i}{\partial t} + \frac{\partial\, {}^{\mathrm{f}}\!\rho\, {}^{\mathrm{f}}v_i}{\partial x_j} w_j + {}^{\mathrm{f}}\!\rho\, {}^{\mathrm{f}}v_i \frac{\partial w_k}{\partial x_k}\Big)\mathrm{d}v &= -\int_{\partial\Omega} {}^{\mathrm{f}}\!\rho v_i({}^{\mathrm{f}}v_j - w_j)n_j \mathrm{d}a + \\ &\quad + \int_{\partial\Omega} {}^{\mathrm{f}}\sigma_{ji} n_j \mathrm{d}a + \int_\Omega {}^{\mathrm{f}}\!\rho f_i \mathrm{d}v \;.\end{aligned} \tag{1.305}$$

The first boundary integrals are called *convective* terms. The second boundary integral in the momentum balance is a *non-convective* term.[94] For boundary integrals we apply GAUSS's theorem and obtain the local forms of the balance equations:

[94] In some textbooks this term is called a diffusive term, however, a diffusion is a motion of mass. Since a traction on the boundary implies a momentum flow without mass diffusion, we refrain from naming traction as a diffusive term.

$$
\begin{aligned}
&\frac{\partial {}^{\mathrm{f}}\!\rho}{\partial t} + \frac{\partial {}^{\mathrm{f}}\!\rho}{\partial x_i} w_i + {}^{\mathrm{f}}\!\rho \frac{\partial w_k}{\partial x_k} + \frac{\partial}{\partial x_i}\Big({}^{\mathrm{f}}\!\rho({}^{\mathrm{f}}\!v_i - w_i)\Big) = 0\,, \\
&\frac{\partial {}^{\mathrm{f}}\!\rho\, {}^{\mathrm{f}}\!v_i}{\partial t} + \frac{\partial {}^{\mathrm{f}}\!\rho\, {}^{\mathrm{f}}\!v_i}{\partial x_j} w_j + {}^{\mathrm{f}}\!\rho\, {}^{\mathrm{f}}\!v_i \frac{\partial w_k}{\partial x_k} + \frac{\partial}{\partial x_j}\Big({}^{\mathrm{f}}\!\rho v_i({}^{\mathrm{f}}\!v_j - w_j) - {}^{\mathrm{f}}\!\sigma_{ji}\Big) - {}^{\mathrm{f}}\!\rho f_i = 0\,.
\end{aligned}
\tag{1.306}
$$

Formally, we can eliminate all terms including w_i and obtain Eqs. (1.235). Therefore, we realize that the domain velocity, w_i, can be chosen arbitrarily. The underlying physics will not be affected by the choice of the domain velocity. For computational reasons we skip the elimination of w_i and solve Eqs. (1.306). In order to complement the equations, we need a constitutive relation for the *true* or CAUCHY stress tensor at the current time instant, ${}^{\mathrm{f}}\!\sigma_{ji}$. The simplest relation is the linear material equation for viscous fluids, i.e., the NAVIER–STOKES equation:

$$
{}^{\mathrm{f}}\!\sigma_{ij} = -{}^{\mathrm{f}}\!p\delta_{ij} + {}^{\mathrm{f}}\!\tau_{ij}\,,\quad {}^{\mathrm{f}}\!\tau_{ij} = {}^{\mathrm{f}}\!\lambda \frac{\partial\, {}^{\mathrm{f}}\!v_k}{\partial x_k}\delta_{ij} + {}^{\mathrm{f}}\!\mu\Big(\frac{\partial\, {}^{\mathrm{f}}\!v_i}{\partial x_j} + \frac{\partial\, {}^{\mathrm{f}}\!v_j}{\partial x_i}\Big)\,. \tag{1.307}
$$

For water this choice is appropriate. The pressure, ${}^{\mathrm{f}}\!p$, is unknown and there are two ways to close the above equations. Either, we find a constitutive equation for pressure by means of mass density and velocity, or, we assume the mass density as being constant (an incompressible flow) and solve the pressure distribution by employing the mass balance. The latter option will be used herein. By using incompressibility, the weak form in Eq. (1.306)$_1$ reads (in the unit of power)

$$
\overset{\mathrm{f}}{\mathrm{F}}_p = \int_\Omega \Big(\frac{\partial\, {}^{\mathrm{f}}\!v_i}{\partial x_i}\,\delta p + \frac{\partial w_k}{\partial x_k}\,\delta p - ({}^{\mathrm{f}}\!v_i - w_i)\frac{\partial\, \delta p}{\partial x_i}\Big)\mathrm{d}v + \int_{\partial\Omega^{\mathrm{N}}} ({}^{\mathrm{f}}\!v_i - w_i)\,\delta p n_i \mathrm{d}a\,. \tag{1.308}
$$

Moreover, by employing Eq. (1.307)$_1$ into Eq. (1.306)$_2$, by using the incompressibility, and by implementing the time discretization, we acquire the weak form for the (fluid) velocity again in the unit of power

$$
\begin{aligned}
\overset{\mathrm{f}}{\mathrm{F}}_v = &\int_\Omega \Big({}^{\mathrm{f}}\!\rho \frac{{}^{\mathrm{f}}\!v_i - {}^{\mathrm{f}}\!v_i^0}{\Delta t}\,\delta v_i + {}^{\mathrm{f}}\!\rho \frac{\partial\, {}^{\mathrm{f}}\!v_i}{\partial x_j} w_j\,\delta v_i + {}^{\mathrm{f}}\!\rho\, {}^{\mathrm{f}}\!v_i \frac{\partial w_k}{\partial x_k}\,\delta v_i + \\
&+ \frac{\partial p}{\partial x_i}\,\delta v_i - {}^{\mathrm{f}}\!\rho v_i({}^{\mathrm{f}}\!v_j - w_j)\frac{\partial\, \delta v_i}{\partial x_j} + {}^{\mathrm{f}}\!\tau_{ji}\frac{\partial\, \delta v_i}{\partial x_j} - {}^{\mathrm{f}}\!\rho f_i\,\delta v_i\Big)\mathrm{d}v + \\
&+ \int_{\partial\Omega^{\mathrm{N}}} \big({}^{\mathrm{f}}\!\rho v_i({}^{\mathrm{f}}\!v_j - w_j)\,\delta v_i n_j - {}^{\mathrm{f}}\!\tau_{ji} n_j\,\delta v_i\big)\mathrm{d}a\,.
\end{aligned}
\tag{1.309}
$$

For the fluid problem we know the velocity on the interaction boundary. We suppose that the fluid particles adhere on the structure so that the computed velocity of the structure is also the velocity of fluid particles being on the interaction boundary.

The domain velocity, w_i, can be chosen arbitrarily. It is easier to comprehend, if we visualize this velocity as the velocity of the underlying mesh. We will compute w_i and deform the mesh, where the computation of fluid motion takes place. From a computational point of view, the choice of the mesh velocity makes a difference.

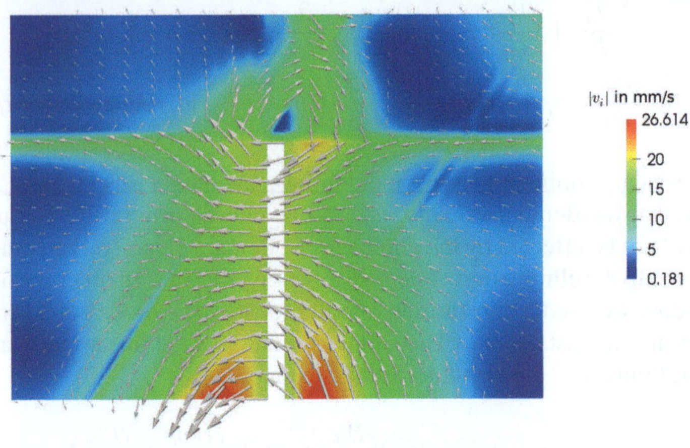

Fig. 1.20 Velocity distribution at $t = 0.84$ s for the water flow due to the embedded beam of steel moving to the left after 8 cycles with period of 0.1 s. Only the domain of fluid is shown. The *colors* indicate the magnitude of the fluid velocity drawn as *arrows*

The solution of ${}^{\mathrm{f}}v_i$ converges more easily if the mesh deforms close to the particles of fluid. In other words, the simplest (trivial) solution is to set $v_i \equiv w_i$. However, this fails to be the best choice since the mesh deformation has some limitations. The connectivity of elements shall be conserved. A fluid velocity with vortices[95] would deform the mesh such that the connectivity of the elements gets distorted. In order to restrict the motion of mesh, we solve a model problem for the mesh velocity:

$$\overset{\mathrm{m}}{\mathrm{F}} = \int_\Omega a \frac{\partial w_{(i}}{\partial x_{j)}} \frac{\partial \delta w_i}{\partial x_j} \mathrm{d}v \,. \tag{1.310}$$

The mesh velocity is implemented by moving the coordinates of the mesh. The parameter a in the latter variational form can be visualized as an artificial viscosity of the mesh. Its value does not affect the velocity of the fluid, since w_i can be chosen arbitrarily. Hence, any value of a allows us to accomplish a computation, this value can be seen as the artificial viscosity of the mesh.

Consider a beam of steel embedded in water as depicted in Fig. 1.19. The system stands still at the beginning under normal atmospheric pressure. The bottom of the beam is moved harmonically with a frequency of 10 Hz such that the motion of the beam accelerates the fluid. After 8 cycles the velocity profile can be seen in Fig. 1.20 where the beam is moving to the left. Basically, we solve first the structure, then the mesh velocity, and then the fluid pressure and velocity. This cycle is repeated three times for minimizing the errors obtained by using a sequential solving method. All the parameters and boundary conditions can be found in the code given below.

---

[95] A vortex means a circulation in the fluid.

```python
"""Computational reality 09, fluid structure interaction"""
__author__ = "B. Emek Abali"
__license__ = "GNU GPL Version 3.0 or later"
#This code underlies the GNU General Public License, http://www.gnu.org/licenses/gpl-3.0.en.html

from fenics import *
parameters["allow_extrapolation"]=True
set_log_level(ERROR)

xlength = 10.0  # in mm
ylength = 150.0 # in mm
mesh = RectangleMesh(Point(0.0, 0.0), Point(31*xlength, 1.5*ylength), 31*2,15*2)
Dim = mesh.topology().dim()

structure = CompiledSubDomain(\
'x[0] >= 15.0*xl && x[0] <= 16.0*xl && x[1] <= yl',xl=xlength,yl=ylength)

interaction = CompiledSubDomain(\
'(near(x[0],15.0*xl) && x[1] <= yl) || (near(x[0],16.0*xl) && x[1] <= yl) || (x[0] >= 15.0*xl && x[0] <= 16.0*xl && near(x[1], yl) )', xl=xlength, yl=ylength )

bottom = CompiledSubDomain('near(x[1],0.0)')
boundaries_all = CompiledSubDomain('on_boundary')
sub_domains = CellFunction('size_t',mesh)
sub_domains.set_all(0)
structure.mark(sub_domains, 1)
mesh_f = SubMesh(mesh, sub_domains, 0)
mesh_s = SubMesh(mesh, sub_domains, 1)
#plot(mesh_f,interactive=True)
#plot(mesh_s,interactive=True)

#facets for solid: 0 for interior, 1 on bottom, 2 on interaction
facets_s = FacetFunction('size_t', mesh_s)
facets_s.set_all(0)
interaction.mark(facets_s,2)
bottom.mark(facets_s,1)
#facets for fluid: 0 for interior, 1 on boundary, 2 on interaction
facets_f = FacetFunction('size_t', mesh_f)
facets_f.set_all(0)
boundaries_all.mark(facets_f,1)
interaction.mark(facets_f,2)

cells_s = CellFunction('size_t', mesh_s)
cells_f = CellFunction('size_t', mesh_f)

dA = Measure('ds', domain=mesh_s, subdomain_data=facets_s)
dV = Measure('dx', domain=mesh_s, subdomain_data=cells_s)
da = Measure('ds', domain=mesh_f, subdomain_data=facets_f)
```

```
dv = Measure('dx', domain=mesh_f, subdomain_data=cells_f)

N = FacetNormal(mesh_s)
n = FacetNormal(mesh_f)

S_s_space = FunctionSpace(mesh_s, 'P', 1)
V_s_space = VectorFunctionSpace(mesh_s, 'P', 1)
T_s_space = TensorFunctionSpace(mesh_s, 'P', 1)

S_f_space = FunctionSpace(mesh_f, 'P', 1)
V_f_space = VectorFunctionSpace(mesh_f, 'P', 1)
T_f_space = TensorFunctionSpace(mesh_f, 'P', 1)
SV_f_space = MixedFunctionSpace([S_f_space, V_f_space])

t=0.0 # in s
dt=0.01 # in s
t_end=5.0 # in s

pwd = '/calcul/CR09/'

file_u_s = File(pwd + 'u_s.pvd')
file_v_f = File(pwd + 'v_f.pvd')
file_p_f = File(pwd + 'p.pvd')

u_s_ = Function(V_s_space, name='u')
v_f_ = Function(V_f_space, name='v')
p_f_ = Function(S_f_space, name='p')

i,j,k,l,m = indices(5)
delta = Identity(Dim)
f= Constant((0.0,0.0))

def s_solve(u0,u00,sigma_f,t):
    rho_s = 8.3E-9 #tonne/mm^3
    nu_s = 0.3
    E_s = 200E3 #in MPa
    mu_s = E_s/(2.*(1.+nu_s))
    lam_s = 2.*mu_s*nu_s/(1.-2.*nu_s)
    sigma_s = project(sigma_f, T_s_space, solver_type="mumps",\
         form_compiler_parameters={"cpp_optimize": True, "representation": "quadrature", "quadrature_degree": 2})
    #plot(facets_s, interactive=True)
    displ_s_bottom = Expression(('A*sin(2.*pi*nu*t)','0.0'),\
     A=0.50, nu=10., t=t)
    bc_s = [DirichletBC(V_s_space, displ_s_bottom, facets_s, 1)]
    u_s = Function(V_s_space)
    del_u = TestFunction(V_s_space)
    du = TrialFunction(V_s_space)

    F_s = as_tensor( u_s[k].dx(i) + delta[k,i], (k,i) )
    J_s = det(F_s)
```

```
    C_s = as_tensor( F_s[k,i]*F_s[k,j], (i,j) )
    E_s = as_tensor( 1./2.*(C_s[i,j]-delta[i,j]), (i,j) )
    S_s = as_tensor( lam_s*E_s[k,k]*delta[i,j] + \
        2.*mu_s*E_s[i,j], (i,j) )
    P_s = as_tensor( F_s[i,j]*S_s[k,j], (k,i) )

    t_hat = as_tensor( J_s*inv(F_s)[k,j]*sigma_s[j,i]*N[k] ,
        ↪ (i,) )

    Form_s = ( rho_s*(u_s -2.*u0_s+u00_s)[i]/(dt*dt)*del_u[i]
        ↪ \
        + P_s[k,i]*del_u[i].dx(k) - rho_s*f[i]*del_u[i] )*dV
            ↪ \
        - t_hat[i]*del_u[i]*dA(2)
    Gain_s = derivative(Form_s, u_s, du)

    solve(Form_s==0, u_s, bc_s, J=Gain_s, \
        solver_parameters={"newton_solver":{"linear_solver":
            ↪ "mumps", "relative_tolerance": 1e-3} }, \
        form_compiler_parameters={"cpp_optimize": True, "
            ↪ representation": "quadrature", "
            ↪ quadrature_degree": 2})

    v_s = project( (u_s-u0_s)/dt, V_s_space, solver_type="
        ↪ mumps",\
        form_compiler_parameters={"cpp_optimize": True, "
            ↪ representation": "quadrature", "
            ↪ quadrature_degree": 2})
    #plot(v_s, interactive=True)

    return u_s, v_s

def m_solve(v_s):
    # artificial viscosity
    a = 1.0E-11 # MPa/s
    del_w = TestFunction(V_f_space)
    w = Function(V_f_space)
    dw = TrialFunction(V_f_space)
    bc_m = [DirichletBC(V_f_space, v_s, facets_f,2)]

    Form_m = a*sym(grad(w))[i,j]*del_w[i].dx(j)*dv
    Gain_m = derivative(Form_m, w, dw)

    solve(Form_m==0, w, bc_m, J=Gain_m, \
        solver_parameters={"newton_solver":{"linear_solver":
            ↪ "mumps", "relative_tolerance": 1e-3} }, \
        form_compiler_parameters={"cpp_optimize": True, "
            ↪ representation": "quadrature", "
            ↪ quadrature_degree": 2})

    return w

def f_solve(w, v0_f):
    rho_f = 998.21E-12  # in tonne / mm^3
```

```
    mu_f = 1001.6E-12    # in N s / mm^2
    lam_f = 1.0E-2       # in N s / mm^2
    bc1 = DirichletBC(SV_f_space.sub(0),Constant(0.1),
        ↪ facets_f,1)
    bc2 = DirichletBC(SV_f_space.sub(1),w, facets_f,2)
    bc_f = [bc1,bc2]

    u_f = Function(SV_f_space)
    u0_f = Function(SV_f_space)
    del_u_f = TestFunction(SV_f_space)
    du_f = TrialFunction(SV_f_space)

    p, v_f = split(u_f)
    del_p, del_v = split(del_u_f)

    tau_f = as_tensor(lam_f*v_f[k].dx(k)*delta[i,j] \
    + mu_f*sym(grad(v_f))[i,j], (i,j))

    Form_f_p = (v_f[i].dx(i)*del_p + w[i].dx(i)*del_p \
    - (v_f-w)[i]*del_p.dx(i) )*dv + (v_f-w)[i]*del_p*n[i]*da
        ↪ (2)
    Form_f_v = (rho_f*(v_f-v0_f)[i]/dt*del_v[i]
    + rho_f*v_f[i].dx(j)*w[j]*del_v[i] + rho_f*v_f[i]*w[k].dx
        ↪ (k)*del_v[i] \
    + p.dx(i)*del_v[i] - rho_f*v_f[i]*(v_f-w)[j]*del_v[i].dx(
        ↪ j) \
    + tau_f[j,i]*del_v[i].dx(j) - rho_f*f[i]*del_v[i])*dv \
    + (rho_f*v_f[i]*(v_f-w)[j] - tau_f[j,i])*del_v[i]*n[j]*da
        ↪ (1)

    Form_f = Form_f_p + Form_f_v
    Gain_f = derivative(Form_f, u_f, du_f)

    solve(Form_f==0, u_f, bc_f, J=Gain_f, \
        solver_parameters={"newton_solver":{"linear_solver":
            ↪ "mumps", "relative_tolerance": 1e-3} }, \
        form_compiler_parameters={"cpp_optimize": True, "
            ↪ representation": "quadrature", "
            ↪ quadrature_degree": 2})

    p, v_f = u_f.split(deepcopy=True)
    sigma_f = project(-p*delta+tau_f, T_f_space, solver_type=
        ↪ "mumps",\
        form_compiler_parameters={"cpp_optimize": True, "
            ↪ representation": "quadrature", "
            ↪ quadrature_degree": 2})

    return p, v_f, sigma_f


u0_s = Function(V_s_space)
u00_s = Function(V_s_space)
sigma_f = Function(T_f_space)
v0_f = Function(V_f_space)
```

```
while t<t_end:
    t += dt
    print "time: ", t
    # inner loop for convergence between fluid and structure
    for ii in range(3):
        u_s, v_s = s_solve(u0_s,u00_s,sigma_f,t)
        w = m_solve(v_s)
        p, v_f, sigma_f = f_solve(w,v0_f)

    u00_s.assign(u0_s)
    u0_s.assign(u_s)
    v0_f.assign(v_f)
    for x in mesh_f.coordinates(): x[:] += dt*w(x)[:]
    u_s_.assign(u_s)
    file_u_s << (u_s_,t)
    v_f_.assign(v_f)
    file_v_f << (v_f_,t)
    p_f_.assign(p)
    file_p_f << (p_f_,t)
```

To-do

A beam embedded in water has been implemented.

- Change the mesh viscosity and investigate the convergence of the linearization.
- Implement a softer beam, for example out of polyethylene.
- Instead of water we can also implement air as a fluid. In that case, the incompressibility would not hold. Search for a constitutive equation for the pressure. Start by looking for an *equation of state* by assuming that air is an ideal gas.

References

1. Abali, B.E., Müller, W.H., Georgievskii, D.V.: A discrete-mechanical approach for computation of three-dimensional flows. ZAMM-J. Appl. Math. Mech./Zeitschrift für Angewandte Mathematik und Mechanik **93**(12), 868–881 (2013)
2. Alnaes, M.S., Mardal, K.A.: On the efficiency of symbolic computations combined with code generation for finite element methods. ACM Trans. Math. Softw. **37**(1) (2010)
3. Alnaes, M.S., Mardal, K.A.: Automated solution of differential equations by the finite element method, the FEniCS book. In: Syfi and sfc: symbolic finite elements and form compilation, Chap. 15. Springer (2012)
4. Bauschinger, J.: Über die Veränderung der Elastizitätsgrenze und der Festigkeit des Eisens und Stahls durch Strecken und Quetschen, durch Erwärmen und Abkühlen und durch oftmals wiederholte Beanspruchung. Mitteilungen des mechanisch-technischen Laboratoriums der Königlich Technischen Hochschule München **13**, 1 (1886)
5. Bronstein, I.N., Semendjajew, K.A., Musiol, G., Mühlig, H.: Taschenbuch der Mathematik. Deutsch (2001)
6. FEniCS Project: Development of tools for automated scientific computing, 2001–2016. http://fenicsproject.org (2016)
7. Flügge, W.: Tensor Analysis and Continuum Mechanics. Springer (1972)

8. Franca, L.P., Hauke, G., Masud, A.: Revisiting stabilized finite element methods for the advective-diffusive equation. Comput. Meth. Appl. Mech. Eng. **195**(13), 1560–1572 (2006)
9. Friedrich, C.: Mechanical stress relaxation in polymers: fractional integral model versus fractional differential model. J. Non-Newton. Fluid Mech. **46**(2–3), 307–314 (1993)
10. Handge, U.A., Zeiler, R., Dijkstra, D.J., Meyer, H., Altstädt, V.: On the determination of elastic properties of composites of polycarbonate and multi-wall carbon nanotubes in the melt. Rheologica acta **50**(5–6), 503–518 (2011)
11. Loverro, A.: Fractional calculus: history, definitions and applications for the engineer. Report, Department of Aerospace and Mechanical Engineering, Notre Dame, IN **46556**, (2004)
12. MatWeb: Material Property Data, 1996–2016. http://matweb.com (2016)
13. Melan, E.: Zur Plastizität des räumlichen Kontinuums. Arch. Appl. Mechan. **9**(2), 116–126 (1938)
14. Morton, K.W.: Finite element methods for non-self-adjoint problems. In: Turner, P. (ed.) Topics in Numerical Analysis, pp. 113-148. Springer, Berlin (1982)
15. Müller, W.H.: An Excursion to Continuum Mechanics. Springer (2014)
16. Odqvist, F.K.G.: Die Verfestigung von flußeisenähnlichen Körpern. ein Beitrag zur Plastizitätstheorie. ZAMM-J. Appl. Math. Mech./Zeitschrift für Angewandte Mathematik und Mechanik **13**(5), 360–363 (1933)
17. Paraview: Parallel visualization application. http://paraview.org/ (2011)
18. Podlubny, I.: Fractional Differential Equations: An Introduction to Fractional Derivatives, Fractional Differential Equations, to Methods of Their Solution and Some of Their Applications. Academic Press, London (1999)
19. Prager, W.: The theory of plasticity: a survey of recent achievements. Proc. Inst. Mech. Eng. **169**(1), 41–57 (1955)
20. Prandtl, L.: Spannungsverteilung in plastischen Körpern. In: Proceedings of the 1st International Congress on Applied Mechanics, Delft, pp. 43–54 (1924)
21. Reddy, J.N., Gartling, D.K.: The Finite Element Method in Heat Transfer and Fluid Dynamics. CRC press (2010)
22. Reuss, A.: Berücksichtigung der elastischen Formänderung in der Plastizitätstheorie. ZAMM-J. Appl. Math. Mech./Zeitschrift für Angewandte Mathematik und Mechanik **10**(3), 266–274 (1930)
23. Shield, R.T., Ziegler, H.: On Prager's hardening rule. Zeitschrift für angewandte Mathematik und Physik ZAMP **9**(3), 260–276 (1958)
24. Simo, J.C., Hughes, T.J.: Computational Inelasticity, vol. 7. Springer Science & Business Media (2006)
25. Spencer, A.J.M.: Theory of Invariants, chap. Part III, pp. 239–352. Academic Press Inc. London (1971)
26. Synge, J.L., Schild, A.: Tensor Calculus. Dover Publications Inc, New York (1969)
27. Taylor, C., Hood, P.: A numerical solution of the navier-stokes equations using the finite element technique. Comput. Fluids **1**(1), 73–100 (1973)
28. Truesdell, C., Toupin, R.A.: Principles of classical mechanics and field theory. In: Flügge, S. (ed.) Handbuch der Physik, vol. III/1 (1960)
29. VDI Gesellschaft (ed.): VDI Wärmeatlas, 10. Auflage. Springer (2006)
30. Ziegler, H.: An Introduction to Thermomechanics. North Holland, Amsterdam (1977)
31. Ziegler, H., Wehrli, C.: The derivation of constitutive relations from the free energy and the dissipation function. Adv. Appl. Mech. **25**, 183–238 (1987)

Chapter 2
Thermodynamics

Thermodynamics includes a *theoretical* and an *applied* part. The applied thermodynamics has its roots at the end of 19th century and it is used to calculate the temperature distribution in a continuum body. This aim is fulfilled by employing the balance of internal energy. We will study this approach in Sect. 2.1 for macroscopic systems and in Sect. 2.2 for microscopic systems. The difference between macroscopic and microscopic systems relies on the used constitutive equation.

The theoretical thermodynamics has started around 1950s. It has the goal of defining constitutive (material) equations that close the balance equations. By using thermodynamics we will derive the constitutive equations necessary in the computational reality. In Chap. 1 we have employed many constitutive equations with an ad-hoc method. In this chapter we will answer the question of how to derive these equations in a thermodynamically consistent manner. Concretely, in Sect. 2.3 we will analyze such an approach and derive the NAVIER–STOKES–FOURIER equations for a viscous fluid. Unfortunately, there are various methods in the literature for the thermodynamically consistent derivation of constitutive equations—we will not discuss the pros and cons of these different approaches. Herein, we present an engineering approach suitable for many simple material models. Although the method fails to cover some processes, it is general enough for determining all of the constitutive equations necessary for the simulated engineering applications in this book. Moreover, the necessary mathematical knowledge is fairly low. After having studied the method in Sect. 2.3 we will employ it in Sect. 2.4 for viscoelastic materials and in Sect. 2.5 for plastic deformations. Much use of the method will be made in the next chapter, too.

B.E. Abali, *Computational Reality*, Advanced Structured Materials 55,
DOI 10.1007/978-981-10-2444-3_2

2.1 Temperature Distribution in Macromechanics

A *conductive* material possesses the ability of transferring thermal energy, *heat*, without mass transport. In other words, heat travels between particles where particles remain at their positions. If we hold one end of a steel spoon in hot water, heat conducts to the other end without any deformation of the spoon itself. Of course there is a small expansion due to the temperature change in the spoon, however, in this section we neglect this reversible deformation and assume the body as *rigid* throughout the simulation. The balance equations can be introduced in a material or open system. The local forms of the equations are identical in both systems.

We motivate the governing equations for a solid body by using a material system. Mass and momentum balances in current frame read

$$\Big(\int_{\mathcal{B}} \rho \,\mathrm{d}v\Big)^{\bullet} = 0\,, \quad \Big(\int_{\mathcal{B}} \rho v_i \,\mathrm{d}v\Big)^{\bullet} = \int_{\partial\mathcal{B}} \sigma_{ji} n_j \,\mathrm{d}a + \int_{\mathcal{B}} \rho f_i \,\mathrm{d}v\,. \tag{2.1}$$

From the mass balance we calculate the mass density (or pressure) and from the momentum balance we acquire the displacement (or velocity). For the temperature calculation we will use another balance equation. In order to obtain this equation we start off with the balance of *total* energy in the current frame:

$$\Big(\int_{\mathcal{B}} \rho e \,\mathrm{d}v\Big)^{\bullet} = \int_{\partial\mathcal{B}} F_j n_j \,\mathrm{d}a + \int_{\mathcal{B}} \rho s \,\mathrm{d}v\,, \tag{2.2}$$

where ρ, e, F_i, and s denote the mass density,[1] the specific[2] total energy, the energy flux, and the specific energy supply, respectively. The total energy is a conserved quantity like mass and momentum; the balance of total energy lacks a production term. We can decompose total energy density:

$$\rho e = \rho u + \frac{1}{2}\rho \boldsymbol{v}^2\,, \tag{2.3}$$

where the first term is the so-called *internal* energy density with the specific internal energy, u, and the second term is the *kinetic* energy density due to the velocity, $\boldsymbol{v}$. Now, by inserting the mass balance into the balance of energy as well as into the balance of momentum, we obtain

$$\begin{aligned} \int_{\mathcal{B}} \rho v_i^{\bullet} \,\mathrm{d}v &= \int_{\partial\mathcal{B}} \sigma_{ji} n_j \,\mathrm{d}a + \int_{\mathcal{B}} \rho f_i \,\mathrm{d}v\,, \\ \int_{\mathcal{B}} \rho e^{\bullet} \,\mathrm{d}v &= \int_{\partial\mathcal{B}} F_j n_j \,\mathrm{d}a + \int_{\mathcal{B}} \rho s \,\mathrm{d}v\,. \end{aligned} \tag{2.4}$$

---

[1] Density means per volume.

[2] Specific means per mass.

After applying GAUSS's law, we acquire the local forms:

$$\rho v_i^{\bullet} - \frac{\partial \sigma_{ji}}{\partial x_j} - \rho f_i = 0 \,, \quad \rho e^{\bullet} - \frac{\partial F_j}{\partial x_j} - \rho s = 0 \,. \tag{2.5}$$

We observe a clear structure in the balance equations. The first terms indicate which term is balanced. The second terms are divergence of fluxes. The third terms are supply terms. On the right-hand side the production terms are written—momentum and total energy are conserved quantities. Equation (2.3) implies that the rate of specific energy can be rewritten by making use of the momentum balance:

$$\begin{aligned} e^{\bullet} &= \Big(u + \frac{1}{2} v_i v_i\Big)^{\bullet} = u^{\bullet} + v_i^{\bullet} v_i \,, \\ \rho e^{\bullet} &= \rho u^{\bullet} + v_i \Big(\frac{\partial \sigma_{ji}}{\partial x_j} + \rho f_i\Big) \,. \end{aligned} \tag{2.6}$$

By using the latter in the balance of total energy we obtain

$$\begin{aligned} &\rho u^{\bullet} + v_i \frac{\partial \sigma_{ji}}{\partial x_j} - \frac{\partial F_j}{\partial x_j} + v_i \rho f_i - \rho s = 0 \,, \\ &\rho u^{\bullet} - \frac{\partial}{\partial x_j}\big(F_j - \sigma_{ji} v_i\big) - \rho (s - f_i v_i) = \sigma_{ji} \frac{\partial v_i}{\partial x_j} \,. \end{aligned} \tag{2.7}$$

This equation has a structure of a balance equation. The first term denotes that the internal energy is balanced. The second term is a divergence of the internal energy flux and the third term is the internal energy supply. These terms are often abbreviated as

$$-q_j = F_j - \sigma_{ji} v_i \,, \quad r = s - f_i v_i \,, \tag{2.8}$$

where the so-called *heat flux*, q_i, and the *supply term*, r, needs to be defined or given. The minus sign in front of the heat flux is due to the convention that the heat pumped into the system has been seen as a positive quantity. The first power generators were using coal to burn and they did produce mechanical energy. Heat added into the system as well as the mechanical work taken out of the system were seen as *positive* quantities. We keep herein this convention and define the flux term of the internal energy as $-q_i$. The balance of internal energy reads

$$\rho u^{\bullet} + \frac{\partial q_j}{\partial x_j} - \rho r = \sigma_{ji} \frac{\partial v_i}{\partial x_j} \,, \tag{2.9}$$

or in global form (after using the balance of mass and GAUSS's law)

$$\Big(\int_{\mathcal{B}} \rho u \,\mathrm{d}v\Big)^{\bullet} = -\int_{\partial \mathcal{B}} q_j n_j \,\mathrm{d}a + \int_{\mathcal{B}} \rho r \,\mathrm{d}v + \int_{\mathcal{B}} \sigma_{ji} \frac{\partial v_i}{\partial x_j} \,\mathrm{d}v \,. \tag{2.10}$$

Here we see again the effect of the minus sign in front of the flux term. The heat flux is defined as the rate of energy transported *into* the system across the boundary. Since the plane normal is directed outward, we need a minus sign in order to describe a transport into the system. In other words, the heat fluxes into the body against the direction of the plane normal, $\boldsymbol{n}$. Therefore, a minus sign is necessary to heat the system up, if the heat flux is positive. The second term on the right-hand side is called a radiation term, r. Actually, the name radiation might be misleading; this is not a thermal radiation, for example, radiation from the sun cannot be modeled with this term. This term is a specific (per mass) heat supply, r, used in the microwave oven or in a laser welding. We will call the term *internal heating* or heat supply in the following. The last term on the right-hand side denotes a production term. When a deformation occurs, this production term will alter the internal energy. We cannot simply neglect this term. As long as there is a deformation in the continuum body, internal energy will be produced. The internal energy is related to the temperature such that an increasing internal energy will imply a temperature increase. Therefore, for any process where a deformation occurs, there will be a change in temperature. The production term is also called an *internal friction*. In many systems this production term may be small, especially for slow deformations, such that we can assume that the process is isothermal. This justification has been used in the Chap. 1.

In this section we restrict the model for rigid bodies, $v_i = 0$, no deformations are allowed. Balances of mass and momentum are satisfied identically and the balance of internal energy simplifies to

$$\rho u^{\bullet} + \frac{\partial q_j}{\partial x_j} - \rho r = 0 \,. \tag{2.11}$$

For a rigid body the internal energy depends only on temperature,

$$u = u(T) \,. \tag{2.12}$$

The internal heating or heat supply, r, increases the temperature by affecting all particles together. This term is a volumetric power; the food in the microwave oven heats up in each of its particles at the same time. In the example of a spoon in hot water, heat is transferred across the surface and then fluxes from one end to the other. If we put the spoon in a microwave oven, all of its particles heat up together due to the supplied heat supply, r. The same holds in case of a laser beam.[3] A laser beam supplies energy on a focused location. Suppose that a laser beam irradiates on one end of the spoon. All particles on that end are irradiated and they all heat up simultaneously. The laser radiates heat in such a way that it increases the temperature volumetrically. We model a laser welding via specific heat supply term, r.

We want to implement a laser welding process and choose a plate as geometry: A steel plate is welded with a laser beam. The energy supply comes as a laser beam in

[3]A laser (Light Amplification by Stimulated Emission of Radiation) generates a focused beam of photons in the same wavelength (coherent).

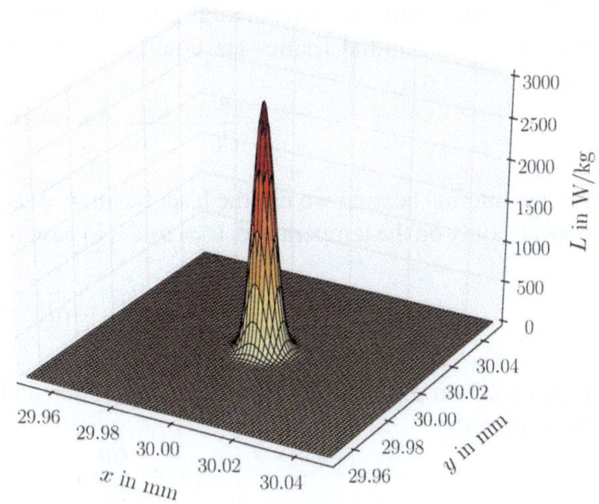

Fig. 2.1 Laser beam distribution as the GAUSSian bell shape in xy-plane

a concentrated manner and heats up the whole thickness of the plate at once, roughly alike cylinder, but instead a circle cross section, a GAUSSian bell shape[4] should be modeled. Any circular GAUSSian bell shape in x_1x_2-plane can be expressed as

$$r = P\exp\Big(-\big((x_1-\hat{x}_1)^2+(x_2-\hat{x}_2)^2\big)\Big)\,, \tag{2.13}$$

such that the laser beam reaches its maximum value P at the position $(\hat{x}_1, \hat{x}_2)$, since $r = P\exp(0) = P$. The power becomes asymptotically zero for coordinates away from $(\hat{x}_1, \hat{x}_2)$ owing to the minus sign. We want to simulate a laser beam evolving in time. The power (energy rate) of the laser reads

$$L(x_i,t) = P\exp\Bigg(-50000\bigg(\Big(x_1-\frac{l}{2}\big(1+\frac{1}{2}\sin(2\pi\tau)\big)\Big)^2+\big(x_2-v_Lt\big)^2\bigg)\Bigg)\,, \tag{2.14}$$

where the laser beam moves with a time parameter $\tau = t/t_{\text{end}} = [0, \dots 1]$ in x_1 sinusoidally and along x_2 linearly with a constant speed v_L. Of course, this is a model problem. In reality, the path is already defined in design and programmed into a NC (Numeric Controller) laser welding machine. Here we want to implement a complicated path description in order to present how to implement such a process into the code. The power of the laser beam, P, can be found in the data sheet of the laser welding machine. In order to visualize the implemented laser beam distribution, we plot the distribution for $\hat{x} = \hat{y} = 30\,\text{mm}$ and $P = 3000\,\text{W/kg}$ in Fig. 2.1. Since we multiplied L in Eq. (2.14) with a huge number of 50 000, the laser beam affects only locally, as expected from a focused light.

[4]It is named for Carl Friedrich Gauß.

For a rigid body the deformation gradient equals to the KRONECKER delta such that current and initial frames are equal, $x_i = X_i$. The balance of internal energy reads

$$\rho \frac{\partial u}{\partial t} + \frac{\partial q_i}{\partial x_i} - \rho r = 0 \,. \tag{2.15}$$

For the internal heating we use the laser beam, $r = L$. Moreover, the internal energy depends only on the temperature, $u = u(T)$, in case of a rigid body. Hence we obtain

$$\rho \frac{\partial u}{\partial T} \frac{\partial T}{\partial t} + \frac{\partial q_i}{\partial x_i} - \rho L = 0 \,. \tag{2.16}$$

In Sect. 2.3 on p. 126 we will formally introduce and discuss the so-called *specific heat capacity*:

$$c = \frac{\partial u}{\partial T} \,, \tag{2.17}$$

and explain how to measure this material parameter. The specific heat capacity is constant for many engineering materials. We also need a constitutive (material) equation for the heat flux, q_i. The simplest relation is given by FOURIER's law:[5]

$$q_i = -\kappa \frac{\partial T}{\partial x_i} \,, \tag{2.18}$$

where the material parameter, κ, is referred to as *thermal conductivity*. We assume it being constant. The minus sign denotes that the heat flux conducts in the direction of decreasing temperature gradient. Hence, heat fluxes from higher to lower temperature. This phenomenon is known intuitively and its validity is a subject of theoretical thermodynamics. For the moment we take it for granted and insert FOURIER's law into the balance of internal energy

$$\rho c \frac{\partial T}{\partial t} - \kappa \frac{\partial^2 T}{\partial x_i \partial x_i} - \rho L = 0 \,. \tag{2.19}$$

This differential equation is called the *field* equation for T, where its solution leads to the temperature distribution in a rigid body. We acquire the variational form of the latter differential equation by discretizing in time and by using integration by parts. First we utilize a backward difference scheme for temporal discretization, as usual,

$$\frac{\partial T}{\partial t} = \frac{T(t) - T(t - \Delta t)}{\Delta t} = \frac{T - T^0}{\Delta t} \,. \tag{2.20}$$

[5]The constitutive equation is named after Jean-Baptiste Joseph Fourier.

Secondly, we multiply the field equation by the test function, δT, for spatial discretization

$$\int_{\mathcal{B}} \left(\frac{\rho c}{\Delta t}(T - T^0) - \kappa \frac{\partial^2 T}{\partial x_i \partial x_i} - \rho L \right) \delta T \, \mathrm{d}v = 0 \, . \tag{2.21}$$

The second term in the integral form possesses a second derivative in space, whereas the multiplied test function has no derivatives. Thus, the continuity conditions of T and δT are different. In the GALERKIN procedure we utilize the same function space for the primitive variable T and its test function δT. Hence it has to belong to class C^2 at least. We can integrate by parts and acquire a form where both terms are differentiated once such that the continuity condition of T, δT is weakened and they need to belong to C^1. After integrating by parts we acquire the weak form:

$$\mathrm{F}_T = \int_{\mathcal{B}} \left(\frac{\rho c}{\Delta t}(T - T^0)\, \delta T + \kappa \frac{\partial T}{\partial x_i} \frac{\partial \, \delta T}{\partial x_i} - \rho L \, \delta T \right) \mathrm{d}v - \int_{\partial \mathcal{B}} \kappa \frac{\partial T}{\partial x_i} \, \delta T n_i \, \mathrm{d}a \, . \tag{2.22}$$

The integrand on the boundaries $\partial \mathcal{B}$ shall be given. In Chap. 1 we have seen DIRICHLET boundary conditions, where the solution itself is given, and NEUMANN boundary conditions, where the gradient of the solution in the surface outward normal, n_i, is defined. For the case of the energy equation the heat flux, q_i, projected into the surface normal, $q_i n_i$, defines a NEUMANN condition. This condition is the heat exchange from the surface of the plate to the surrounding medium, probably air or some special kind of gases like Argon or Helium for a better welding. The heat exchange is due to the temperature difference between the surface and the surroundings. Surface becomes hot as a consequence of the energy delivered by the laser beam. The temperature of the surroundings might be set constant[6] to an ambient temperature, T_{amb}. Thus, we obtain the DIRICHLET condition $T = T_{\mathrm{amb}}$ on the boundary. Both conditions can be mixed together in order to create another type of boundary condition. We introduce a mixed boundary condition, to which is referred as a ROBIN boundary condition,[7] for the whole surface $\partial \mathcal{B}$ as follows

$$q_i n_i = h(T - T_{\mathrm{amb}}) \; \forall x_i \in \partial \mathcal{B} \, , \tag{2.23}$$

where a (positive) *convective heat transfer coefficient*, h in W/(m^2 K), is introduced that depends on the material and state of the ambient. If the body is embedded in fluid the convection heat transfer coefficient is higher than in air. In case of a moving

[6] Constant temperature of the surroundings is a *warm bath idealization*. Suppose that there is so much water in a bath; a heat exchange with the body within the bath does not affect the temperature of the bath, at all. The water on the surface of the body remains at the same constant temperature all the time.

[7] It is named after Victor Gustave Robin.

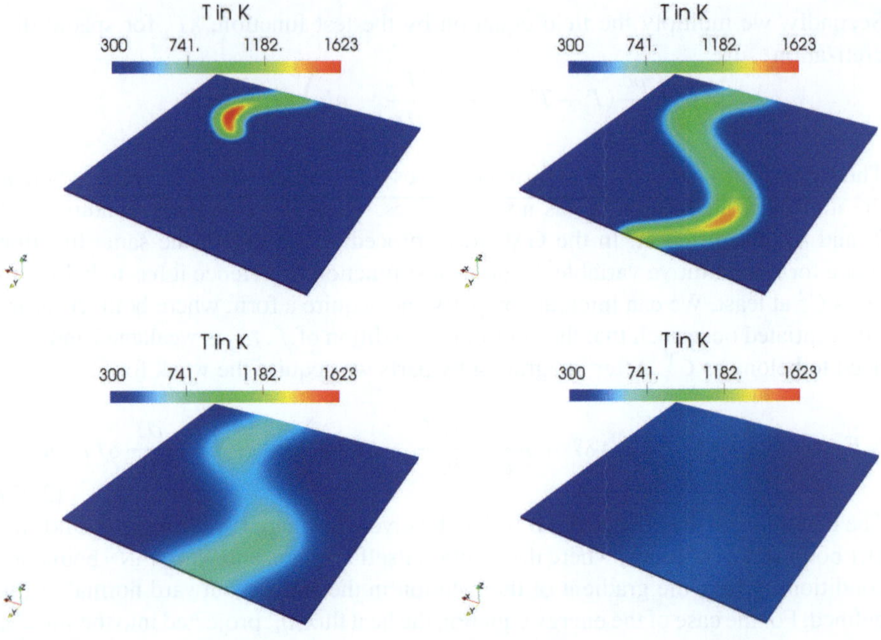

Fig. 2.2 Temperature distribution at 2, 5, 15, and 50 s in the steel plate during the laser welding

fluid that surrounds the body, the coefficient is even higher. This so-called natural convection in Eq. (2.23) provides a positive energy flux into the body if the ambient temperature is higher than the surface temperature,

$$-q_i n_i = h(T_{\text{amb}} - T) > 0 \text{ when } T_{\text{amb}} > T \,. \tag{2.24}$$

Additionally, for the heat flux we readily employ FOURIER's law in Eq. (2.18). Now the *linear* variational form reads

$$\begin{aligned} \mathrm{F}_T = & \int_{\mathcal{B}} \left(\frac{\rho c}{\Delta t}(T - T^0)\, \delta T + \kappa \frac{\partial T}{\partial x_i} \frac{\partial\, \delta T}{\partial x_i} - \rho L\, \delta T \right) \mathrm{d}v + \\ & + \int_{\partial \mathcal{B}} h(T - T_{\text{amb}})\, \delta T\, \mathrm{d}a \,. \end{aligned} \tag{2.25}$$

The code below computes the temperature distribution transiently, which can be seen in Fig. 2.2.

```
"""Computational reality 10, temperature distribution in a macroscopic body"""
__author__ = "B. Emek Abali"
__license__  = "GNU GPL Version 3.0 or later"
#This code underlies the GNU General Public License, http://www.gnu.org/licenses/gpl-3.0.en.html

from fenics import *
import numpy
parameters["allow_extrapolation"]=True
parameters["form_compiler"]["cpp_optimize"] = True
set_log_level(ERROR)

rho=7860.0      # mass density of steel in kg/m^3
c=624.0         # heat capacity in J/(kg K)
kappa=30.1 # thermal conductivity in W/(m K)
h=18.0     # heat convection out of the surface into ambient in W/(m^2 K)
Ta=300.0     # ambient temperature in K
l=0.1        # length in x and y directions in m
thickness=0.001 # thickness of the plate in m
P=3.0e6      # laser power in W/kg
speed=0.02  # laser speed in m/s
# Create mesh and define function space
mesh = BoxMesh(Point(0.0,0.0,0.0), Point(l,l,thickness), 200,200,2)
Space = FunctionSpace(mesh, 'P', 1)
cells = CellFunction('size_t', mesh)
facets = FacetFunction('size_t', mesh)
da = Measure('ds', domain=mesh, subdomain_data=facets)
dv = Measure('dx', domain=mesh, subdomain_data=cells)
t=0.0
t_end=50.0
Dt=0.1
initial_T = Expression("Tini", Tini=Ta)
T0 = interpolate(initial_T, Space)
Laser = Expression("P*exp(-50000.0*(pow(x[0]-0.5*l*(1+0.5*sin(2*pi*t/t_e)), 2)+pow(x[1]-velo*t, 2)))",P=P,t=0,t_e=t_end/10.,l=l,velo=speed)
T = TrialFunction(Space)
del_T = TestFunction(Space)
Form = (rho*c/Dt*(T-T0)*del_T \
    + kappa*T.dx(i)*del_T.dx(i) \
    - rho*Laser*del_T ) * dv \
    + h*(T-Ta)*del_T*da

left=lhs(Form)
right=rhs(Form)

A = assemble(left)  # non-changing by time stepping
b = None          # dynamically assembled acc. to time
T = Function(Space)
file_T = File("/calcul/CR10/T.pvd")
```

```
for t in numpy.arange(0,t_end,Dt):
    print "Time ",t
    Laser.t=t
    b=assemble(right, tensor=b)
    solve(A, T.vector(), b, 'cg')
    if t==int(t): file_T << (T,t)
    T0.assign(T)
```

To-do

Temperature distribution in a macroscopic rigid body has been computed. In order to deepen the understanding of the implementation try to do the following steps:

- Since the integral form is linear we may have implemented as in the previous sections by using "Form" and "Gain." Try to implement in this way, the results have to be identical.
- Change the boundary condition to a weak[8] DIRICHLET condition and then to adiabatic[9] boundaries by manipulating the parameter h.
- Search for approximate values of h for *natural* and *forced* convection. Which one has been established in the given code?
- Since heat escapes over the boundaries, sooner or later the temperature becomes the ambient temperature homogeneously in all body. Find the material parameters for a polymer and apply the same laser power. Is the body out of steel or polymer will reach the ambient temperature quicker?

2.2 Heat Transfer in Micromechanics

Heat propagation in a rigid body has been described by a parabolic differential equation in the last section. A mathematician recognizes the differential equations under two classes: *diffusion* and *wave* problems. A parabolic differential equation models a diffusion problem and a hyperbolic differential equation models a wave propagation. We refrain from using this terminology and point on the famous "paradox" due to the characteristics of the parabolic differential equation used in the heat transfer.[10] Consider FOURIER's law describing the heat flux:

$$q_i(\boldsymbol{x}, t) = -\kappa G_i(\boldsymbol{x}, t) , \quad G_i = \frac{\partial T}{\partial x_i} . \tag{2.26}$$

[8] We apply a DIRICHLET condition *strongly* by exchanging the solution with the given solution by using "DirichletBC" in the code. Instead of this method, we can satisfy the condition *weakly* by writing it under the boundary integral.

[9] An adiabatic boundary prevents heat transfer across boundaries.

[10] See [16] for some interesting explanations on the characteristics of the heat propagation.

Flux is the rate of energy exchange, in other words, it describes how quickly the heat energy travels between neighboring particles. Since the flux depends only on the temperature gradient, as long as there exists a temperature gradient the exchange may happen as quick as possible. Therefore, heat flux depending only on the temperature gradient results in an infinite propagation of the information. A typical example is a long bar excited on one end by a laser pulse. As expected, the temperature changes on the excited end. According to Eq. (2.26) at the very moment changed the temperature the neighboring element feels this change. The temperature change implies a heat flux instantaneously, thus whole bar "knows" this temperature change. In other words, the temperature starts changing in the whole bar instantaneously. For a bar in a macroscopic length scale, the temperature change is insignificant. Hence, FOURIER's law is quite accurate for many engineering problems in macroscale. However, in a microscopic length scale and laser pulses in femtoseconds the accuracy of FOURIER's law is inappropriate. There are even measurements in these scales suggesting a more sophisticated definition of the heat flux[11] than FOURIER's law. In this section we will generalize the heat flux by adding a rate dependency, similar to ideas used in Sect. 1.4, and simulate in microscale.

The generalization of the heat flux can be introduced in many different ways. We are interested in applied thermodynamics; so we use an argumentation that the flux and temperature gradient occur in different time instants. It is challenging to choose cause and effect, thus, we basically introduce a time lag both to the heat flux and temperature gradient:

$$q_i(\boldsymbol{x}, t + \tau_q) = -\kappa G_i(\boldsymbol{x}, t + \tau_T) \,. \tag{2.27}$$

The parameters τ_q and τ_T defines the time-delay or relaxation times between the flux and its response, i.e., the temperature gradient. By setting $\tau_T = 0$ we attain CATTANEO–VERNOTTE's heat flux[12] Since we want to evaluate the heat flux in the current time, t, we expand the left and right sides of Eq. (2.27) around the current time by using a TAYLOR series with linear terms:

$$q_i(\boldsymbol{x}, t) + \tau_q q_i^{\bullet}(\boldsymbol{x}, t) = -\kappa G_i(\boldsymbol{x}, t) - \kappa \tau_T G_i^{\bullet}(\boldsymbol{x}, t) \,. \tag{2.28}$$

By suppressing the arguments the heat flux for micromechanics reads

$$q_i = -\tau_q q_i^{\bullet} - \kappa G_i - \kappa \tau_T G_i^{\bullet} \,. \tag{2.29}$$

[11]See, for example, [18].

[12]It is named for Carlo Cattáneo and Pierre Vernotte.

Since the body is rigid, there is no difference between the total and partial time rate. Time discretization delivers

$$\begin{aligned} q_i &= -\tau_q \frac{q_i - q_i^0}{\Delta t} - \kappa G_i - \kappa \tau_T \frac{G_i - G_i^0}{\Delta t} \ , \\ q_i &= \frac{\Delta t}{\Delta t + \tau_q}\Big(\frac{\tau_q}{\Delta t} q_i^0 - \kappa\Big(1 + \frac{\tau_T}{\Delta t}\Big) G_i + \frac{\kappa \tau_T}{\Delta t} G_i^0 \Big) \ . \end{aligned} \tag{2.30}$$

As we have established the field equation for a rigid body in the last section

$$\rho c \frac{\partial T}{\partial t} + \frac{\partial q_i}{\partial x_i} - \rho L = 0 \ , \tag{2.31}$$

we can now implement the heat flux with relaxation times. After making the first term discrete in time, multiplying by the test function, δT, and then applying an integration by parts, we obtain the weak form:

$$\int_{\mathcal{B}} \Big(\frac{\rho c}{\Delta t} (T - T_0)\, \delta T - q_i \frac{\partial\, \delta T}{\partial x_i} - \rho L\, \delta T \Big) \,\mathrm{d}v + \int_{\partial \mathcal{B}} \hat{q}\, \delta T \,\mathrm{d}a \ , \tag{2.32}$$

where $\hat{q} = q_i n_i$ is the given boundary condition.

Every body consists of electronic particles and thus emits energy as a result of the changes in the electronic configurations of the atoms. This phenomenon happens above the absolute temperature, 0 K, for all times in form of radiation. Radiation is a volumetric supply and every particle of the body emits energy in the form of electromagnetic waves (photons), however, they are again absorbed from the neighboring particles. Only the particles building the surface emits energy such that this type of radiation is modeled as a surface phenomenon.[13] The maximum rate of energy emitted from a rigid body into a vacuum is given by the STEFAN–BOLTZMANN law:[14]

$$\tilde{q} = \sigma T^4 \ , \tag{2.33}$$

where the STEFAN–BOLTZMANN constant, $\sigma = 5.670 \cdot 10^{-8}\,\mathrm{W/m^2}$, is a universal constant. In the vacuum, the radiation waves propagate with the speed of light, $c_0 = 2.998 \cdot 10^8\,\mathrm{m/s}$. In air, they propagate with $c_{\mathrm{air}} = c_0/n$, where the *refractive index* is $n = 1.0008$ for air. The energy rate to the air is

$$\tilde{q} = n^2 \sigma T^4 \ , \tag{2.34}$$

in other words, it is only 0.16 % more than into the vacuum; this difference is negligible. The aforementioned relation holds for the idealized surface referred to as a

[13] For some intuitive explanations and examples of the thermal radiation, see [2, Chap. II, Sect. 9–4].

[14] The law is named after Josef Stefan and Ludwig Boltzmann.

blackbody. Real surfaces emit less energy than the blackbody and this rate is measured by the *emissivity*, ε, of the surface:

$$\tilde{q} = \varepsilon\sigma T^4 \ . \tag{2.35}$$

The emissivity is between zero and one. Even in vacuum this type of energy exchange occurs. If the body is surrounded by air at T_{amb} then the energy rate emitted from the body reads

$$\tilde{q} = \varepsilon\sigma(T^4 - T_{\text{amb}}^4) \ . \tag{2.36}$$

We can now combine the natural convection and the surface radiation in order to find out the boundary condition as follows

$$\hat{q} = q_i n_i = h(T - T_{\text{amb}}) + \varepsilon\sigma(T^4 - T_{\text{amb}}^4) \ , \tag{2.37}$$

for a body embedded in resting air. If the air possesses a velocity, as in case of a forced convection, the value of h increases and the heat exchange via convection dominates; the surface radiation can be completely neglected.

We acquire the following weak form for the computation:

$$\begin{aligned} \text{Form} = &\int_{\mathcal{B}} \left(\frac{\rho c}{\Delta t}(T - T_0)\, \delta T - q_i \frac{\partial\, \delta T}{\partial x_i} - \rho L\, \delta T \right) \mathrm{d}v + \\ &+ \int_{\partial\mathcal{B}} \left(h(T - T_{\text{amb}}) + \varepsilon\sigma(T^4 - T_{\text{amb}}^4) \right) \delta T \, \mathrm{d}a \ , \end{aligned} \tag{2.38}$$

with the heat flux, q_i, as in Eq. (2.30). Obviously, the weak form is nonlinear due to the thermal radiation in the boundary conditions.

Suppose we have a tiny beam of $100 \times 5 \times 5\,\mu\text{m}$ excited on one end with a laser beam. Taken from [18] time lag parameters are given below in picoseconds:

$$\begin{aligned} \text{for Cu}\ , \quad \tau_T &= 70.833\,\text{ps}\ , \quad \tau_q = 0.4648\,\text{ps}\ , \\ \text{for Au or Ag}\ , \quad \tau_T &= 89.286\,\text{ps}\ , \quad \tau_q = 0.7438\,\text{ps}\ , \\ \text{for Pb}\ , \quad \tau_T &= 12.097\,\text{ps}\ , \quad \tau_q = 0.1720\,\text{ps}\ . \end{aligned} \tag{2.39}$$

We choose an appropriate unit system:

$$1\,\text{ps} = 10^{-12}\,\text{s}\ , \quad 1\,\mu\text{g} = 10^{-9}\,\text{kg}\ , \quad 1\,\mu\text{m} = 10^{-6}\,\text{m}\ , \quad 1\,\text{nN} = 10^{-9}\,\text{N}\ , \tag{2.40}$$

where energy is in femtoJoule, $1\,\text{nN} \times 1\,\mu\text{m} = 10^{-15}\,\text{N}\,\text{m} = 1\,\text{fJ}$, and temperature in K as usual. A short pulse of laser on one end ignites a heat transport to the surface and toward to other end, see Fig. 2.3. The implementation with consistent units has been realized by the code below.

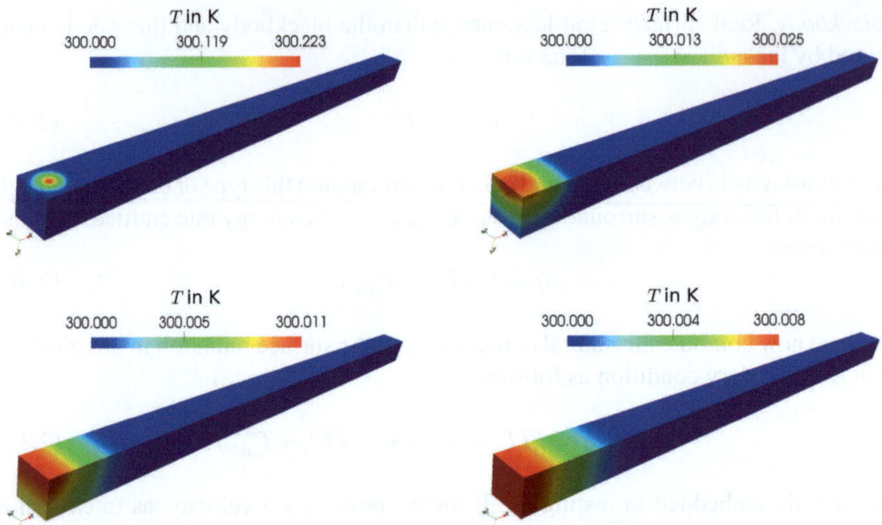

Fig. 2.3 Temperature distribution at 1, 15, 50, and 100 ns in the gold bar, consider the change of the temperature scale for a better visualization

```
"""Computational reality 11, temperature distribution in a microscopic body"""
__author__ = "B. Emek Abali"
__license__ = "GNU GPL Version 3.0 or later"
#This code underlies the GNU General Public License, http://www.gnu.org/licenses/gpl-3.0.en.html

from fenics import *
import numpy
parameters["allow_extrapolation"]=True
parameters["form_compiler"]["cpp_optimize"] = True
set_log_level(ERROR)

rho=19.3E-6      # mass density of gold (Au) in mug/mum^3]
c=129.0E6  # heat capacity in fJ/(mug K)
kappa=318.0E-3  # thermal conductivity in fJ/(mum K ps)
tau_T = 89.286  # in ps
tau_q = 0.7438  # in ps
h=18.0E-9  # natural convection coefficient in fJ/(ps mum^2 K)]
emis=0.47  # emissivity of gold not polished
sigma=5.670E-17 # Stefan-Boltzmann constant in fJ/(ps mum**2)
Ta=300.0  # ambient temperature in K
l=100.0  # length in mum
th=5.0  # thickness in mum
P=30.0E3  # laser power in fJ/(mum ps)

mesh = BoxMesh(Point(0.0,0.0,0.0), Point(l,th,th), 200,10,10)
Space = FunctionSpace(mesh, 'P', 1)
VectorSpace = VectorFunctionSpace(mesh, 'P', 1)
cells = CellFunction('size_t', mesh)
```

```
facets = FacetFunction('size_t', mesh)
da = Measure('ds', domain=mesh, subdomain_data=facets)
dv = Measure('dx', domain=mesh, subdomain_data=cells)
t=0.0
t_end=100000.0 #0.1 mus
Dt=1000.0
initial_T = Expression("Tini", Tini=Ta)
T0 = interpolate(initial_T, Space)
Laser = Expression("P*exp(-1.0*(pow(x[0]-2, 2)+pow(x[1], 2)+pow(x[2]-2.5, 2)))",P=P)
T = Function(Space)
del_T = TestFunction(Space)
dT = TrialFunction(Space)
q0 = Function(VectorSpace)
G = as_tensor(T.dx(i), (i,))
G0 = as_tensor(T0.dx(i), (i,))
q = as_tensor(Dt/(Dt+tau_q)*(tau_q/Dt*q0[i]-kappa*(1+tau_T/Dt)*G[i]+kappa*tau_T/Dt*G0[i]), (i,))
Form = (rho*c/Dt*(T-T0)*del_T \
    - q[i]*del_T.dx(i) \
    - rho*Laser*del_T ) * dv \
    + (h*(T-Ta) \
    + emis*sigma*(T**4-Ta**4) )*del_T*da
Gain = derivative(Form, T, dT)

file_T = File("/calcul/CR11/T.pvd")
for t in numpy.arange(0,t_end,Dt):
    print "Time ",t
    if t>=2000.: Laser.P=0
    solve(Form==0, T, [], J=Gain, \
        solver_parameters={"newton_solver":{"linear_solver": "mumps", "relative_tolerance": 1e-3} }, \
        form_compiler_parameters={"cpp_optimize": True, "representation": "quadrature", "quadrature_degree": 2} )

    if t==int(t): file_T << (T,t)
    q0 = project(q,VectorSpace)
    T0.assign(T)
```

To-do

Temperature distribution with a time lag has been implemented.

- Change the boundary condition to a DIRICHLET boundary without radiation.
- Simulate the same problem with FOURIER's law.
- Produce a 2D plot with temperature versus x-coordinates. Plot the results with CATTANEO–VERNOTTE's and FOURIER's law on top of each other and analyse the difference.

2.3 Thermodynamics in a Nutshell

Theoretical thermodynamics concerns derivation of the constitutive equations, which we have already been using in the former sections. For some viscous fluids, for example water, we are certain that the NAVIER–STOKES equation is an adequate model to describe the flow behavior. The material model given by the constitutive equation has been attained phenomenologically (by using empirical research). In the early 1940s the concept of continuum mechanics has been redesigned under the name *rational mechanics*. Different scientific branches of mechanics: solid body mechanics, fluid mechanics, and applied thermodynamics, have been fused by using this concept. Such an abstraction of different studies has lead to theoretical thermodynamics[15] used to derive the constitutive equations.

Theoretical thermodynamics is a non-unique approach. At least four prominent methodologies can be listed: the COLEMAN–NOLL procedure, MULLER's rational thermodynamics, and non-equilibrium thermodynamics.[16] They deliver the basic equations like NAVIER–STOKES's equation such that we believe that all methodologies are correct. We omit an introduction and discussion of different methodologies and design a method using elements from all of them. The output is again the well-known equations for simple viscous fluids such that we can convince ourselves that the procedure is acceptable. The proposed method possesses some limitations that we will remark by presenting and applying the procedure in the following. Although the method has some weak points, it is relatively simple and allows to be generalized easily in order to involve electromagnetic interactions in Chap. 3.

We have introduced and implemented the following three balance equations in their local forms: the balance of mass, the balance of linear momentum, and the balance of internal energy, respectively:

$$\rho^{\boldsymbol{\cdot}} + \rho \frac{\partial v_i}{\partial x_i} = 0 \,, \quad \rho v_i^{\boldsymbol{\cdot}} - \frac{\partial \sigma_{ji}}{\partial x_j} - \rho f_i = 0 \,, \quad \rho u^{\boldsymbol{\cdot}} + \frac{\partial q_i}{\partial x_i} - \rho r = \sigma_{ij} \frac{\partial v_j}{\partial x_i} \,, \tag{2.41}$$

in the current frame expressed in a Cartesian coordinate system. The first two has zero sources, i.e., zero right-hand sides and the balance of internal energy has a non-zero source. The source is a production term. We cannot eliminate or "shut off" the production term. For example the production term of the internal energy—known as the internal friction—alters the internal energy, as long as the material undergoes a motion. Internal energy fails to be a conserved quantity. Mass and linear momentum are conserved quantities, since they lack a production term.

From the balance equations we want to solve the mass density, ρ, the velocity, v_i, and the temperature, T. First we need to close the balance equations by defining the constitutive equations for the CAUCHY stress, σ_{ji}, for the specific internal energy, u,

---

[15] Thermodynamics of irreversible processes started with papers of Carl Eckart, see [4], [5], [6], [7].

[16] The COLEMAN-NOLL procedure is named after Bernard D. Coleman and Walter Noll, see [3]. MULLER's rational thermodynamics is named for Ingo Müller, see [11]. The non-equilibrium thermodynamics is introduced by Sybren Ruurds de Groot and Peter Mazur, see [8].

and for the heat flux, q_i. After having found the definitions, the formulation will be complete such that we can generate a weak form and solve the balance equations augmented with the constitutive equations. Theoretical thermodynamics has the aim of determining the constitutive equations. In this section we will perform the necessary steps leading to the adequate constitutive equations modeling a viscous fluid.

We want to compute (ρ, v_i, T) for a fluid in a EULERian frame, (x_i, t). The unknowns (ρ, v_i, T) are referred to as *primitive* variables: their mathematical existence is accepted without any further investigation. Our goal is to compute the primitive variables by satisfying the balance equations. We can solve the balance equations if they are closed: All constitutive equations for σ_{ij}, q_i, and u are given by functions depending on the primitive variables or their time and space derivatives.

The first assumption is that the specific total energy, e, is additive so that the specific internal energy, u, and specific kinetic energy, $e^{\text{kin.}}$, are separable and independent. Hence, the internal energy fails to depend on the velocity. By following [14] we assume that the change of the internal energy[17] is recoverable. The kinetic energy is irreversible. There are many different formulations in the literature and none of them is wrong, because we cannot measure different parts of the energy separately; they are just various models approximating the behavior of the material with different accuracies. The accuracy of any formalism can only be tested by measurements. In the end, the primitive variables are calculated with adequate accuracy, if the material modeling is appropriate. Without discussing its limitations, we assume that the internal energy possesses only terms occurring a recoverable change. This assumption is of paramount importance and leads to a useful methodology described in the following.

We introduce mass density, $\rho(x_i, t)$, velocity, $v_i(x_j, t)$, and temperature $T(x_i, t)$ as primitive variables. We axiomatically assume that they are independent functions.[18] Velocity fails to be an *objective* variable. If we perform a EUCLIDian coordinate transformation, velocity transforms other than a tensor of rank one. Fortunately, symmetric velocity gradient, $d_{ij} = \partial v_{(i}/\partial x_{j)}$, is an objective variable, it is a tensor of rank two.[19] Constitutive equations shall depend on objective variables.

We start the formulation by proposing an equation for the internal energy. Actually, this equation can be derived in various ways. We present here a method based on the balance of internal energy since we will make much use of it in the next chapter.[20] Consider a material of having the following constitutive equation:

$$\sigma_{ji} = {}^{\text{r}}\sigma_{ji} + {}^{\text{d}}\sigma_{ji} , \tag{2.42}$$

[17]Under the assumption that no phase changes occur.

[18]Of course they are coupled, however, independent. We can hold the temperature fixed and move the body or restrict any motion and change the temperature.

[19]The formulation holds for fluids with elasticity, too. Therefore, we need to introduce, $d_{ij} = \dot{\varepsilon}_{ij}$, for a fixed frame, $w_i = 0$. The proof of this identity is out of scope, therefore, we explain it in Appendix A.4 on p. 301.

[20]For another, more conventional derivation, see [1, Sect. 3].

where the reversible (recoverable) change is given by the first and irreversible (dissipative) change is described by the second term. The assumption of separating stress tensor additively is justified by the energy assumption, where we have also separated the reversible (recoverable) and irreversible (dissipative) terms into an internal and kinetic energy, respectively. For a fluid without elasticity, we introduce the reversible term by using pressure, p, as follows

$$ {}^{\mathrm{r}}\sigma_{ji} = -p\delta_{ji} \,. \tag{2.43} $$

The dissipative term will be a function of d_{ij} leading to a flow in the system. The stress consists of the reversible term, if the fluid rests (zero velocity),[21] this state is called a mechanical *equilibrium*. In equilibrium, the pressure enables a reversible momentum transport, for example the sound waves in a fluid are transported by the pressure, p. This process is reversible since the motion of fluid particles are recovered after the sound wave has passed by. This small motion of fluid particles are neglected at all, we only calculate the velocity leading to a convection of the fluid. If the fluid rests, pressure still transports sound waves reversibly. Hence, we can call the pressure as a *hydrostatic* pressure since it is responsible for a momentum transport in the static or equilibrium state. In the mechanical equilibrium the stress reads

$$ \sigma_{ji}\Big|_{\text{eq.}} = -p\delta_{ji} \,. \tag{2.44} $$

Formally, the decomposition in Eq. (2.42) is correct since we still have not defined ${}^{\mathrm{d}}\sigma_{ij}$. Instead of searching for a definition of σ_{ij} we now have to search for p and ${}^{\mathrm{d}}\sigma_{ji}$.

For a thermal equilibrium we introduce a quantity called a specific *entropy*, η, as

$$ -\frac{\partial q_i}{\partial x_i}\Big|_{\text{eq.}} = q^{\bullet}\Big|_{\text{eq.}} \,, \quad \frac{1}{T} q^{\bullet}\Big|_{\text{eq.}} = \rho\eta^{\bullet} \,, \tag{2.45} $$

by following [8, Chap. XIV, Sect. 2]. The entropy is responsible for a reversible transport of energy, i.e., the process is in a thermal equilibrium. Thermal equilibrium is often explained as a slow temperature change, actually, it is just the reversible part of the process without any dissipation. Indeed a slow temperature change is undertaken by the reversible part. Similar to the previous case, we now have to determine a constitutive equation for η and q_i. The balance of internal energy:

$$ \rho u^{\bullet} + \frac{\partial q_i}{\partial x_i} - \rho r = \sigma_{ij} \frac{\partial v_j}{\partial x_i} \,, \tag{2.46} $$

reads at thermal and mechanical equilibrium

[21] Since we use a EUCLIDean transformation to test the objectivity, a constant velocity is accepted, too. Consider a rigid body moving with a constant velocity, it actually *rests* in a coordinate system moving with this velocity. Hence, we can always introduce a constantly moving coordinate system, which is allowed in the EUCLIDean transformation, where the body rests.

$$\rho u^{\cdot} - \rho T \eta^{\cdot} = -p \frac{\partial v_i}{\partial x_i} , \tag{2.47}$$

since by existing internal heating (supply term), equilibrium cannot take place. We recall that the internal energy is fully recoverable. By introducing a specific volume, $v = 1/\rho$, a volume per mass, we can rewrite the balance of mass as follows

$$\begin{aligned}
(v^{-1})^{\cdot} v &= -\frac{\partial v_i}{\partial x_i} , \\
\frac{v^{\cdot}}{v} &= \frac{\partial v_i}{\partial x_i} , \\
\rho v^{\cdot} &= \frac{\partial v_i}{\partial x_i} .
\end{aligned} \tag{2.48}$$

Now by inserting the latter into the balance of internal energy we obtain

$$\begin{aligned}
\rho u^{\cdot} - \rho T \eta^{\cdot} &= -\rho p v^{\cdot} , \\
u^{\cdot} &= T \eta^{\cdot} - p v^{\cdot} .
\end{aligned} \tag{2.49}$$

Furthermore from the latter expression, we acquire so-called GIBBS's equation:

$$\mathrm{d}u = T \, \mathrm{d}\eta - p \, \mathrm{d}v , \tag{2.50}$$

under the condition that u has a first integral:

$$u = \int \mathrm{d}u . \tag{2.51}$$

Often, this condition is referred to as the 1st law of thermodynamics.[22] The integration is between two states. Suppose we start from the state, $\{T = T_0, v = v_0\}$, and end up in a state at another temperature and mass density (thus specific volume), $\{T, v\}$. Since the internal energy is a total differential we can obtain the energy by integrating from the state one to state two

$$u = \int_{(T_0, v_0)}^{(T, v)} \mathrm{d}u = u(T, v) - u(T_0, v_0) . \tag{2.52}$$

Only the first and last states are important, not the states in between. This condition is a limitation and the methodology herein with this limitation is called the *equilibrium* thermodynamics. Exactly this assumption is a deficiency on the way to a general

[22]For instance in [14] the internal energy is introduced as a full recoverable quantity such that the first integral is automatically justified. Either we can accept the axiom of existence of $\mathrm{d}u$ as the 1st law, or the assumption that the internal energy is fully recoverable as the 1st law.

theory.[23] In other words, for some processes this assumption may lead to constitutive equations not capable of describing the process accurately. The fact that the total energy has a dissipative term only due to the kinetic energy might be too restrictive. The methodology presented here would not suffice for describing a process where temperature (or its rate) plays a dissipative role in the total energy. However, at least for all processes presented in this book, this restriction is admissible and the implemented method is reliable.

Usually GIBBS's equation is an axiomatic start; its validity is taken for granted. Herein, we have motivated it by using the balance of internal energy at the equilibrium. Since the internal energy is assumed to be recoverable, the differential relation holds for the non-equilibrium, too. By considering Eq. (2.50) we realize that the internal energy is a function of entropy and specific volume,

$$u = u(\eta, v)\ . \tag{2.53}$$

We have an inconsistency by defining energy depending on a variable, η, which is not yet defined. Better we shall find a constitutive equation for energy depending on the primitive variables or their derivatives, i.e., on the so-called *state* variables.[24] In this section the state or primary variables are $\{T, v\}$. Thus, we need an energy definition depending only on the primary variables. In order to obtain such an energy we introduce the so-called specific HELMHOLTZ free energy:

$$\psi = u - T\eta\ . \tag{2.54}$$

Its total differential is assumed to exist[25]

$$\mathrm{d}\psi = \mathrm{d}u - \eta\,\mathrm{d}T - T\,\mathrm{d}\eta\ . \tag{2.55}$$

By inserting Eq. (2.50) into the latter we obtain

$$\mathrm{d}\psi = -\eta\,\mathrm{d}T - p\,\mathrm{d}v\ . \tag{2.56}$$

From this differential form we realize that the free energy depends on the primary variables

$$\psi = \psi(T, v)\ . \tag{2.57}$$

---

[23] For a brief explanation of this deficiency, we refer to [13, Chap. 2].

[24] Since state variables are derived from the primitive variables we can also name them as primary variables.

[25] This assumption is another weak point in the methodology, the 1st law of thermodynamics only states that the internal energy has a perfect differential, $\mathrm{d}u$, but not the free energy. Concretely, we have to make sure that $\mathrm{d}(T\eta)$ exists.

Obviously, we have the following relations:

$$\mathrm{d}\psi = \frac{\partial \psi}{\partial T}\,\mathrm{d}T + \frac{\partial \psi}{\partial v}\,\mathrm{d}v \ , \quad \eta = -\frac{\partial \psi}{\partial T} \ , \quad p = -\frac{\partial \psi}{\partial v} \ . \tag{2.58}$$

Our goal is to determine the *dual* variables depending on the primary variables[26]

$$\eta = \eta(T, v) \ , \quad p = p(T, v) \ , \tag{2.59}$$

leading to the following differentials:

$$\begin{aligned} \mathrm{d}\eta &= \frac{\partial \eta}{\partial T}\,\mathrm{d}T + \frac{\partial \eta}{\partial v}\,\mathrm{d}v = A\,\mathrm{d}T + B\,\mathrm{d}v \ , \\ \mathrm{d}p &= \frac{\partial p}{\partial T}\,\mathrm{d}T + \frac{\partial p}{\partial v}\,\mathrm{d}v = C\,\mathrm{d}T + D\,\mathrm{d}v \ . \end{aligned} \tag{2.60}$$

We need to determine the material coefficients A, B, C, D as functions depending on the state variables. The partial derivatives are taken by holding the other arguments fixed. We skip an explicit notation about the dependency, since it is superfluous. The small change $\mathrm{d}(\cdot)$ is simply how we shall undertake the measurements.

We cannot measure the (specific) entropy, η, directly. Instead, heat (thermal energy) is measured in a calorimeter, $\delta q = T\,\mathrm{d}\eta$. The notational difference, δq, is only due to the fact that we cannot form a total differential of the heat. In other words, it is necessary to integrate over the whole process that is an evolution, the knowledge of the start and end states is not sufficient for δq. Technically, we just measure the heat energy supplied to the system. By holding the specific volume constant, $\mathrm{d}v = 0$, and by varying the temperature, $\mathrm{d}T$, we obtain a change of heat, δq, which is measured as

$$\delta q = c\,\mathrm{d}T \Rightarrow A = \frac{c}{T} \ , \tag{2.61}$$

where c is called the *specific heat capacity* by holding volume constant. It may depend on temperature and have to be measured for different specific volumes. Of course for different specific volumes, the numerical value of c may vary, too. Hence, $c = c(T, v)$, at least in principle. These measurements are difficult such that either heat capacity is assumed to be constant or solely the dependence on the temperature is found in the literature.

We can easily measure the coefficient D by fixing the temperature, $\mathrm{d}T = 0$; varying the specific volume, $\mathrm{d}v$; and measuring the pressure change, $\mathrm{d}p$. It is possible to change the temperature and measure the pressure for a fixed (specific) volume for determining C. However, it is rather difficult to set the temperature constant and measure the heat exchange due to the variation in the specific volume. The problem is

[26] Although this is mathematically obvious that the dual variables have to depend on the same set of arguments of the free energy, namely on the state variables, this condition is called the *equipresence* principle, see [17, Sect. 293.η].

that we normally measure the heat change over the temperature measurement. There is a MULLER-calorimeter for this purpose but it is not appropriate for all materials.[27]

Fortunately, we can skip measurements for determining C coefficient because $C = B$. In order to see this relation, we write the dual variables in a matrix notation:

$$\begin{pmatrix} \mathrm{d}\eta \\ \mathrm{d}p \end{pmatrix} = \begin{pmatrix} c/T & B \\ C & D \end{pmatrix} \begin{pmatrix} \mathrm{d}T \\ \mathrm{d}v \end{pmatrix} . \tag{2.62}$$

The condition $B = C$ is a *symmetry* condition in the matrix of coefficients. This condition can be seen readily by using the free energy depending on T and v,

$$B = \frac{\partial \eta}{\partial v} = -\frac{\partial^2 \psi}{\partial v \partial T} = -\frac{\partial^2 \psi}{\partial T \partial v} = \frac{\partial p}{\partial T} = C . \tag{2.63}$$

This condition is known as MAXWELL's reciprocal relation.[28] For a viscous fluid we can integrate and find out the constitutive equations:

$$\eta = \int \mathrm{d}\eta = \int \frac{c}{T}\,\mathrm{d}T + \int B\,\mathrm{d}v , \quad p = \int \mathrm{d}p = \int B\,\mathrm{d}T + \int D\,\mathrm{d}v , \tag{2.64}$$

from a reference state $\{T_{\text{ref.}}, v_{\text{ref.}}\}$ to the current state $\{T, v\}$. For a *linear* material the coefficients are constants such that we readily obtain

$$\begin{aligned} \eta &= c \ln\left(\frac{T}{T_{\text{ref.}}}\right) + B(v - v_{\text{ref.}}) , \\ p &= B(T - T_{\text{ref.}}) + D(v - v_{\text{ref.}}) . \end{aligned} \tag{2.65}$$

For an incompressible material $v = v_{\text{ref.}}$ such that B and D fails to be measured by a variation of the specific volume, since $\mathrm{d}v = 0$. GIBBS's equation becomes $\mathrm{d}u = c\,\mathrm{d}T$ and the specific entropy reads

$$\eta = c \ln\left(\frac{T}{T_{\text{ref.}}}\right) . \tag{2.66}$$

We calculate p from the balance of mass.

In order to derive the heat flux, q_i, and the dissipative part of the stress tensor, ${}^{\mathrm{d}}\sigma_{ij}$, we continue the methodology in the following. Since we have defined the internal energy, we can insert Eq. $(2.49)_2$:

$$u^{\bullet} = T\eta^{\bullet} - pv^{\bullet} \tag{2.67}$$

[27] See [10] for a detailed explanation about the MULLER-calorimeter named after F. Horst Müller.

[28] It is named after James Clerk Maxwell.

into the balance of internal energy in Eq. (2.41)$_3$ and obtain

$$
\begin{aligned}
T\rho\eta^{\bullet} - p\rho v^{\bullet} + \frac{\partial q_i}{\partial x_i} - \rho r &= \sigma_{ij}\frac{\partial v_j}{\partial x_i} \,, \\
T\rho\eta^{\bullet} + \frac{\partial q_i}{\partial x_i} - \rho r &= {}^{\mathrm{d}}\sigma_{ij}\frac{\partial v_j}{\partial x_i} \,,
\end{aligned}
\tag{2.68}
$$

after having used the balance of mass as in Eq. (2.48). We can rewrite the latter equation further,

$$
\begin{aligned}
\rho\eta^{\bullet} + \frac{1}{T}\frac{\partial q_i}{\partial x_i} - \rho\frac{r}{T} &= \frac{1}{T}\,{}^{\mathrm{d}}\sigma_{ij}\frac{\partial v_j}{\partial x_i} \,, \\
\rho\eta^{\bullet} + \frac{\partial}{\partial x_i}\Big(\frac{q_i}{T}\Big) - q_i\frac{\partial}{\partial x_i}\Big(\frac{1}{T}\Big) - \rho\frac{r}{T} &= \frac{1}{T}\,{}^{\mathrm{d}}\sigma_{ij}\frac{\partial v_j}{\partial x_i} \,,
\end{aligned}
\tag{2.69}
$$

in order to acquire a balance equation:

$$
\rho\eta^{\bullet} + \frac{\partial}{\partial x_i}\Big(\frac{q_i}{T}\Big) - \rho\frac{r}{T} = -\frac{q_i}{T^2}\frac{\partial T}{\partial x_i} + \frac{1}{T}\,{}^{\mathrm{d}}\sigma_{ij}\frac{\partial v_j}{\partial x_i} \,. \tag{2.70}
$$

This balance equation is the balance of entropy:

$$
\rho\eta^{\bullet} + \frac{\partial \Phi_i}{\partial x_i} - \rho\frac{r}{T} = \Sigma \,, \tag{2.71}
$$

with the flux term, Φ_i, and the production term, Σ, as follows

$$
\Phi_i = \frac{q_i}{T} \,, \quad \Sigma = -\frac{q_i}{T^2}\frac{\partial T}{\partial x_i} + \frac{1}{T}\,{}^{\mathrm{d}}\sigma_{ij}\frac{\partial v_j}{\partial x_i} \,. \tag{2.72}
$$

The 2nd law of thermodynamics asserts that any process attains a positive entropy production:

$$
\Sigma \geq 0 \,. \tag{2.73}
$$

This law restricts the possible constitutive relations. In other words, q_i, σ_{ji} have to be such that $\Sigma \geq 0$ is assured for every possible processes. This restriction leads to the constitutive equations for q_i and σ_{ij} as presented in the following in three steps. First we introduce the following notation:

$$
G_i = \frac{\partial T}{\partial x_i} \,, \quad d_{ij} = \frac{\partial v_{(i}}{\partial x_{j)}} \,, \quad d_{|ij|} = d_{ij} - \frac{1}{3}d_{kk}\delta_{ij} \,. \tag{2.74}
$$

Moreover, in non-polar materials the dissipative stress is symmetric, ${}^{\mathrm{d}}\sigma_{ij} = {}^{\mathrm{d}}\sigma_{ji}$. We exclude polar materials in this book.[29] Secondly, we rewrite the entropy production:

[29]Nematic fluids used in LCD (Liquid Crystal Display) is a prominent polar material.

$$\Sigma = -\frac{1}{T^2} q_i G_i + \frac{1}{3T} {}^{\mathrm{d}}\sigma_{ii} d_{jj} + \frac{1}{T} {}^{\mathrm{d}}\sigma_{|ij|} d_{|ij|} \geq 0 \,. \tag{2.75}$$

This multiplication can be seen as *thermodynamical fluxes*:

$$\mathcal{F}^\alpha = \Big\{ -q_i \,,\; {}^{\mathrm{d}}\sigma_{ii} \,,\; {}^{\mathrm{d}}\sigma_{|ij|} \Big\} \,, \tag{2.76}$$

multiplied (by an inner product) with *thermodynamical forces*:

$$\mathcal{K}^\alpha = \Big\{ \frac{G_i}{T^2} \,,\; \frac{1}{3T} d_{jj} \,,\; \frac{1}{T} d_{|ij|} \Big\} \,, \tag{2.77}$$

as follows

$$\Sigma = \mathcal{F}^\alpha \cdot \mathcal{K}^\alpha \geq 0 \,, \quad \alpha = 1, 2, 3 \,. \tag{2.78}$$

The thermodynamical forces are independent among each other. Thirdly, each thermodynamical force is of another rank. For an isotropic material a thermodynamical flux may depend only on the same rank of the thermodynamical forces. Since otherwise under a coordinate transformation different rank tensors transform differently such that the dependency of one flux component on the force component changes. Hence, for isotropic materials the thermodynamical fluxes depend only on the thermodynamical forces of the same rank:

$$\mathcal{F}^1 = \mathcal{F}^1(\mathcal{K}^1) \,, \quad \mathcal{F}^2 = \mathcal{F}^2(\mathcal{K}^2) \,, \quad \mathcal{F}^3 = \mathcal{F}^3(\mathcal{K}^3) \,. \tag{2.79}$$

This condition is known as the CURIE symmetry principle.[30] Consider the following relations:

$$-q_i = \bar{a} \frac{1}{T^2} G_i \,, \quad {}^{\mathrm{d}}\sigma_{ii} = \bar{b} \frac{1}{3T} d_{jj} \,, \quad {}^{\mathrm{d}}\sigma_{|ij|} = \bar{c} \frac{1}{T} d_{|ij|} \,, \tag{2.80}$$

where the material coefficients may depend on the thermodynamical forces.[31] If we insert the constitutive relations into the entropy production

$$\Sigma = \bar{a} \frac{1}{T^4} G_i G_i + \bar{b} \frac{1}{9T^2} d_{ii} d_{jj} + \bar{c} \frac{1}{T^2} d_{|ij|} d_{|ij|} \geq 0 \,, \tag{2.81}$$

has to hold. We know that every term is a multiplication between different types of fluxes and forces, i.e., every term is independent. In order to demand $\Sigma \geq 0$ for any process, the coefficients have to be

$$\bar{a} \geq 0 \,, \quad \bar{b} \geq 0 \,, \quad \bar{c} \geq 0 \,, \tag{2.82}$$

---

[30] It is named after Pierre Curie.

[31] For the sake of clarity, the coefficients are functions of the invariants of thermodynamical forces.

since $T > 0$ in K(elvin). This conclusion is quite general and the coefficients may be scalar functions of the thermodynamical forces, i.e., the thermodynamical fluxes may depend on the thermodynamical forces nonlinearly. If we rename the coefficients:

$$\bar{a}\frac{1}{T^2} = \kappa \, , \ \bar{b}\frac{1}{3T} = 3\lambda + 2\mu \, , \ \bar{c}\frac{1}{T} = 2\mu \, , \tag{2.83}$$

and simplify to *linear* materials by assuming that κ, λ, and μ are constant, then we end up in a NAVIER–STOKES–FOURIER fluid:

$$\begin{aligned} q_i &= -\kappa G_i \, , \\ \sigma_{ij} = -p\delta_{ij} + {}^{\mathrm{d}}\sigma_{ij} &= -p\delta_{ij} + \frac{1}{3}(3\lambda + 2\mu) d_{kk}\delta_{ij} + 2\mu d_{|ij|} = \\ &= (-p + \lambda d_{kk})\delta_{ij} + 2\mu d_{ij} \, . \end{aligned} \tag{2.84}$$

Obviously the material constants are $\kappa \geq 0$, $\mu \geq 0$, and $3\lambda + 2\mu \geq 0$, in order to fulfill the 2nd law. Since $d_{ii} = \rho v^{\bullet}$ for an incompressible fluid, commonly $3\lambda + 2\mu = 0$ is used, which is known as STOKES's hypothesis. However, this hypothesis cannot be verified experimentally. Since $d_{ii} = 0$ holds for an incompressible fluid flow, it is not possible to detect the value of $3\lambda + 2\mu$. We will never use this hypothesis. Certainly, by utilizing $d_{ii} = 0$ we assume that the flow is incompressible. This interpretation is correct, incompressibility is not a material property; even water flows compressible under great pressure and temperature conditions.[32] By using the incompressibility we spare the mass balance for the computation of p. For a numerical stability λ shall be a large number.

For a non-linear fluid we can now quickly generalize the constitutive equations and propose

$$\bar{c}\frac{1}{T} = 2\left(\mu_0 + \frac{k}{\pi\sqrt{II}} \arctan\left(\frac{\sqrt{II}}{B}\right)\right) \, , \ II = \frac{1}{2} d_{ij} d_{ij} \, , \tag{2.85}$$

as already employed in Sect. 1.8. For the thermodynamical consistency, material parameters, μ_0, k, B, have positive values.

We want to compute the same channel problem as in Sect. 1.7, this time by computing not only the pressure and velocity but also the temperature change. We choose water, a viscous linear fluid and model with FOURIER–NAVIER–STOKES constitutive equations:

$$\sigma_{ij} = -p\delta_{ij} + {}^{\mathrm{d}}\sigma_{ij} \, , \ {}^{\mathrm{d}}\sigma_{ij} = \lambda d_{kk}\delta_{ij} + 2\mu d_{ij} \, , \ q_i = -\kappa \frac{\partial T}{\partial x_i} \, . \tag{2.86}$$

[32] However, if we use $3\lambda + 2\mu = 0$ then we assume that the incompressibility is a property of the material. This is definitely not the case, see [9] and also [15].

Since water can be assumed as incompressible, $\mathrm{d}v = 0$, we have the rate of internal energy from GIBBS's equation:

$$\rho u^{\bullet} = \rho T \eta^{\bullet} = \rho c T^{\bullet} \,. \tag{2.87}$$

For a fixed[33] domain, Ω, we obtain a weak form from Eq (2.41)$_1$ for computing the pressure:

$$\mathrm{F}_p = \int_\Omega \Big(\frac{\partial \rho}{\partial t} + v_i \frac{\partial \rho}{\partial x_i} + \rho \frac{\partial v_i}{\partial x_i} \Big) \frac{\delta p}{\rho} \, \mathrm{d}v \,. \tag{2.88}$$

This integral form is in the unit of power. For an incompressible flow the latter reduces to

$$\mathrm{F}_p = \int_\Omega \frac{\partial v_i}{\partial x_i} \, \delta p \, \mathrm{d}v \,. \tag{2.89}$$

Analogously the balance of linear momentum in Eq (2.41)$_2$ leads to a weak form for computing the velocity:

$$\begin{aligned} \mathrm{F}_v = \int_\Omega \Big(\rho \frac{\partial v_i}{\partial t} \, \delta v_i + \rho v_j \frac{\partial v_i}{\partial x_j} \, \delta v_i + \frac{\partial p}{\partial x_i} \, \delta v_i + {}^{\mathrm{d}}\sigma_{ji} \frac{\partial \, \delta v_i}{\partial x_j} - \rho f_i \, \delta v_i \Big) \, \mathrm{d}v - \\ - \int_{\partial\Omega} {}^{\mathrm{d}}\sigma_{ji} n_j \, \delta v_i \, \mathrm{d}a \,. \end{aligned} \tag{2.90}$$

This integral form is also in the unit of power. If we want to apply a NEUMANN boundary by defining a mechanical pressure applied on the boundary, then the traction vector:

$$\hat{t}_i = \sigma_{ji} n_j = -p n_i + {}^{\mathrm{d}}\sigma_{ji} n_j \,, \tag{2.91}$$

is necessary. We can apply a mechanical pressure on left and right openings, $\hat{p}$. The mechanical pressure on a boundary reads

$$-\frac{1}{3} \sigma_{kk} = \hat{p} = -\frac{1}{3} \big(-p \delta_{kk} + {}^{\mathrm{d}}\sigma_{kk} \big) \,. \tag{2.92}$$

For applying this pressure we need the traction vector inward the domain

$$\hat{t}_i = -n_i \hat{p} = \frac{n_i}{3} \big(-p \delta_{kk} + {}^{\mathrm{d}}\sigma_{kk} \big) \,. \tag{2.93}$$

For an incompressible flow, $d_{kk} = 0$, the spherical dissipative stress vanishes such that we obtain

$$\hat{t}_i = -p n_i \,, \tag{2.94}$$

[33] In a fixed domain we simply write $\frac{\mathrm{d}(\cdot)}{\mathrm{d}t}$ instead of $(\cdot)^{\bullet}$ and obtain the balance equations for open systems.

leading to

$$ {}^{\mathrm{d}}\sigma_{ji} n_j = \hat{t}_i + p n_i = 0 \,. \tag{2.95} $$

Due to the latter the boundary terms vanish in Eq. (2.91), i.e., in general for incompressible flows, the boundary terms in the weak form of velocity vanish.

In order to compute the temperature, we need a weak form in the unit of power. Either we generate it from the balance of internal energy as in Eq. (2.41)$_3$ by dividing by T and multiplying with δT, or from the balance of entropy as in Eq. (2.70) by multiplying with δT. For the first option we need to use $\rho u^\bullet = \rho c T^\bullet$ and for the second option $\eta^\bullet = T^\bullet c/T$. The result is the same. We use the entropy balance and obtain

$$ \int_\Omega \Big(\frac{\rho c}{T} \frac{\partial T}{\partial t} + \frac{\rho c}{T} v_i \frac{\partial T}{\partial x_i} + \frac{\partial}{\partial x_i} \Big(\frac{q_i}{T} \Big) - \frac{\rho r}{T} - \frac{1}{T} \sigma_{ij} \frac{\partial v_j}{\partial x_i} + \frac{1}{T^2} q_i \frac{\partial T}{\partial x_i} \Big) \, \delta T \, \mathrm{d}v = 0 \,. \tag{2.96} $$

The term with the heat flux needs to be integrated by parts, since it consists a second gradient of temperature. Hence we acquire the weak form

$$ \begin{aligned} \mathrm{F}_T = \int_\Omega \Big(& \frac{\rho c}{T} \frac{\partial T}{\partial t} \, \delta T + \frac{\rho c}{T} v_i \frac{\partial T}{\partial x_i} \, \delta T - \frac{q_i}{T} \frac{\partial \, \delta T}{\partial x_i} - \frac{\rho r}{T} \, \delta T - \\ & - \frac{1}{T} \sigma_{ij} \frac{\partial v_j}{\partial x_i} \, \delta T + \frac{1}{T^2} q_i \frac{\partial T}{\partial x_i} \, \delta T \Big) \, \mathrm{d}v + \int_{\partial\Omega} \frac{1}{T} \hat{q} \, \delta T \, \mathrm{d}a \,. \end{aligned} \tag{2.97} $$

After utilizing the time discretization we can sum up all integral forms since they are all in the unit of power

$$ \begin{aligned} \text{Form} = \mathrm{F}_p + \mathrm{F}_v + \mathrm{F}_T = & \int_\Omega \Big(v_{i,i} \, \delta p + \rho \frac{v_i - v_i^0}{\Delta t} \, \delta v_i + \rho v_j v_{i,j} \, \delta v_i + \\ + p_{,i} \, \delta v_i + {}^{\mathrm{d}}\sigma_{ji} \, \delta v_{i,j} - \rho f_i \, \delta v_i + & \frac{\rho c}{T} \frac{T - T^0}{\Delta t} \, \delta T + \frac{\rho c}{T} v_i T_{,i} \, \delta T - \frac{q_i}{T} \, \delta T_{,i} - \\ - \frac{\rho r}{T} \, \delta T - \frac{1}{T} \sigma_{ij} v_{j,i} \, \delta T + & \frac{1}{T^2} q_i T_{,i} \, \delta T \Big) \, \mathrm{d}v + \int_{\partial\Omega} \frac{1}{T} \hat{q} \, \delta T \, \mathrm{d}a \,, \end{aligned} \tag{2.98} $$

where we have employed the usual comma notation for a partial space derivative. This form can be solved by applying appropriate boundary conditions. In a 2D channel filled with water, the primitive variables, $\{p, v_i, T\}$, are computed. On top and bottom walls fluid rests at 300 K temperature. Due to the greater pressure on the left side than on the right hand side, water flows from left to right. The pressure on the left rises linearly in time such that the viscous fluid moves faster in time, with the typical parabolic flow profile. The production term increases the temperature because of the viscous flow. We keep on top and bottom at 300 K and on the left and right we implement adiabatic boundaries. In Figs. 2.4, 2.5 and 2.6 we present distributions of

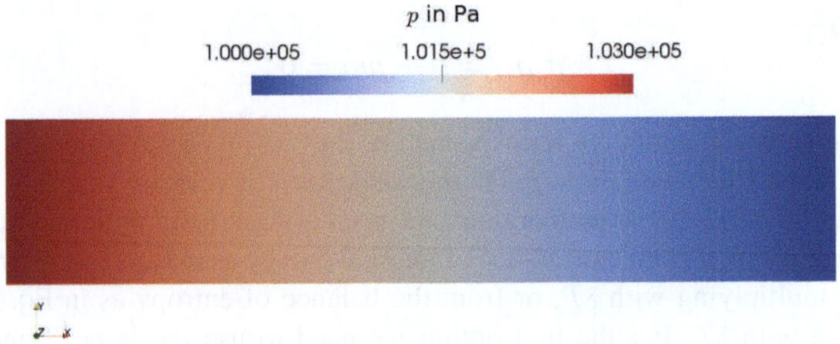

Fig. 2.4 Pressure distribution in channel flow at 300 s

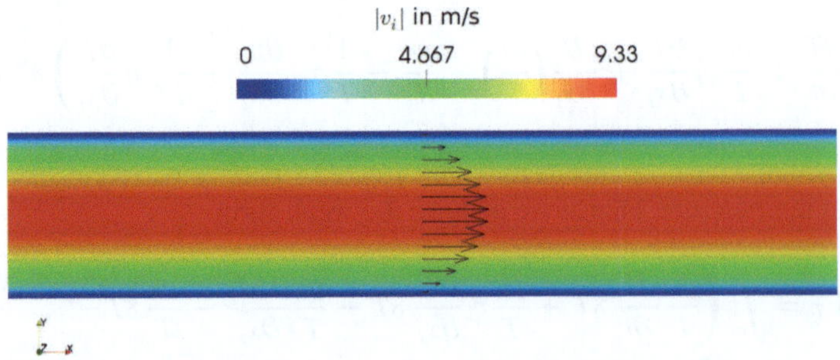

Fig. 2.5 Velocity distribution in channel flow at 300 s. Colors denote the magnitude of velocity

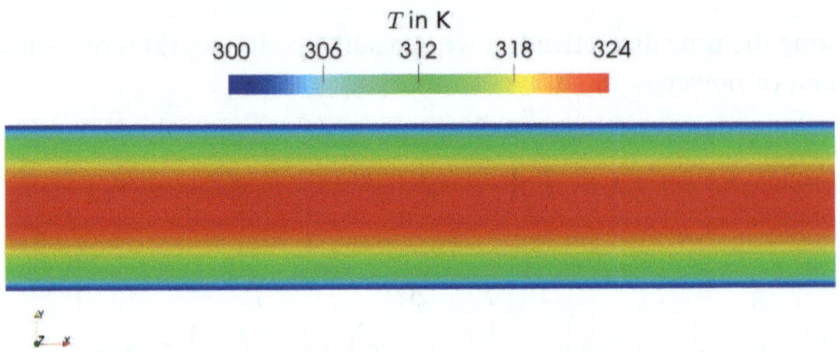

Fig. 2.6 Distribution of temperature in channel flow at 300 s. Temperature increases due to the internal friction. It is forced to be 300 K on *top* and *bottom* boundaries

primitive variables after 5 min obtained with the code below. The viscous flow with a relatively high velocity causes a significant temperature change. We have used the code given below. By using a mixed function space for primitive variables we have computed all unknowns at once, i.e., monolithically. This method is necessary since we have inserted the balance equations in each other by obtaining the governing equations. Moreover, the field equations are coupled and nonlinear, so we are not allowed to solve them separately.

```python
"""Computational reality 12, channel flow of Navier-Stokes-Fourier fluid"""
__author__ = "B. Emek Abali"
__license__ = "GNU GPL Version 3.0 or later"
#This code underlies the GNU General Public License, http://www.gnu.org/licenses/gpl-3.0.en.html

from fenics import *
import numpy
set_log_level(ERROR)

xlength = 0.5 # m
ylength = 0.1 # m
mesh=RectangleMesh(Point(0.0,-ylength/2.0), Point(xlength, ylength/2.0), 200,40)

TensorSpace = TensorFunctionSpace(mesh, 'P', 1)
VectorSpace = VectorFunctionSpace(mesh, 'P', 1)
ScalarSpace = FunctionSpace(mesh, 'P', 1)
# p, v, T
Space = MixedFunctionSpace([ScalarSpace,VectorSpace, ScalarSpace])

facets = FacetFunction('size_t',mesh)
cells = CellFunction('size_t',mesh)
da = Measure('ds', domain=mesh, subdomain_data=facets)
dv = Measure('dx', domain=mesh, subdomain_data=cells)

left = CompiledSubDomain('near(x[0],0) && on_boundary')
right = CompiledSubDomain('near(x[0],l) && on_boundary',l=xlength)
bottom = CompiledSubDomain('near(x[1],-yl/2.0)',yl=ylength)
top = CompiledSubDomain('near(x[1],yl/2.0)',yl=ylength)

v_noslip = Constant((0.0, 0.0))
pL = Expression('100000.0+10*t',t=0)
pR = Constant(100000.0)
Tini = 300. #K
bc1=DirichletBC(Space.sub(0), pL, left)
bc2=DirichletBC(Space.sub(0), pR, right)
bc3=DirichletBC(Space.sub(1), v_noslip, bottom)
bc4=DirichletBC(Space.sub(1), v_noslip, top)
bc5=DirichletBC(Space.sub(1).sub(1), 0.0, left)
bc6=DirichletBC(Space.sub(1).sub(1), 0.0, right)
bc7=DirichletBC(Space.sub(2), Tini, top)
bc8=DirichletBC(Space.sub(2), Tini, bottom)

bc=[bc1,bc2,bc3,bc4,bc5,bc6,bc7,bc8]
u_init = Expression(('p0','0.0','0.0','T0'), p0=100000.0,T0=Tini)

i, j, k, l =indices(4)
n = FacetNormal(mesh)
```

```
t = 0.0
Dt = 50.
t_end = 500.

test = TestFunction(Space)
du = TrialFunction(Space)
u0 = Function(Space)
u = Function(Space)
u0 = interpolate(u_init ,Space)
u = interpolate(u_init ,Space)
p0, v0, T0 = split(u0)
p, v, T = split(u)
delp, delv, delT = split(test)
delta = Identity(2)

#water approx. at 300 K
rho = 995.7 #kg/m^3
mu = 0.8 #Ns/m^2 = kg /s /m
lambada = mu*1E5
c = 4180. #J/(kgK)
kappa = 0.58 #W/(m K)
h=18.0 #W/(m^2 K)
Tamb = Tini

d = sym(grad(v))
dsigma = as_tensor( lambada*d[k,k]*delta[i,j] + 2.0*mu*d[i,j], (i,j))
sigma = as_tensor( -p*delta[i,j] + dsigma[i,j], (i,j) )
q = as_tensor( -kappa*T.dx(i), (i,))
f = Constant((0,0))
r = Constant(0)

Form = (  v[i].dx(i)*delp \
     + rho*(v-v0)[i]/Dt*delv[i] \
     + rho*v[j]*v[i].dx(j)*delv[i] \
     + p.dx(i)*delv[i] \
     + dsigma[j,i]*delv[i].dx(j) \
     - rho*f[i]*delv[i]
     + rho*c/T*(T-T0)/Dt*delT \
     + rho*c/T*v[i]*T.dx(i)*delT \
     - q[i]/T*delT.dx(i) \
     - rho*r/T*delT \
     - 1./T*sigma[i,j]*v[j].dx(i)*delT \
     + 1./T**2*q[i]*T.dx(i)*delT \
     )*dv

Gain = derivative(Form, u, du)

pwd='/calcul/CR12/'
file_p = File(pwd+'pressure.pvd')
file_v = File(pwd+'velocity.pvd')
file_T = File(pwd+'temperature.pvd')
```

```
for t in numpy.arange(0,t_end,Dt):
    print 'time:' ,t
    pL.t = t
    solve(Form==0, u, bc, J=Gain, \
        solver_parameters={"newton_solver":{"linear_solver":
            ↪ "mumps", "relative_tolerance": 1e-5} }, \
        form_compiler_parameters={"cpp_optimize": True, "
            ↪ representation": "quadrature", "
            ↪ quadrature_degree": 2} )
    file_p << (u.split()[0], t)
    file_v << (u.split()[1], t)
    file_T << (u.split()[2], t)
    u0.assign(u)
```

To-do

We have implemented the same channel flow as in Sect. 1.7 by incorporating the temperature distribution caused by the viscous flow. For such high velocities, the temperature increase is significant. This outcome is partly due to the implemented adiabatic boundaries, which are actually not very realistic.

- Implement ROBIN boundary conditions for heat flux on all boundaries.
- Implement the code by using a material with a higher viscosity (search for properties of a polymer melt).

2.4 Thermoviscoelasticity

By considering the principles of thermodynamics in a EULERian frame, we have derived all of the necessary constitutive equations for a viscous fluid in the last section. For fluids we use an open system. In this section we will derive the constitutive equations for a deformable solid in a LAGRANGEan (reference) frame expressed in Cartesian coordinates. A material system is utilized for solids. As the reference frame we choose the initial frame, where the positions (coordinates) of particles are known. We start first by transforming the balance equations from the current frame to the initial frame. The following identities in a Cartesian coordinate system:

$$dv = J\,dV\ ,\quad n_j\,da = (\boldsymbol{F}^{-1})_{kj} J N_k\,dA\ , \tag{2.99}$$

have been derived in Sect. 1.4 for arbitrary coordinate systems. The balance equations of mass, momentum, and internal energy in the current frame for a material system:[34]

[34] A material system is a closed system possessing the same particles over time. In a material system no (mass) convection is allowed.

$$
\begin{aligned}
\Big(\int_{\mathcal{B}} \rho \,\mathrm{d}v\Big)^{\boldsymbol{\cdot}} &= 0 \,, \\
\Big(\int_{\mathcal{B}} \rho v_i \,\mathrm{d}v\Big)^{\boldsymbol{\cdot}} &= \int_{\partial\mathcal{B}} \sigma_{ji} n_j \,\mathrm{d}a + \int_{\mathcal{B}} \rho f_i \,\mathrm{d}v \,, \\
\Big(\int_{\mathcal{B}} \rho u \,\mathrm{d}v\Big)^{\boldsymbol{\cdot}} &= -\int_{\partial\mathcal{B}} q_j n_j \,\mathrm{d}a + \int_{\mathcal{B}} \Big(\rho r + \sigma_{ji}\frac{\partial v_i}{\partial x_j}\Big) \,\mathrm{d}v \,,
\end{aligned}
\tag{2.100}
$$

are transformed into the initial frame

$$
\begin{aligned}
\Big(\int_{\mathcal{B}_0} \rho J \,\mathrm{d}V\Big)^{\boldsymbol{\cdot}} &= 0 \,, \\
\Big(\int_{\mathcal{B}_0} \rho v_i J \,\mathrm{d}V\Big)^{\boldsymbol{\cdot}} &= \int_{\partial\mathcal{B}_0} \sigma_{ji} (\boldsymbol{F}^{-1})_{kj} J N_k \,\mathrm{d}A + \int_{\mathcal{B}_0} \rho f_i J \,\mathrm{d}V \,, \\
\Big(\int_{\mathcal{B}_0} \rho u J \,\mathrm{d}V\Big)^{\boldsymbol{\cdot}} &= -\int_{\partial\mathcal{B}_0} q_j (\boldsymbol{F}^{-1})_{kj} J N_k \,\mathrm{d}A + \int_{\mathcal{B}_0} \Big(\rho r + \sigma_{ji}\frac{\partial v_i}{\partial x_j}\Big) J \,\mathrm{d}V \,.
\end{aligned}
\tag{2.101}
$$

Initial frame is constant in time, $(\mathrm{d}V)^{\boldsymbol{\cdot}} = 0$, thus, the balance of mass in the initial frame reads

$$
\rho_0 = \rho J \,. \tag{2.102}
$$

The mass density in the initial state, ρ_0, is of course constant in time, $\rho_0^{\boldsymbol{\cdot}} = 0$. By introducing fluxes in the initial frame:

$$
P_{ki} = \sigma_{ji} (\boldsymbol{F}^{-1})_{kj} J \,, \quad Q_k = q_j (\boldsymbol{F}^{-1})_{kj} J \,, \tag{2.103}
$$

and inserting the mass balance into the momentum balance and internal energy balance, we acquire

$$
\begin{aligned}
\int_{\mathcal{B}_0} \rho_0 v_i^{\boldsymbol{\cdot}} \,\mathrm{d}V &= \int_{\partial\mathcal{B}_0} P_{ki} \,\mathrm{d}A + \int_{\mathcal{B}_0} \rho f_i J \,\mathrm{d}V \,, \\
\int_{\mathcal{B}_0} \rho_0 u^{\boldsymbol{\cdot}} \,\mathrm{d}V &= -\int_{\partial\mathcal{B}_0} Q_k N_k \,\mathrm{d}A + \int_{\mathcal{B}_0} \Big(\rho_0 r + J\sigma_{ji}\frac{\partial v_i}{\partial x_j}\Big) \,\mathrm{d}V \,.
\end{aligned}
\tag{2.104}
$$

After utilizing GAUSS's law on the boundary integrals, we write the balance equations in their local forms:

$$
\rho_0 v_i^{\boldsymbol{\cdot}} - \frac{\partial P_{ki}}{\partial X_k} - \rho_0 f_i = 0 \,, \quad \rho_0 u^{\boldsymbol{\cdot}} + \frac{\partial Q_k}{\partial X_k} - \rho_0 r = J\sigma_{ji}\frac{\partial v_i}{\partial x_j} \,. \tag{2.105}
$$

We have written the production terms on the right-hand side. Since the formulation is in the initial frame, the partial derivative with respect to x_i needs to be reformulated as a differentiation in X_i. The velocity gradient in the current frame reads

$$\frac{\partial v_i}{\partial x_j} = \frac{\partial v_i}{\partial X_k}\frac{\partial X_k}{\partial x_j} = \frac{\partial v_i}{\partial X_k}(\boldsymbol{F}^{-1})_{kj} , \tag{2.106}$$

hence, we obtain

$$J\sigma_{ji}\frac{\partial v_i}{\partial x_j} = J\sigma_{ji}\frac{\partial v_i}{\partial X_k}(\boldsymbol{F}^{-1})_{kj} = P_{ki}\frac{\partial v_i}{\partial X_k} . \tag{2.107}$$

The second PIOLA–KIRCHHOFF stress tensor:

$$S_{ij} = (\boldsymbol{F}^{-1})_{jk}P_{ik} = (\boldsymbol{F}^{-1})_{jk}\sigma_{lk}(\boldsymbol{F}^{-1})_{il}J , \tag{2.108}$$

is more beneficial by obtaining constitutive equations. From the latter the nominal stress becomes

$$P_{ij} = F_{jl}S_{il} . \tag{2.109}$$

We further rewrite the production term. By starting with the right CAUCHY–GREEN deformation tensor, $C_{ij} = F_{ki}F_{kj}$, and its corresponding GREEN–LAGRANGE strain tensor, $2E_{ij} = (C_{ij} - \delta_{ij})$, we obtain

$$\begin{aligned} C_{ij} &= F_{ki}F_{kj} = F_{kj}F_{ki} = C_{ji} , \\ 2E^{\bullet}_{ij} &= C^{\bullet}_{ij} = 2F^{\bullet}_{k(i}F_{kj)} . \end{aligned} \tag{2.110}$$

In the initial frame we have the following identity:

$$\frac{\partial v_j}{\partial X_i} = \frac{\partial^2 x_j}{\partial X_i \partial t} = \frac{\partial^2 x_j}{\partial t \partial X_i} = F^{\bullet}_{ji} , \tag{2.111}$$

since $x_i = x_i(t, X_j)$. By using the aforementioned relations we acquire the following version of the production term:

$$P_{ij}\frac{\partial v_j}{\partial X_i} = F_{jl}S_{il}\frac{\partial v_j}{\partial X_i} = F_{jl}S_{il}F^{\bullet}_{ji} = F_{j(l}S_{il}F^{\bullet}_{ji)} = S_{il}E^{\bullet}_{il} , \tag{2.112}$$

for a symmetric stress tensor, $S_{ij} = S_{ji}$. In case of non-polar materials, the CAUCHY stress tensor is symmetric, leading to the symmetric second PIOLA–KIRCHHOFF stress tensor given in Eq. (2.108). For non-polar materials the balance of internal energy in the initial frame reads

$$\rho_0 u^{\bullet} + \frac{\partial Q_k}{\partial X_k} - \rho_0 r = S_{ij}E^{\bullet}_{ij} . \tag{2.113}$$

At equilibrium the balance of internal energy is

$$\rho_0 u^{\bullet} - \rho_0 T\eta^{\bullet} = {}^{\mathrm{e}}S_{ij}E^{\bullet}_{ij} , \tag{2.114}$$

since the internal energy is fully recoverable and the stress tensor is decomposed into an elastic (reversible) term, ${}^{\mathrm{e}}S_{ij}$, and into a dissipative (irreversible) term, ${}^{\mathrm{d}}S_{ij}$, such that

$$S_{ij} = {}^{\mathrm{e}}S_{ij} + {}^{\mathrm{d}}S_{ij} \ . \tag{2.115}$$

We need constitutive equations for the specific entropy, η, for the heat flux, Q_i, and for the elastic and dissipative stress tensors, ${}^{\mathrm{e}}S_{ij}$, ${}^{\mathrm{d}}S_{ij}$. By using the 1st law of thermodynamics we can rewrite the rate of internal energy as a differential form:

$$\mathrm{d}u = T\,\mathrm{d}\eta + {}^{\mathrm{e}}S_{ij} v\,\mathrm{d}E_{ij} \ , \tag{2.116}$$

where the specific volume, $v = 1/\rho_0$, is a known quantity. The latter differential form is often introduced as GIBBS's equation.[35]

In Eq. (2.116) the internal energy is given as a function of η and E_{ij}. Having a function of the strains is adequate since the strains are given by the primitive variables (displacement). However, we have just introduced a variable called entropy, η, we lack a definition for it. We simply want to exchange the dependency from entropy to the temperature, which is one of the primitive variables. We transform[36] the differential form in Eq. (2.116) by introducing a free energy:

$$\psi = u - T\eta \ , \tag{2.117}$$

into the following form:

$$\mathrm{d}\psi = \mathrm{d}u - \eta\,\mathrm{d}T - T\,\mathrm{d}\eta = -\eta\,\mathrm{d}T + {}^{\mathrm{e}}S_{ij} v\,\mathrm{d}E_{ij} \ . \tag{2.118}$$

This differential form implies an energy depending on the temperature and strain,

$$\psi = \psi(T, E_{ij}) \ , \tag{2.119}$$

such that

$$-\eta = \frac{\partial \psi}{\partial T} \ , \quad {}^{\mathrm{e}}S_{ij} v = \frac{\partial \psi}{\partial E_{ij}} \ . \tag{2.120}$$

The temperature and strain are called the primary or state variables. Since the energy depends on the primary variables, its derivatives depend on the same set of variables, too. So the derived, dual, or conjugate variables, η, ${}^{\mathrm{e}}S_{ij}$, depend on the primary variables

$$\begin{aligned} \mathrm{d}\eta &= A\,\mathrm{d}T + \bar{p}_{ij}\,\mathrm{d}E_{ij} \ , \\ \mathrm{d}\,{}^{\mathrm{e}}S_{ij} &= p_{ij}\,\mathrm{d}T + C_{ijkl}\,\mathrm{d}E_{kl} \ . \end{aligned} \tag{2.121}$$

---

[35] For an alternative derivation of GIBBS's equation we refer to [12, Chap. 8].

[36] Mathematicians call this transformation a LEGENDRE transformation named after Adrien-Marie Legendre.

We can readily apply the MAXWELL symmetry condition (reciprocal relation):

$$\bar{p}_{ij} = \frac{\partial \eta}{\partial E_{ij}} = -\frac{\partial^2 \psi}{\partial E_{ij} \partial T} = -\frac{\partial^2 \psi}{\partial T \partial E_{ij}} = -\frac{\partial\, {}^{e}S_{ij} v}{\partial T} = -v \frac{\partial\, {}^{e}S_{ij}}{\partial T} = -v p_{ij} \ . \tag{2.122}$$

The specific volume is a given function in space for heterogeneous materials or a constant value for homogeneous materials. It is coupled to the temperature through constitutive equations, however, it is independent on T so we have taken it out in the differentiation with respect to the temperature. The dual variables read

$$\begin{aligned} \mathrm{d}\eta &= A\, \mathrm{d}T - p_{ij} v\, \mathrm{d}E_{ij} \ , \\ \mathrm{d}\, {}^{e}S_{ij} &= p_{ij}\, \mathrm{d}T + C_{ijkl}\, \mathrm{d}E_{kl} \ . \end{aligned} \tag{2.123}$$

As in the previous section $A = c/T$, where the specific heat capacity, c, is measured by varying the temperature and recording the change of heat by fixed strains, $\mathrm{d}E_{ij} = 0$. In other words, all of the boundaries are clamped and the temperature is varied. The *stiffness tensor* C_{ijkl} is measured on a constant temperature, $\mathrm{d}T = 0$, by varying the strains $\mathrm{d}E_{ij}$ and recording the stress changes $\mathrm{d}\, {}^{e}S_{ij}$. Since C_{ijkl} consists of many coefficients, we also need to establish various measurements. One of such measurements is the prominent tensile test. Throughout the experiment, the temperature is fixed such that the components of C_{ijkl} are valid for a specific temperature. One needs to redo the experiments in different temperatures for determining components as a function in T. The thermal pressure p_{ij} is the pressure occurring due to temperature variation by fixed strains, $\mathrm{d}E_{ij} = 0$. The body tries to expand or shrink and applies a pressure on the clamped boundaries holding the strains fixed.

The values for the thermal pressure are difficult to find in the literature. Therefore, we introduce the coefficients of thermal expansion, α_{ij}, which are measured by varying the temperature and measuring the strain change

$$\mathrm{d}E_{ij} = \alpha_{ij}\, \mathrm{d}T \ , \tag{2.124}$$

for a specific stress. Since such a measurement is realized by fixed stress, $\mathrm{d}\, {}^{e}S_{ij} = 0$, we can observe from Eq. (2.123)$_2$

$$\begin{aligned} 0 = p_{ij}\, \mathrm{d}T + C_{ijkl}\, \mathrm{d}E_{kl} \ , \quad p_{ij}\, \mathrm{d}T &= -C_{ijkl} \alpha_{kl}\, \mathrm{d}T \\ \Rightarrow p_{ij} &= -C_{ijkl} \alpha_{kl} \ . \end{aligned} \tag{2.125}$$

Now, the dual variables become

$$\begin{aligned} \mathrm{d}\eta &= \frac{c}{T}\, \mathrm{d}T + C_{ijkl} \alpha_{kl} v\, \mathrm{d}E_{ij} \ , \\ \mathrm{d}\, {}^{e}S_{ij} &= -C_{ijkl} \alpha_{kl}\, \mathrm{d}T + C_{ijkl}\, \mathrm{d}E_{kl} \ . \end{aligned} \tag{2.126}$$

For non-polar materials the stress tensor is symmetric, we assume that the elastic part is also symmetric, ${}^{e}S_{ij} = {}^{e}S_{ji}$. We restrict the formalism for linear materials such that the stiffness tensor, C_{ijkl}, the coefficients of thermal expansion, α_{ij}, and the specific heat capacity, c, are constants and we acquire the dual variables by integrating from the reference state, $T = T_{\text{ref.}}$, $E_{ij} = 0$, to the current state

$$\begin{aligned} \eta &= c \ln\left(\frac{T}{T_{\text{ref.}}}\right) + C_{ijkl}\alpha_{kl} v E_{ij} \ , \\ {}^{\text{e}}S_{ij} &= -C_{ijkl}\alpha_{kl}(T - T_{\text{ref.}}) + C_{ijkl}E_{kl} \ . \end{aligned} \tag{2.127}$$

Often, thermal strains are introduced

$${}^{\text{th}}E_{kl} = \alpha_{kl}(T - T_{\text{ref.}}) \ , \tag{2.128}$$

such that the elastic stress is written as

$${}^{\text{e}}S_{ij} = C_{ijkl}\big(E_{kl} - {}^{\text{th}}E_{kl}\big) \ . \tag{2.129}$$

Finally, we have determined the GIBBS equation:

$$\begin{aligned} \mathrm{d}u &= T\,\mathrm{d}\eta + {}^{\text{e}}S_{ij} v \,\mathrm{d}E_{ij} = \\ &= c\,\mathrm{d}T + T C_{ijkl}\alpha_{kl} v \,\mathrm{d}E_{ij} + C_{ijkl}\big(E_{kl} - \alpha_{kl}(T - T_{\text{ref.}})\big) v \,\mathrm{d}E_{ij} = \\ &= c\,\mathrm{d}T + v C_{ijkl}\alpha_{kl} T_{\text{ref.}} \,\mathrm{d}E_{ij} + v C_{ijkl} E_{kl} \,\mathrm{d}E_{ij} \ , \end{aligned} \tag{2.130}$$

solely depending on the temperature and displacement (over the known relation between strain and displacement). For a linear thermoelastic isotropic body, the material parameters reduce to

$$C_{ijkl} = \lambda\delta_{ij}\delta_{kl} + \mu\delta_{ik}\delta_{jl} + \mu\delta_{il}\delta_{jk} \ , \quad \alpha_{ij} = \alpha\delta_{ij} \ , \tag{2.131}$$

thus, the internal energy rate reads

$$u^{\bullet} = cT^{\bullet} + v(3\lambda + 2\mu)\alpha T_{\text{ref.}} E^{\bullet}_{ii} + v(\lambda\delta_{ij}E_{kk} + 2\mu E_{ij})E^{\bullet}_{ij} \ . \tag{2.132}$$

For deriving the heat flux, Q_i, and the dissipative stress, ${}^{\text{d}}S_{ij}$, we start with Eq. (2.116) in the following form:

$$\rho_0 u^{\bullet} = \rho_0 T \eta^{\bullet} + \big(S_{ij} - {}^{\text{d}}S_{ij}\big)E^{\bullet}_{ij} \ , \tag{2.133}$$

and insert it into Eq. (2.113) in order to acquire the balance of entropy in the reference frame:

$$\begin{aligned} \rho_0\eta^{\bullet} + \frac{\partial}{\partial X_i}\left(\frac{Q_i}{T}\right) - \rho_0\frac{r}{T} &= \frac{1}{T}\,{}^{\text{d}}S_{ij}E^{\bullet}_{ij} + Q_i\frac{\partial}{\partial X_i}\left(\frac{1}{T}\right) , \\ \rho_0\eta^{\bullet} + \frac{\partial}{\partial X_i}\left(\frac{Q_i}{T}\right) - \rho_0\frac{r}{T} &= \frac{1}{T}\,{}^{\text{d}}S_{ij}E^{\bullet}_{ij} - \frac{1}{T^2}Q_i\frac{\partial T}{\partial X_i} \ . \end{aligned} \tag{2.134}$$

The right-hand side is the production term and it has to be positive according to the 2nd law of thermodynamics:

$$\Sigma = \frac{1}{T}\,{}^{\mathrm{d}}S_{ij}E^{\bullet}_{ij} - \frac{1}{T^2}Q_i G_i \geq 0 \,, \tag{2.135}$$

where again for simplicity we have used the following notation:

$$G_i = \frac{\partial T}{\partial X_i} \,. \tag{2.136}$$

The stress tensor is symmetric for non-polar materials; we have employed a symmetric reversible term, the dissipative term has to be symmetric, too. A symmetric tensor of rank two can be decomposed into a spherical (volumetric) term and a deviatoric term. Multiplication of a volumetric with a deviatoric term vanishes such that the entropy production reads

$$\Sigma = \frac{1}{3T}\,{}^{\mathrm{d}}S_{ii}E^{\bullet}_{jj} + \frac{1}{T}\,{}^{\mathrm{d}}S_{|ij|}E^{\bullet}_{|ij|} - \frac{1}{T^2}Q_i G_i \geq 0 \,. \tag{2.137}$$

By introducing thermodynamical fluxes:

$$\mathcal{F}^{\alpha} = \Big\{ Q_i \,,\ {}^{\mathrm{d}}S_{ii} \,,\ {}^{\mathrm{d}}S_{|ij|} \Big\} \,, \tag{2.138}$$

and thermodynamical forces:

$$\mathcal{K}^{\alpha} = \Big\{ -\frac{G_i}{T^2} \,,\ \frac{1}{3T}E^{\bullet}_{jj} \,,\ \frac{1}{T}E^{\bullet}_{|ij|} \Big\} \,, \tag{2.139}$$

we can rewrite the 2nd law:

$$\Sigma = \mathcal{F}^{\alpha} \cdot \mathcal{K}^{\alpha} \,, \quad \alpha = 1, 2, 3 \,. \tag{2.140}$$

All of thermodynamical fluxes are of different type (tensors of different ranks). According to the CURIE principle thermodynamical fluxes depend only on their corresponding thermodynamical forces of the same rank such that we obtain

$$\mathcal{F}^1 = \mathcal{F}^1(\mathcal{K}^1) \,,\ \mathcal{F}^2 = \mathcal{F}^2(\mathcal{K}^2) \,,\ \mathcal{F}^3 = \mathcal{F}^3(\mathcal{K}^3) \,. \tag{2.141}$$

We can readily propose linear constitutive equations:

$${}^{\mathrm{d}}S_{ii} = \mu_1 E^{\bullet}_{ii} \,,\ {}^{\mathrm{d}}S_{|ij|} = \mu_2 E^{\bullet}_{|ij|} \,,\ Q_i = -\kappa G_i \,, \tag{2.142}$$

where μ_1, μ_2, and κ are all positive constants such that $\Sigma \geq 0$. The viscous part of the stress reads

$$
\begin{aligned}
{}^{\mathrm{d}}S_{ij} &= \frac{1}{3}\,{}^{\mathrm{d}}S_{kk}\delta_{ij} + {}^{\mathrm{d}}S_{|ij|} = \frac{\mu_1}{3}E^{\bullet}_{kk}\delta_{ij} + \mu_2\Big(E^{\bullet}_{ij} - \frac{1}{3}E^{\bullet}_{kk}\delta_{ij}\Big) = \\
&= \frac{\mu_1 - \mu_2}{3}E^{\bullet}_{kk}\delta_{ij} + \mu_2 E^{\bullet}_{ij}\,.
\end{aligned}
\tag{2.143}
$$

Then by using the obtained elastic stress we acquire a linear thermoviscoelastic material model:

$$
S_{ij} = C_{ijkl}\big(E_{kl} - \alpha_{ij}(T - T_{\text{ref.}})\big) + \frac{\mu_1 - \mu_2}{3}E^{\bullet}_{kk}\delta_{ij} + \mu_2 E^{\bullet}_{ij}\,. \tag{2.144}
$$

For a constant κ the constitutive equation:

$$
Q_i = -\kappa\frac{\partial T}{\partial X_i}\,, \tag{2.145}
$$

is called FOURIER's law in the LAGRANGEan frame.

In order to compute the displacement and temperature in a linear thermoviscoelastic body, we employ the balance of momentum and the balance of entropy:

$$
\begin{aligned}
\rho_0 u^{\bullet\bullet}_i - \frac{\partial P_{ji}}{\partial X_j} - \rho_0 f_i &= 0\,, \\
\rho_0\eta^{\bullet} + \frac{\partial}{\partial X_i}\Big(\frac{Q_i}{T}\Big) - \rho_0\frac{r}{T} &= \frac{1}{T}\,{}^{\mathrm{d}}S_{ij}E^{\bullet}_{ij} - \frac{1}{T^2}Q_i\frac{\partial T}{\partial X_i}\,.
\end{aligned}
\tag{2.146}
$$

The primitive variables are displacement, u_i, and temperature, T. Hence we multiply the balance of linear momentum with δu_i and integrate over the continuum body for generating a form in the unit of energy. By multiplying the balance of entropy with δT and integrating over the body, we obtain a form in the unit of power. After discretizing in time, we can multiply the equation with Δt in order to acquire both forms in the unit of energy. Having forms in the same unit, we can sum them up. Furthermore, we apply GAUSS's law in order to weaken the forms and acquire

$$
\begin{aligned}
\text{Form} = \int_{\mathcal{B}_0}\Big(&\rho_0\frac{u_i - 2u_i^0 + u_i^{00}}{\Delta t\,\Delta t}\,\delta u_i + P_{ji}\,\delta u_{i,j} - \rho_0 f_i\,\delta u_i + \\
&+\frac{\rho_0}{T}(\eta - \eta^0)\,\delta T - \Delta t\frac{1}{T}Q_i\,\delta T_{,i} - \Delta t\frac{\rho_0 r}{T}\,\delta T - \frac{1}{T}\,{}^{\mathrm{d}}S_{ij}(E_{ij} - E^0_{ij})\,\delta T + \\
&+\Delta t\frac{1}{T^2}Q_i T_{,i}\,\delta T\Big)\,\mathrm{d}V + \int_{\partial\mathcal{B}_0}\Big(\Delta t\frac{1}{T}\hat{Q}\,\delta T - \hat{t}_i\,\delta u_i\Big)\,\mathrm{d}A\,,
\end{aligned}
\tag{2.147}
$$

where the comma notation has been used for a partial space derivative in X_i. We summarize the necessary relations:

$$
F_{ij} = \frac{\partial u_i}{\partial X_j} + \delta_{ij}\,,\quad C_{ij} = F_{ki}F_{kj}\,,\quad E_{ij} = \frac{1}{2}(C_{ij} - \delta_{ij})\,,
$$

$$
\begin{aligned}
{}^{\text{e}}S_{ij} &= -C_{ijkl}\alpha_{kl}(T - T_{\text{ref.}}) + C_{ijkl}E_{kl} \ , \quad \eta = c \ln\left(\frac{T}{T_{\text{ref.}}}\right) + C_{ijkl}\alpha_{kl} v E_{ij} \ , \\
{}^{\text{d}}S_{ij} &= \frac{\mu_1 - \mu_2}{3} E^{\bullet}_{kk}\delta_{ij} + \mu_2 E^{\bullet}_{ij} \ , \quad Q_i = -\kappa \frac{\partial T}{\partial X_i} \ , \\
S_{ij} &= {}^{\text{e}}S_{ij} + {}^{\text{d}}S_{ij} \ , \quad P_{ij} = F_{jl} S_{il} \ . \qquad (2.148)
\end{aligned}
$$

For an isotropic body the stiffness tensor and coefficients of thermal expansion are

$$
C_{ijkl} = \lambda\delta_{ij}\delta_{kl} + \mu\delta_{ik}\delta_{jl} + \mu\delta_{il}\delta_{jk} \ , \quad \alpha_{ij} = \alpha\delta_{ij} \ . \qquad (2.149)
$$

Therefore, in case of an isotropic body we need seven material parameters, viz., λ, μ, α, μ_1, μ_2, κ, and c.

In a tensile testing we normally assume that the process is isothermal. By computing the reality where heat is produced due to the entropy production, we can validate this engineering assumption. The geometry is a beam along X_1 and we use a ROBIN boundary condition for the heat flux over all boundaries:

$$
\hat{Q} = h(T - T_{\text{amb}}) \ . \qquad (2.150)
$$

On the left side we hold the beam fixed and on the right side we pull with the force given by the traction vector $\hat{t}_i = (800t, 0, 0)$ MPa linearly in time, t. The traction (force per area) is the controlled parameter, i.e., the machine is steered by the force. The tip displacement is measured, it is an observed quantity. Conveniently we plot stress versus strain, where the stress (on the right tip) is the (axial) traction and the (normal axial) strain, E_{11}, is the displacement divided by the initial length. The traction vector, $\hat{t} = N_j P_{ji}$, is given by the nominal or engineering stress, P_{ji}. The strain, E_{ij}, is called the engineering strain; we have introduced it as the GREEN–LAGRANGE strain measure.

We apply a mechanical load and measure the temperature in the middle of the beam as well as the stress and strain on the tip. In Fig. 2.7 the temperature change can be seen, it is clearly negligible. This is good news, because we measure the elasticity

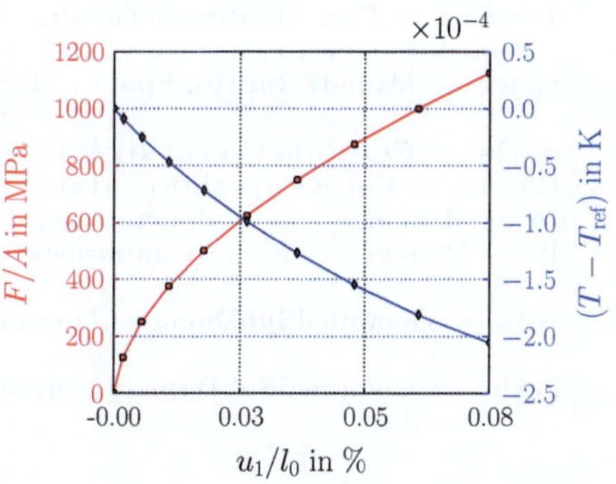

Fig. 2.7 Tensile testing and temperature change due to the deformation

components, λ, ν for isotropic materials by using a tensile testing and assume that the temperature remains constant. The code for the computation is given below.

```
"""Computational Reality 13, thermoviscoelasticity"""
__author__ = "B. Emek Abali"
__license__ = "GNU GPL Version 3.0 or later"
#This code underlies the GNU General Public License, http://www.gnu.org/licenses/gpl-3.0.en.html

from fenics import *
import numpy
set_log_level(ERROR)
#units: mm, 1000 kg=ton, s, MPa, mJ, K
delta = Identity(3)
f = Constant((0.0, 0.0, -9810.))
r = 0.
Tref = 293.15 #in K
Tamb=Tref

# Material data of P265GH (St 45.8) from VDI Waermeatlas, at 293.15 K
rho0 = 7850.0E-9 #in kg / mm^3
kappa = 57.0 #in mJ / (s mm K)
capacity = 430.0E6 #in mJ / (ton K)
alpha = 12.2E-6 #in 1/K at 373.15 K
EModul = 211.E+3 #in MPa
nu = 0.28
h = 10.E-3 #in mJ / (s m^2 K)
mu1 = 1.E+6 #in MPa / s
mu2 = 3.E+6 #in MPa / s

tMax = 5.0
Dt = 0.5
t = 0.0

xMin, xMax, xElements = 0.0, 100.0, 10
yMin, yMax, yElements = -10., +10., 10
zMin, zMax, zElements = +10., -10., 10
mesh = BoxMesh(Point(xMin,yMin,zMin), Point(xMax,yMax,zMax), xElements,yElements,zElements)
N = FacetNormal(mesh)
length = abs(xMax-xMin)

T_Space = FunctionSpace(mesh, 'P', 1)
u_Space = VectorFunctionSpace(mesh, 'P', 1)
Space = MixedFunctionSpace([T_Space, u_Space])

cells = CellFunction('size_t', mesh)
facets = FacetFunction('size_t', mesh)
dA = Measure('ds', domain=mesh, subdomain_data=facets)
dV = Measure('dx', domain=mesh, subdomain_data=cells)

left = CompiledSubDomain('near(x[0],l) && on_boundary',l=xMin)
right = CompiledSubDomain('near(x[0],l) && on_boundary',l=xMax)
```

```
back = CompiledSubDomain('near(x[1],l) && on_boundary',l=yMin
    )
front = CompiledSubDomain('near(x[1],l) && on_boundary',l=
    yMax)
bottom = CompiledSubDomain('near(x[2],l) && on_boundary',l=
    zMin)
top = CompiledSubDomain('near(x[2],l) && on_boundary',l=zMax)

facets.set_all(0)
right.mark(facets, 1)
tHat = Expression(('A*t','0.','0.'), A=250., t=0.)
bc = [DirichletBC(Space.sub(1), Constant((0.0, 0.0, 0.0)),
    left),\
DirichletBC(Space.sub(1).sub(1), Constant(0.0), right),\
DirichletBC(Space.sub(1).sub(2), Constant(0.0), right)]

dunkn = TrialFunction(Space)
test = TestFunction(Space)
delT, delu = split(test)

unkn = Function(Space)
unkn0 = Function(Space)
unkn00 = Function(Space)

unkn_init = Expression(('T_ini','0','0','0'),T_ini=Tref)
unkn = interpolate(unkn_init,Space)
unkn0.assign(unkn)
unkn00.assign(unkn0)

T, u = split(unkn)
T0, u0 = split(unkn0)
T00, u00 = split(unkn0)

i, j, k, l = indices(4)
delta = Identity(3)
F = as_tensor(u[i].dx(j)+delta[i,j], (i,j))
F0 = as_tensor(u0[i].dx(j)+delta[i,j], (i,j))
C = as_tensor(F[k,i]*F[k,j], (i,j))
C0 = as_tensor(F0[k,i]*F0[k,j], (i,j))
E = as_tensor(1./2.*(C[i,j]-delta[i,j]), (i,j))
E0 = as_tensor(1./2.*(C0[i,j]-delta[i,j]), (i,j))
lambada = EModul * nu / (1. + nu) / (1. - 2. * nu)
mu = 0.5 * EModul / (1. + nu)
C_ = as_tensor(lambada*delta[i,j]*delta[k,l]+mu*delta[i,k]*
    delta[j,l]+mu*delta[i,l]*delta[j,k], (i,j,k,l))
alp = as_tensor(alpha*delta[i,j], (i,j))
eStress = as_tensor(-C_[i,j,k,l]*alp[k,l]*(T-Tref) + C_[i,j,k
    ,l]*E[k,l], (i,j))
dStress = as_tensor((mu1-mu2)/3.*(E-E0)[k,k]/Dt*delta[i,j] +
    mu2*(E-E0)[i,j]/Dt, (i,j))
S = as_tensor(eStress[i,j]+dStress[i,j], (i,j))
P = as_tensor(F[j,l]*S[i,l], (i,j))
eta = as_tensor( capacity*ln(T/Tref) + C_[i,j,k,l]*alp[k,l
```

```
    ↪ ]*1./rho0*E[i,j] , ())
eta0 = as_tensor( capacity*ln(T0/Tref) + C_[i,j,k,l]*alp[k,l
    ↪ ]*1./rho0*E0[i,j] , ())
Q = as_tensor(-kappa*T.dx(i), (i,))

Form = (rho0*(u-2.*u0+u00)[i]/Dt/Dt*delu[i] + P[j,i]*delu[i].
    ↪ dx(j) - rho0*f[i]*delu[i] + rho0/T*(eta-eta0)*delT -
    ↪ Dt/T*Q[i]*delT.dx(i) - Dt*rho0*r/T*delT - 1./T*dStress
    ↪ [i,j]*(E-E0)[i,j]*delT + Dt/T**2*Q[i]*T.dx(i)*delT )*
    ↪ dV + Dt/T*h*(T-Tamb)*delT*(dA(0)+dA(1)) - tHat[i]*delu
    ↪ [i]*dA(1)

Gain = derivative(Form, unkn, dunkn)

import matplotlib as mpl
mpl.use('Agg')
import matplotlib.pyplot as pylab
pylab.rc('text', usetex=True )
pylab.rc('font', family='serif', serif='cm', size=30 )
pylab.rc('legend', fontsize=30)
pylab.rc(('xtick.major','ytick.major'), pad=15)
pylab.subplots_adjust(top=0.90)
pylab.subplots_adjust(bottom=0.17)
pylab.subplots_adjust(left=0.20)
pylab.subplots_adjust(right=0.8)

fig = pylab.figure(1, figsize=(14,10))
ax1 = fig.add_subplot(111)
ax1.grid(True, axis='x')
ax1.set_xlabel('$u_1/l_0$ in $\%$')
ax1.set_ylabel('$F/A$ in MPa', color='r')
ax1.tick_params(axis='y', colors='r')
ax1.grid(True, axis='y', color='r')
ax2 = ax1.twinx()
ax2.set_ylabel('$(T-T_\mathrm{ref})$ in K', color='b')
ax2.tick_params(axis='y', colors='b')
ax2.grid(True, axis='y', color='b')
ax2.ticklabel_format(style='sci', axis='y', scilimits=(-2,2))

pwd='/calcul/CR13/'
file_u = File(pwd+'displ.pvd')
file_T = File(pwd+'temp.pvd')
strain,stress,temp =[],[],[]

while t < tMax:
    print 'time: ',t
    tHat.t = t
    solve(Form==0, unkn, bc, J=Gain, \
        solver_parameters={"newton_solver":{"linear_solver":
            ↪ "mumps", "relative_tolerance": 1e-5} }, \
        form_compiler_parameters={"cpp_optimize": True, "
            ↪ representation": "quadrature", "
            ↪ quadrature_degree": 2} )
```

```

        file_T << (unkn.split()[0], t)
        file_u << (unkn.split()[1], t)

        strain.append(unkn.split()[1](xMax,0.,0.)[0]/length*100.)
        stress.append(tHat(xMax,0.,0.)[0])
        temp.append(unkn.split()[0](xMax/2.,0.,0.)-Tref)
        ax1.plot(strain, stress, 'o-', color='r')
        ticks = numpy.linspace(numpy.array(strain).min(), numpy.
            ↪ array(strain).max(), 4)
        ax1.set_xticks(ticks)
        ax1.set_xticklabels(['%1.2f' % i_ticks for i_ticks in
            ↪ ticks])
        ax2.plot(strain,temp, 'd-', color='b')
        fig.savefig(pwd+'CompReal13_tensiletest.pdf')
        unkn00.assign(unkn0)
        unkn0.assign(unkn)
        t += Dt
```

To-do

We have employed the 1st and 2nd laws of thermodynamics, obtained constitutive (material) equations, and computed a coupled thermoviscoelastic problem. In a tensile testing the temperature change is negligible.

- Which term is responsible for the temperature change?
- Implement the code for a thermoelastic problem by setting $\mu_1 = \mu_2 = 0$, thus, ${}^{\mathrm{d}}S_{ij} = 0$. Solve a laser welding application as in Sect. 2.1 and determine the deformations.
- Try to implement a bimetal and apply a thermal loading. Guess and inspect the occurring deformation.

2.5 Thermoplasticity

We have seen a methodology for deriving material equations from thermodynamical restrictions called the 1st and 2nd laws. Unfortunately, it is rather difficult to utilize this procedure for plasticity. There are numerous different suggestions but none of them is accepted by all communities. We present here a more or less widely accepted methodology—it is used in many commercial codes.[37] Within its derivation there occur many assumptions, hence, the method fails to rely on a sound thermodynamical formulation. From a pragmatic point of view, however, it works!

[37] We present a monolithic approach; however, many commercial codes still use a staggered schema. A staggered schema solves the field equations subsequently such that the results from each solution are used in the subsequent field equation. Such an approach is used in Sect. 1.9, where the balance equations are solved subsequently.

Balance equations of mass, momentum, and internal energy has been motivated in the last section. By neglecting big deformations with $F_{ij} \approx \delta_{ij}$ they result in[38]

$$\rho_0 = \rho\,, \quad \rho u_i^{\bullet\bullet} - \sigma_{ji,j} - \rho f_i = 0\,, \quad \rho u^{\bullet} + q_{i,i} - \rho_0 r = \sigma_{ij}\varepsilon_{ij}^{\bullet}\,. \tag{2.151}$$

We axiomatically assume that the production of internal energy consists of two parts: a reversible part including elastic and thermal strain, and an irreversible part due to the plastic strain:

$$\sigma_{ij}\varepsilon_{ij}^{\bullet} = \sigma_{ij}\big({}^{\mathrm{r}}\varepsilon_{ij}^{\bullet} + {}^{\mathrm{p}}\varepsilon_{ij}^{\bullet}\big)\,. \tag{2.152}$$

This assumption is by no means more restrictive than the assumption of decomposing the stress in the last sections. Again we assume that the reversible part remains at equilibrium such that we obtain from the balance of internal energy at equilibrium

$$\rho u^{\bullet} - T\eta^{\bullet} = \sigma_{ij}\,{}^{\mathrm{r}}\varepsilon_{ij}^{\bullet}\,. \tag{2.153}$$

In this setting, GIBBS's equation reads

$$\mathrm{d}u = T\,\mathrm{d}\eta + v\sigma_{ij}\,\mathrm{d}\,{}^{\mathrm{r}}\varepsilon_{ij}\,, \tag{2.154}$$

again with $v = 1/\rho$ as a known quantity. The latter equation allows us to generate the material equations for the dual variables. However, this time ${}^{\mathrm{r}}\varepsilon_{ij}$ is not known. Hence, the chosen primary variables should be $\{T, \sigma_{ij}\}$. We use the same mathematical trick in order to transform the energy into a quantity depending on the primary variables by introducing the so-called specific GIBBS free energy:

$$g = u - T\eta - v\sigma_{ij}\,{}^{\mathrm{r}}\varepsilon_{ij}\,, \tag{2.155}$$

with its differential:

$$\mathrm{d}g = \mathrm{d}u - \eta\,\mathrm{d}T - T\,\mathrm{d}\eta - v\,{}^{\mathrm{r}}\varepsilon_{ij}\,\mathrm{d}\sigma_{ij} - v\sigma_{ij}\,\mathrm{d}\,{}^{\mathrm{r}}\varepsilon_{ij}\,, \tag{2.156}$$

and by inserting the latter in Eq. (2.154)

$$\mathrm{d}g = -\eta\,\mathrm{d}T - v\,{}^{\mathrm{r}}\varepsilon_{ij}\,\mathrm{d}\sigma_{ij}\,. \tag{2.157}$$

The assumption that the free energy possesses a first integral:

$$g = \int \mathrm{d}g\,, \tag{2.158}$$

[38] We also use a linear strain measure, ε_{ij}, instead of E_{ij} in order to attain an identical formulation for plasticity as given in the literature.

is a weakness in the formulation. We take the latter as granted; under this assumption it is obvious that we can write

$$g = g(T, \sigma_{ij}) \, , \quad -\eta = \frac{\partial g}{\partial T} \, , \quad -v\,{}^{\mathrm{r}}\varepsilon_{ij} = \frac{\partial g}{\partial \sigma_{ij}} \, . \tag{2.159}$$

GIBBS's free energy depends on the primary variables, viz., on T and σ_{ij}. The conjugated or dual variables, η and ${}^{\mathrm{r}}\varepsilon_{ij}$, depend on the same set of arguments as the energy,

$$\begin{aligned} \mathrm{d}\eta &= \frac{c}{T}\,\mathrm{d}T + \bar{\alpha}_{ij}\,\mathrm{d}\sigma_{ij} \, , \\ \mathrm{d}\,{}^{\mathrm{r}}\varepsilon_{ij} &= \alpha_{ij}\,\mathrm{d}T + S_{ijkl}\,\mathrm{d}\sigma_{kl} \, . \end{aligned} \tag{2.160}$$

Again we employ the MAXWELL relation:

$$\bar{\alpha}_{ij} = \frac{\partial \eta}{\partial \sigma_{ij}} = -\frac{\partial^2 g}{\partial \sigma_{ij} \partial T} = -\frac{\partial^2 g}{\partial T \partial \sigma_{ij}} = v\frac{\partial\,{}^{\mathrm{r}}\varepsilon_{ij}}{\partial T} = v\alpha_{ij} \, , \tag{2.161}$$

since $v = 1/\rho$ is a function of X_i but not of the temperature. For a *linear* material model the coefficients, c, α_{ij}, S_{ijkl} are all constants. For a linear model we obtain the dual variables by integrating from the *ground* state $\{T = T_{\text{ref.}}, \sigma_{ij} = 0\}$ without strain and entropy to the current state and obtain

$$\begin{aligned} \eta &= c \ln\left(\frac{T}{T_{\text{ref.}}}\right) + v\alpha_{ij}\sigma_{ij} \, , \\ {}^{\mathrm{r}}\varepsilon_{ij} &= \alpha_{ij}(T - T_{\text{ref.}}) + S_{ijkl}\sigma_{kl} \, . \end{aligned} \tag{2.162}$$

The first term can be seen as thermal strains and the second term as elastic strains:

$$\begin{aligned} {}^{\mathrm{th}}\varepsilon_{ij} = \alpha_{ij}(T - T_{\text{ref.}}) \, , \quad {}^{\mathrm{e}}\varepsilon_{ij} = S_{ijkl}\sigma_{kl} \, , \\ {}^{\mathrm{r}}\varepsilon_{ij} = {}^{\mathrm{th}}\varepsilon_{ij} + {}^{\mathrm{e}}\varepsilon_{ij} \, . \end{aligned} \tag{2.163}$$

Then the so-called HOOKE's law with DUHAMEL–NEUMANN supplemental term[39] for thermal strains can be deduced

$$\sigma_{kl} = C_{klij}\left({}^{\mathrm{r}}\varepsilon_{ij} - {}^{\mathrm{th}}\varepsilon_{ij}\right) \, , \tag{2.164}$$

where the stiffness tensor, C_{ijkl}, is the inverse[40] of the compliance tensor, S_{ijkl}. We can even use the assumption already undertaken:

---

[39] It is named after Jean-Marie Constant Duhamel and Franz Ernst Neumann.

[40] For the inverse of a tensor of rank four we need an identity tensor of rank four. This method can be challenging. Instead of that, the inverse is found by using the VOIGT notation. For linear materials we can always rewrite the compliance tensor in the VOIGT notation, which is a 6×6 matrix and its inverse is easy to determine. From the resulting 6×6 matrix in the VOIGT notation, the stiffness tensor is obtained.

$$\varepsilon_{ij} = {}^{\mathrm{r}}\varepsilon_{ij} + {}^{\mathrm{p}}\varepsilon_{ij} , \tag{2.165}$$

in order to rewrite the material equation for stress:

$$\sigma_{ij} = C_{ijkl}\big(\varepsilon_{kl} - {}^{\mathrm{p}}\varepsilon_{kl} - {}^{\mathrm{th}}\varepsilon_{kl}\big) . \tag{2.166}$$

The rate of stress reads

$$\begin{aligned} \sigma^{\bullet}_{ij} &= C_{ijkl}\big(\varepsilon^{\bullet}_{kl} - {}^{\mathrm{p}}\varepsilon^{\bullet}_{kl} - {}^{\mathrm{th}}\varepsilon^{\bullet}_{kl}\big) , \\ \sigma^{\bullet}_{ij} &= C_{ijkl}\big(\varepsilon^{\bullet}_{kl} - {}^{\mathrm{p}}\varepsilon^{\bullet}_{kl} - \alpha_{kl}T^{\bullet}\big) , \end{aligned} \tag{2.167}$$

by using that C_{ijkl}, α_{ij}, as well as $T_{\text{ref.}}$ are constants. From the balance of internal energy in Eq. (2.151)$_3$ augmented by GIBBS's equation (2.154) we obtain

$$\begin{aligned} \rho T\eta^{\bullet} + \sigma_{ij}\,{}^{\mathrm{r}}\varepsilon^{\bullet}_{ij} + q_{i,i} - \rho_0 r &= \sigma_{ij}\varepsilon^{\bullet}_{ij} , \\ \rho T\eta^{\bullet} + q_{i,i} - \rho_0 r &= \sigma_{ij}\,{}^{\mathrm{p}}\varepsilon^{\bullet}_{ij} . \end{aligned} \tag{2.168}$$

Now by using the material equations for dual variables in Eq. (2.160) and in Eq. (2.166) we obtain the field equation for temperature:

$$\begin{aligned} \rho c T^{\bullet} + T\alpha_{ij}\sigma^{\bullet}_{ij} + q_{i,i} - \rho_0 r &= \sigma_{ij}\,{}^{\mathrm{p}}\varepsilon^{\bullet}_{ji} , \\ \rho c T^{\bullet} + T\alpha_{ij}C_{ijkl}\big(\varepsilon^{\bullet}_{kl} - {}^{\mathrm{p}}\varepsilon^{\bullet}_{kl} - \alpha_{kl}T^{\bullet}\big) + q_{i,i} - & \\ -\rho_0 r - C_{ijkl}\big(\varepsilon_{kl} - {}^{\mathrm{p}}\varepsilon_{kl} - \alpha_{kl}(T - T_{\text{ref.}})\big)\,{}^{\mathrm{p}}\varepsilon^{\bullet}_{ji} &= 0 . \end{aligned} \tag{2.169}$$

The field equation for displacement is acquired from Eq. (2.151)$_2$ by augmenting Eq. (2.166) as follows

$$\begin{aligned} \rho u^{\bullet\bullet}_i - \sigma_{ji,j} - \rho f_i &= 0 , \\ \rho u^{\bullet\bullet}_i - C_{jikl}\big(\varepsilon_{kl} - {}^{\mathrm{p}}\varepsilon_{kl} - \alpha_{kl}(T - T_{\text{ref.}})\big)_{,j} - \rho f_i &= 0 . \end{aligned} \tag{2.170}$$

The field equations are nonlinear and coupled. We can solve them after having defined ${}^{\mathrm{p}}\varepsilon_{ij}$, ${}^{\mathrm{p}}\varepsilon^{\bullet}_{ij}$, and q_i.

Plasticity starts with the assumption that we can acquire the rate of plastic strain by using a *dissipation function*, Φ, as follows

$$ {}^{\mathrm{p}}\varepsilon^{\bullet}_{ij} = \Lambda^{\bullet}\frac{\partial\Phi}{\partial\sigma_{ij}} . \tag{2.171}$$

The associated plasticity proposes to use the flow potential, f, for the dissipation function, $f \equiv \Phi$. Modeling kinematic hardening has been discussed in Sect. 1.6.2, we use the same notation and skip the calculations undertaken there. The flow potential:

$$f = \frac{1}{2}(\sigma_{|ij|} - \beta_{ij})(\sigma_{|ij|} - \beta_{ij}) - \frac{1}{3}\sigma_{\mathrm{Y}}^2 , \tag{2.172}$$

results in

$$c\Lambda^{\bullet} = \Gamma^{\bullet} , \quad \Lambda^{\bullet} = \frac{\big(\sigma_{|ij|} - \beta_{ij}\big)\sigma^{\bullet}_{ij}}{\frac{2}{3}c\sigma_{\mathrm{Y}}^2} . \tag{2.173}$$

With the help of Eq. (2.167) we write

$$\Lambda^{\bullet} = \frac{\big(\sigma_{|ij|} - \beta_{ij}\big)C_{ijkl}\big(\varepsilon^{\bullet}_{kl} - {}^{\mathrm{p}}\varepsilon^{\bullet}_{kl} - {}^{\mathrm{th}}\varepsilon^{\bullet}_{kl}\big)}{\frac{2}{3}c\sigma_{\mathrm{Y}}^2} , \tag{2.174}$$

and by inserting the rate of plastic strain:

$${}^{\mathrm{p}}\varepsilon^{\bullet}_{kl} = \Lambda^{\bullet}\frac{\partial f}{\partial \sigma_{kl}} = \Lambda^{\bullet}\big(\sigma_{|kl|} - \beta_{kl}\big) , \tag{2.175}$$

into the multiplier, we obtain

$$\begin{aligned} \Lambda^{\bullet}\Bigg(1 + \frac{\big(\sigma_{|ij|} - \beta_{ij}\big)C_{ijkl}\big(\sigma_{|kl|} - \beta_{kl}\big)}{\frac{2}{3}c\sigma_{\mathrm{Y}}^2}\Bigg) &= \frac{\big(\sigma_{|ij|} - \beta_{ij}\big)C_{ijkl}\big(\varepsilon^{\bullet}_{kl} - {}^{\mathrm{th}}\varepsilon^{\bullet}_{kl}\big)}{\frac{2}{3}c\sigma_{\mathrm{Y}}^2} , \\ \Lambda^{\bullet} &= \frac{\big(\sigma_{|ij|} - \beta_{ij}\big)C_{ijkl}\big(\varepsilon^{\bullet}_{kl} - {}^{\mathrm{th}}\varepsilon^{\bullet}_{kl}\big)}{\frac{4}{9}H\sigma_{\mathrm{Y}}^2 + \big(\sigma_{|ij|} - \beta_{ij}\big)C_{ijkl}\big(\sigma_{|kl|} - \beta_{kl}\big)} , \end{aligned} \tag{2.176}$$

where we have chosen $c = 2/3H$ for an easier association of parameters.[41] By using the conditional parameter $\langle\gamma\rangle$ from Eq. (1.216) we define the plastic strain rate:

$${}^{\mathrm{p}}\varepsilon^{\bullet}_{mn} = \langle\gamma\rangle\frac{\big(\sigma_{|ij|} - \beta_{ij}\big)C_{ijkl}\big(\varepsilon^{\bullet}_{kl} - {}^{\mathrm{th}}\varepsilon^{\bullet}_{kl}\big)}{\frac{4}{9}H\sigma_{\mathrm{Y}}^2 + \big(\sigma_{|ij|} - \beta_{ij}\big)C_{ijkl}\big(\sigma_{|kl|} - \beta_{kl}\big)}\big(\sigma_{|mn|} - \beta_{mn}\big) . \tag{2.177}$$

This equation gives the evolution of the plastic strain, which is accumulated by the rate of plastic strain such that we can acquire it by integration,

$${}^{\mathrm{p}}\varepsilon_{ij} = \int {}^{\mathrm{p}}\varepsilon^{\bullet}_{ij}\,\mathrm{d}t . \tag{2.178}$$

[41] We use H for the plastic modulus instead of h as in Sect. 1.6 since we have started to use h for the convective heat transfer coefficient in the mixed boundary conditions for temperature.

For the heat flux we use FOURIER's law:

$$q_i = -\kappa T_{,i} \,. \tag{2.179}$$

Furthermore, we may test the validity of the evolution equation by employing the 2nd law of thermodynamics. By using Eq. (2.168)$_2$ we reformulate the balance of internal energy into the balance of entropy:

$$\begin{aligned} \rho T \eta^{\bullet} + q_{i,i} - \rho r &= \sigma_{ij}\, {}^{\mathrm{p}}\varepsilon^{\bullet}_{ij} \,, \\ \rho \eta^{\bullet} + \Big(\frac{q_i}{T}\Big)_{,i} - \rho \frac{r}{T} &= -\frac{1}{T^2} q_i T_{,i} + \frac{1}{T}\sigma_{ij}\, {}^{\mathrm{p}}\varepsilon^{\bullet}_{ij} \,, \end{aligned} \tag{2.180}$$

with the production term being positive:

$$\Sigma = -\frac{1}{T^2} q_i T_{,i} + \frac{1}{T}\sigma_{ij}\, {}^{\mathrm{p}}\varepsilon^{\bullet}_{ij} \geq 0 \,. \tag{2.181}$$

Obviously, it is challenging to prove that the rate of plastic strain in Eq. (2.177) is thermodynamically admissible. Therefore, we rather have to "believe in" the formulation than to derive in a thermodynamically compatible way. For the moment a thermodynamically consistent formulation of plasticity is an unresolved issue and still heavily discussed in the literature.

We have obtained two coupled field equations for temperature and displacement from the balance equations of internal energy in Eq. (2.169) and of momentum in Eq. (2.170), respectively. In the initial frame the time derivatives are simply the partial time derivatives. From the balance of momentum and energy we generate the weak forms in the unit of energy. First we apply the usual time discretization. Secondly, we multiply the momentum balance with δu_i and the energy balance with $\Delta t\, \delta T/T$ in order to rectify the unit of energy. Finally, we integrate by parts and obtain

$$\mathrm{F}_{\boldsymbol{u}} = \int_{\mathcal{B}_0} \Big(\rho \frac{u_i - 2u_i^0 + u_i^{00}}{\Delta t^2}\, \delta u_i + \sigma_{ji}\, \delta u_{i,j} - \rho f_i\, \delta u_i \Big)\, \mathrm{d}V - \int_{\partial\mathcal{B}_0} \hat{t}_i\, \delta u_i\, \mathrm{d}A \,, \tag{2.182}$$

as well as

$$\begin{aligned} \mathrm{F}_T = \int_{\mathcal{B}_0} \Big(& \frac{\rho c}{T}(T - T^0)\, \delta T + \Delta t \alpha_{ij} C_{ijkl}\big(\varepsilon^{\bullet}_{kl} - {}^{\mathrm{p}}\varepsilon^{\bullet}_{kl} - {}^{\mathrm{th}}\varepsilon^{\bullet}_{kl} \big)\, \delta T - \\ & - \Delta t q_i \Big(\frac{\delta T}{T}\Big)_{,i} - \Delta t \frac{\rho r}{T}\, \delta T - \frac{\Delta t}{T}\sigma_{ij}\, {}^{\mathrm{p}}\varepsilon^{\bullet}_{ij}\, \delta T + \Delta t \frac{1}{T^2} q_i T_{,i}\, \delta T \Big)\, \mathrm{d}V + \\ & + \int_{\partial\mathcal{B}_0} \Delta t \hat{q} \frac{\delta T}{T}\, \mathrm{d}A \,, \end{aligned} \tag{2.183}$$

with

$$\mathrm{F} = \mathrm{F}_{\boldsymbol{u}} + \mathrm{F}_T \ , \tag{2.184}$$

where we implement the stress, the kinematic hardening, and the plastic strain according to the so-called incremental plasticity,

$$\begin{aligned}
\sigma^{\bullet}_{ij} &= C_{ijkl}\big(\varepsilon^{\bullet}_{kl} - {}^{\mathrm{p}}\varepsilon^{\bullet}_{kl} - {}^{\mathrm{th}}\varepsilon^{\bullet}_{kl}\big) \ , \quad \sigma_{ij} = \sigma^0_{ij} + \Delta t \sigma^{\bullet}_{ij} \ , \\
{}^{\mathrm{th}}\varepsilon^{\bullet}_{kl} &= \alpha_{kl}\frac{T - T^0}{\Delta t} \ , \quad \beta^{\bullet}_{ij} = \frac{(\sigma^0_{|kl|} - \beta^0_{kl})\sigma^{\bullet}_{kl}}{\frac{2}{3}\sigma_{\mathrm{Y}}^2}(\sigma^0_{|ij|} - \beta^0_{ij}) \ , \\
{}^{\mathrm{p}}\varepsilon^{\bullet}_{mn} &= \langle\gamma\rangle \frac{(\sigma^0_{|ij|} - \beta^0_{ij})C_{ijkl}(\varepsilon^{\bullet}_{kl} - {}^{\mathrm{th}}\varepsilon^{\bullet}_{kl})}{\frac{4}{9}H\sigma_{\mathrm{Y}}^2 + (\sigma^0_{|ij|} - \beta^0_{ij})C_{ijkl}\big(\sigma^0_{|kl|} - \beta^0_{kl}\big)}(\sigma^0_{|mn|} - \beta^0_{mn}) \ , \\
\beta_{ij} &= \beta^0_{ij} + \Delta t \beta^{\bullet}_{ij} \ , \quad {}^{\mathrm{p}}\varepsilon_{ij} = {}^{\mathrm{p}}\varepsilon^0_{ij} + \Delta t\, {}^{\mathrm{p}}\varepsilon^{\bullet}_{ij} \ ,
\end{aligned} \tag{2.185}$$

and the heat flux as well as the strain as follows

$$q_i = -\kappa T_{,i} \ , \quad \varepsilon_{ij} = u_{(i,j)} \ , \quad \varepsilon^0_{ij} = u^0_{(i,j)} \ , \quad \varepsilon^{\bullet}_{ij} = \frac{1}{\Delta t}(\varepsilon_{ij} - \varepsilon^0_{ij}) \ . \tag{2.186}$$

Consider again a one-axial tensile testing, as in the previous section. By including plasticity we can capture an effect known from the daily life. If a cyclic loading with plastic deformation is utilized, the structure heats up. This phenomenon can clearly be seen in Fig. 2.8. A part of the energy has been stored such that the temperature decreases and increases. This part of the process is reversible and it is modeled by the entropy, η. Simultaneously, entropy is produced by Σ, which is an irreversible effect increasing the temperature further. In total, after one cycle of deformation, the temperature is increased approximately 1 K. The code is below including all realistic material parameters for a standard steel.

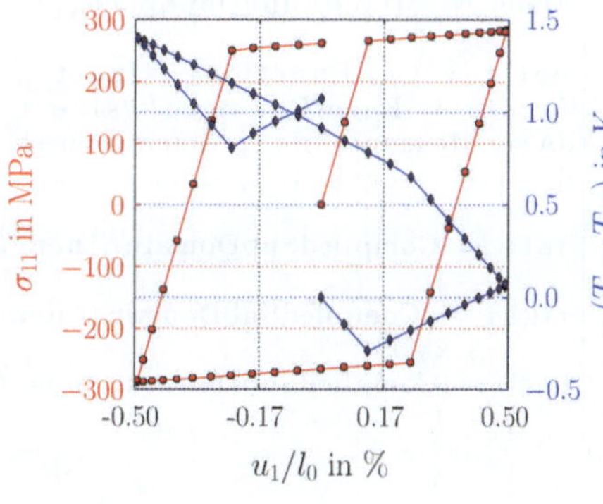

Fig. 2.8 Tensile testing and temperature rise due to the plastic deformation

```
"""Computational reality 14, thermoplasticity"""
__author__ = "B. Emek Abali"
__license__ = "GNU GPL Version 3.0 or later"
#This code underlies the GNU General Public License, http://www.gnu.org/licenses/gpl-3.0.en.html

from fenics import *
import numpy
set_log_level(ERROR)
#units: mm, 1000 kg=ton, s, MPa, mJ, K
Tref = 300. #in K
Tamb=Tref
# Material data of P265GH (St 45.8) from VDI Waermeatlas, at 293.15 K
rho = 7.85E-9 #in tonne / mm^3
kappa = 57.0 #in mJ / (s mm K)
capacity = 430.0E6 #in mJ / (ton K)
alpha = 12.2E-6 #in 1/K at 373.15 K
EModul = 211.0E+3 #in MPa
nu = 0.28
H =  2600. #MPa
h = 10.E-3 #in mJ / (s mm^2 K)
sigmaY = Constant(250.0) #MPa

tMax = 10.0
Dt = 0.2
t = 0.0

xMin, xMax, xElements = 0.0, 100.0, 20
yMin, yMax, yElements = -5., +5., 2
zMin, zMax, zElements = -5., +5., 2
mesh = BoxMesh(Point(xMin,yMin,zMin), Point(xMax,yMax,zMax), xElements,yElements,zElements)
N = FacetNormal(mesh)
length = abs(xMax-xMin)

Scalar = FunctionSpace(mesh, 'P', 1)
Vector = VectorFunctionSpace(mesh, 'P', 1)
Tensor = TensorFunctionSpace(mesh, 'P', 1)
Space = MixedFunctionSpace([Scalar, Vector])

cells = CellFunction('size_t', mesh)
facets = FacetFunction('size_t', mesh)
dA = Measure('ds', domain=mesh, subdomain_data=facets)
dV = Measure('dx', domain=mesh, subdomain_data=cells)

left = CompiledSubDomain('near(x[0],l) && on_boundary',l=xMin)
right = CompiledSubDomain('near(x[0],l) && on_boundary',l=xMax)
back = CompiledSubDomain('near(x[1],l) && on_boundary',l=yMin)
```

```
front = CompiledSubDomain('near(x[1],l) && on_boundary',l=yMax)
bottom = CompiledSubDomain('near(x[2],l) && on_boundary',l=zMin)
top = CompiledSubDomain('near(x[2],l) && on_boundary',l=zMax)

facets.set_all(0)
displ = Expression(('0.5*sin(2.*pi*f*time)','0.0','0.0'),f=0.1,time=0)
bc1 = DirichletBC(Space.sub(1),displ,right)
bc2 = DirichletBC(Space.sub(1),Constant((0.0,0.0,0.0)),left)

bc = [bc1,bc2]

dunkn = TrialFunction(Space)
test = TestFunction(Space)
delT, delu = split(test)

unkn = Function(Space)
unkn0 = Function(Space)
unkn00 = Function(Space)

unkn_init = Expression(('Tini','0','0','0'),Tini=Tref)
unkn = interpolate(unkn_init,Space)
unkn0.assign(unkn)
unkn00.assign(unkn0)

T, u = split(unkn)
T0, u0 = split(unkn0)
T00, u00 = split(unkn0)

i, j, k, l, m, n, o, p, r, s = indices(10)
delta = Identity(3)
lambada = EModul * nu / (1.+nu) / (1.-2.*nu)
mu = 0.5 * EModul / (1.+nu)
C = as_tensor(lambada*delta[i,j]*delta[k,l]+mu*delta[i,k]*delta[j,l]+mu*delta[i,l]*delta[j,k], (i,j,k,l))
alp = alpha*delta

peps0 = Function(Tensor)
sigma0= Function(Tensor)
dev_sigma0 = as_tensor(sigma0[i,j]-1./3.*sigma0[k,k]*delta[i,j], (i,j))
beta0 = Function(Tensor)

eps = sym(grad(u))
eps0 = sym(grad(u0))
Deps = (eps-eps0)/Dt

teps = alp*(T-Tref)
Dteps = alp*(T-T0)/Dt

gamma = Function(Scalar)
```

```
Dpeps = as_tensor(gamma*(dev_sigma0-beta0)[i,j]*C[i,j,k,l]*(Deps-Dteps)[k,l]/(4./9.*H*sigmaY**2+(dev_sigma0-beta0)[m,n]*C[m,n,o,p]*(dev_sigma0-beta0)[o,p])*(dev_sigma0-beta0)[r,s], (r,s))

Dsigma = as_tensor(C[i,j,k,l]*(Deps-Dpeps-Dteps)[k,l], (i,j))
Dbeta = as_tensor( gamma*(dev_sigma0-beta0)[k,l]*Dsigma[k,l]/(2./3.*sigmaY**2)*(dev_sigma0-beta0)[i,j], (i,j))

sigma = sigma0 + Dt*Dsigma
beta = beta0 + Dt*Dbeta
peps = peps0 + Dt*Dpeps

dev_sigma = as_tensor(sigma[i,j]-1./3.*sigma[k,k]*delta[i,j], (i,j))
q = as_tensor(-kappa*T.dx(i), (i,))

f = Constant((0.0,0.0,0.0))
R = Constant(0.0)
qHat = h*(T-Tamb)

F_u = (rho*(u-2.*u0+u00)[i]/Dt/Dt*delu[i] + sigma[j,i]*delu[i].dx(j) - rho*f[i]*delu[i] )*dV
F_T = (rho*capacity/T*(T-T0)*delT+ Dt*alp[i,j]*C[i,j,k,l]*(Deps-Dpeps-Dteps)[k,l]*delT - Dt*q[i]*(delT/T).dx(i) - Dt*rho*R/T*delT - Dt*sigma0[i,j]*Dpeps[j,i]*delT/T )*dV + Dt*qHat*delT/T*dA

Form = F_u + F_T
Gain = derivative(Form, unkn, dunkn)

import matplotlib as mpl
mpl.use('Agg')
import matplotlib.pyplot as pylab
pylab.rc('text', usetex=True )
pylab.rc('font', family='serif', serif='cm', size=30 )
pylab.rc('legend', fontsize=30)
pylab.rc(('xtick.major','ytick.major'), pad=15)
pylab.subplots_adjust(top=0.90)
pylab.subplots_adjust(bottom=0.17)
pylab.subplots_adjust(left=0.20)
pylab.subplots_adjust(right=0.75)

fig = pylab.figure(1, figsize=(14,10))
ax1 = fig.add_subplot(111)
ax1.grid(True, axis='x')
ax1.set_xlabel(r'$u_1/l_0$ in $\%$')
ax1.set_ylabel(r'$\sigma_{11}$ in MPa', color='r')
ax1.tick_params(axis='y', colors='r')
ax1.grid(True, axis='y', color='r')
ax2 = ax1.twinx()
ax2.set_ylabel(r'$(T-T_\mathrm{ref})$ in K', color='b')
ax2.tick_params(axis='y', colors='b')
ax2.grid(True, axis='y', color='b')
```

```
ax2.ticklabel_format(style='sci', axis='y', scilimits=(-2,2))

pwd='/calcul/CR14/'
file_u = File(pwd+'displ.pvd')
file_T = File(pwd+'temp.pvd')
strain,stress,temp =[],[],[]

while t < tMax:
    print 'time: ',t
    displ.time = t
    solve(Form==0, unkn, bc, J=Gain, \
        solver_parameters={"newton_solver":{"linear_solver": "mumps", "relative_tolerance": 1e-5} }, \
        form_compiler_parameters={"cpp_optimize": True, "representation": "quadrature", "quadrature_degree": 2} )

    file_T << (unkn.split()[0], t)
    file_u << (unkn.split()[1], t)

    sigma_ = project(sigma, Tensor, solver_type="mumps",\
        form_compiler_parameters={"cpp_optimize": True, "representation": "quadrature", "quadrature_degree": 2} )
    sigma0.assign(sigma_)
    beta_ = project(beta, Tensor, solver_type="mumps",\
        form_compiler_parameters={"cpp_optimize": True, "representation": "quadrature", "quadrature_degree": 2} )
    beta0.assign(beta_)
    flow_ = project(1./2.*(dev_sigma0-beta0)[i,j]*(dev_sigma0-beta0)[i,j] - 1./3.*sigmaY**2, Scalar, solver_type="mumps",\
        form_compiler_parameters={"cpp_optimize": True, "representation": "quadrature", "quadrature_degree": 2} )
    flow_bool = flow_.vector().array() >= 0.
    direction_ = project((dev_sigma0-beta0)[i,j]*Deps[i,j], Scalar, solver_type="mumps",\
        form_compiler_parameters={"cpp_optimize": True, "representation": "quadrature", "quadrature_degree": 2} )
    direction_bool=1./2.*(numpy.sign(direction_.vector().array())+1.)
    gamma.vector()[:] = numpy.array(flow_bool*direction_bool, dtype=int)

    peps_ = project(peps, Tensor, solver_type="mumps",\
        form_compiler_parameters={"cpp_optimize": True, "representation": "quadrature", "quadrature_degree": 2} )
    peps0.assign(peps_)

    unkn00.assign(unkn0)
```

```
        unkn0.assign(unkn)

        strain.append(unkn0.split()[1](xMax,0.,0.)[0]/length
            ↪ *100.)
        stress.append(sigma0(xMax/2.,0.,0.)[0])
        temp.append(unkn0.split()[0](xMax/2.,0.,0.)-Tref)
        ax1.plot(strain, stress, 'o-', color='r')
        ticks = numpy.linspace(numpy.array(strain).min(), numpy.
            ↪ array(strain).max(), 4)
        ax1.set_xticks(ticks)
        ax1.set_xticklabels(['%1.2f' % i_ticks for i_ticks in
            ↪ ticks] )
        ax2.plot(strain,temp, 'd-', color='b')
        fig.savefig(pwd+'CompReal14_tensiletest.pdf')

        t += Dt
```

To-do

Plastic deformation generates a temperature change in the system.

- Try to change the boundary conditions to adiabatic boundaries. Guess the result before starting the numerical calculation.
- Find out a stress/strain curve for another material and determine the plasticity modulus or hardening parameter H in MPa. What will be the result for a lower or higher H parameter?
- Find out the material behavior of aluminum and explain why the linear hardening model used in this section is inadequate for modeling an aluminum sample.

References

1. Abali, B.E.: Thermodynamically compatible modeling, determination of material parameters, and numerical analysis of nonlinear rheological materials. Ph.D. thesis, Technische Universität. Berlin, Institute of Mechanics (2014)
2. Cengel, Y.A.: Introduction to Thermodynamics and Heat Transfer. McGraw Hill Higher Education Press, New York (2007)
3. Coleman, B., Noll, W.: The thermodynamics of elastic materials with heat conduction and viscosity. Arch. Ration. Mech. Anal. **13**(1), 167–178 (1963)
4. Eckart, C.: The thermodynamics of irreversible processes. i. The simple fluid. Phys. Rev. **58**, 267–269 (1940)
5. Eckart, C.: The thermodynamics of irreversible processes. ii. Fluid mixtures. Phys. Rev. **58**(3), 269–275 (1940)
6. Eckart, C.: The thermodynamics of irreversible processes. iii. Relativistic theory of the simple fluid. Phys. Rev. **58**(10), 919–924 (1940)
7. Eckart, C.: The thermodynamics of irreversible processes. iv. The theory of elasticity and anelasticity. Phys. Rev. **73**, 373–382 (1948)
8. de Groot, S.R., Mazur, P.: Non-Equilibrium Thermodynamics. Dover Publications, New York (1984)
9. Questions in fluid mechanics: Gad-el Hak, M. J. Fluids Eng. **117**, 3–5 (1995)

10. Kilian, H.G., Höhne, G.: The müller-calorimeter and applications. Thermochimica Acta **69**(1), 199–219 (1983)
11. Müller, I.: The coldness, a universal function in thermoelastic bodies. Arch. Ration. Mech. Anal. **41**(5), 319–332 (1971)
12. Müller, I.: Thermodynamics. Pitman Publishing, London (1985)
13. Müller, I., Ruggeri, T.: Rational Extended Thermodynamics. Springer, New York (1998)
14. Pauli, W.: Pauli lectures on physics. In: Thermodynamics and the Kinetic Theory of Gases, vol. 3. Dover (2000) repub. of MIT Press, Cambridge, Massachusetts (1973)
15. Rajagopal, K.R.: A new development and interpretation of the Navier-Stokes fluid which reveals why the "Stokes assumption" is inapt. Int. J. Non-Linear Mechan. **50**, 141–151 (2013)
16. Salazar, A.: Energy propagation of thermal waves. Eur. J. Phys. **27**(6), 1349 (2006)
17. Truesdell, C., Toupin, R.A.: Principles of classical mechanics and field theory. In: Flügge, S. (ed.), Handbuch der Physik, vol. III/1 (1960)
18. Tzou, D.Y.: Experimental support for the lagging behavior in heat propagation. J. Thermophys. Heat Transf. **9**(4), 686–693 (1995)

Chapter 3
Electromagnetism

A continuum body consists of material particles having a mass and an electric charge. In Chaps. 1 and 2 we have analyzed a continuum body made of massive particles, *matter*, without electric charge. We have ignored the electromagnetic interaction completely. In order to include the electric charge and the effects of its motion, we will introduce MAXWELL's equations[1] and solve them. There are different ways of introducing MAXWELL's equations leading to slightly different governing equations.[2] We will follow [28, Chap. 9] for the motivation of MAXWELL's equations from balance equations and use the AMPERE–LORENTZ convention[3] in the rationalized mks system of units,[4] also referred to as GIORGI system.[5] The unit of charge is C(oulomb).[6]

Electromagnetism is the theory of electromagnetic interactions with matter. In this theory there occur various *new* quantities; and this makes a straight-forward introduction of equations challenging. These *new* quantities lead to electromagnetic fields, which can be measured. However, they often lack a clear interpretation. For example, a moving electric charge fails to be understood completely, but an electric current is something we use in our daily lives. Another difficulty arises in the definition of the electromagnetic interactions with matter: we have to redefine all balance equations used in Chaps. 1 and 2. Especially this step is quite confusing owing to various formulations existing in the literature. In order to establish a knowledge of electromagnetism, we will motivate governing equations one-by-one with applications in the following sections. This approach is beneficial for obtaining a familiarity with the electromagnetic fields and their governing equations.

---

[1] They are named after James Clerk Maxwell.

[2] See [30].

[3] It is named for André Marie Ampère and Hendrik Antoon Lorentz.

[4] The abbreviation mks stands for meter, kilogramm, seconds.

[5] It is named after Giovanni Giorgi.

[6] It is named for Charles-Augustin de Coulomb.

B.E. Abali, *Computational Reality*, Advanced Structured Materials 55,
DOI 10.1007/978-981-10-2444-3_3

The purpose of computation is to obtain *five* quantities as functions in space and time within the continuum body: mass density, ρ, velocity, v_i, temperature, T, electric field, E_i, and magnetic flux (area) density, B_i. They are the primitive variables[7] and we need field equations for computing them. The electric field E_i in V(olt)/m and magnetic flux density B_i in T(esla) are[8] acquired by the MAXWELL equations and the mass density, velocity, temperature by the balance equations.

In Sect. 3.1 we consider a conducting wire and investigate how it heats up due to the production term in the balance of internal energy. We introduce electric field and magnetic flux in polarized materials in Sect. 3.2. By using thermodynamical principles we derive the constitutive equations and solve a problem addressing the thermoelectric coupling in Sect. 3.3. We include plasticity in Sect. 3.4. In Sect. 3.5 a piezoelectric sensor is discussed by deriving the constitutive equations in a thermodynamically consistent way. A magnetohydrodynamical problem is presented in Sect. 3.6.

3.1 Conducting Wire

Electric charge is a fundamental quantity like mass. It can neither be destroyed nor supplied, hence, a balance equation in the current frame reads for a material system[9]

$$\Big(\int_{\mathcal{B}} z \mathrm{d}m\Big)^{\bullet} = 0\,, \quad \mathrm{d}m = \rho\, \mathrm{d}v, \tag{3.1}$$

where z denotes a specific electric charge (charge per mass) in C/kg. Instead of massive particles, the continuum body $\mathcal{B}$ consists of charged particles. The coordinates denote charged particles instead of matter and a motion of continuum body means moving charged particles. Electrically charged particles in motion possess a velocity, $v_i^{\mathrm{e}} = x_i^{\bullet}$, measured in a *laboratory* frame. Although the laboratory frame itself may have a velocity, w_i, we ignore it for simplicity and use a fixed frame. For inserting the time rate into the integral we need the following relation:

$$(\mathrm{d}v)^{\bullet} = (\mathrm{d}x_1 \mathrm{d}x_2 \mathrm{d}x_3)^{\bullet} \frac{\mathrm{d}v}{\mathrm{d}x_1 \mathrm{d}x_2 \mathrm{d}x_3} = \Big(\frac{\partial x_1^{\bullet}}{\partial x_1} + \frac{\partial x_2^{\bullet}}{\partial x_2} + \frac{\partial x_3^{\bullet}}{\partial x_3}\Big)\mathrm{d}v = \frac{\partial v_i^{\mathrm{e}}}{\partial x_i}\,. \tag{3.2}$$

---

[7] The electromagnetic theory started by believing that electric field E_i and magnetic field H_i shall be the primitive variables. In 1940s the theoretical thermodynamics introduced the way of starting with E_i and B_i fields. There are still explanations starting different than the approach used herein.

[8] The units are named after Alessandro Volta and Nikola Tesla.

[9] In the case of an open system, there would be an additional convective term.

Now the balance equation reads

$$\begin{aligned}&\Big(\int_{\mathcal{B}} \rho z \mathrm{d}v\Big)^{\boldsymbol{\cdot}} = \int_{\mathcal{B}} \Big((\rho z)^{\boldsymbol{\cdot}} + \rho z \frac{\partial v_i^{\mathrm{e}}}{\partial x_i}\Big)\mathrm{d}v = 0 \,, \\ &\frac{\partial \rho z}{\partial t} + v_i^{\mathrm{e}} \frac{\partial \rho z}{\partial x_i} + \rho z \frac{\partial v_i^{\mathrm{e}}}{\partial x_i} = \frac{\partial \rho z}{\partial t} + \frac{\partial \rho z v_i^{\mathrm{e}}}{\partial x_i} = 0 \,.\end{aligned} \tag{3.3}$$

In a conducting wire electrically charged particles travel or flow with the velocity v_i^{e}, however, it is difficult to measure this velocity for each charge. Instead, the flow of charged particles is observed as the *electric current (area) density*:

$$J_i = \rho z v_i^{\mathrm{e}} \,, \tag{3.4}$$

which is the *conducting* charge per area and per time, i.e., the flux of electric charge.[10] We may move the conducting wire with a velocity, v_i, measured in the (fixed) laboratory frame. This motion is of matter and the charges are carried with that leads to a *convection* current. We measure in a laboratory frame an electric current:

$$J_i = \mathcal{J}_i + \rho z v_i \,, \tag{3.5}$$

caused by a conduction current and a convection current. The conduction current (area density), $\mathcal{J}_i$, is measured on the co-moving continuum, i.e., we "sit" on the moving wire, measure the velocity of flowing particles, and multiply them by their charge density, ρz, in atomic scale. In the macroscopic setting we cannot detect every single moving electron and consider this current as the conduction current. The charge is somehow conducted from one end to the other in the wire. Since this quantity is measured co-movingly, it is *objective*. In any coordinate system we will have the same conduction definition. We need to define a constitutive equation for the electric current due to conduction, $\mathcal{J}_i$. With its definition we can solve the balance of charge:

$$\frac{\partial \rho z}{\partial t} + \frac{\partial}{\partial x_i}(\mathcal{J}_i + \rho z v_i) = 0 \,, \tag{3.6}$$

in order to obtain the charge distribution. For a fixed electric wire, $v_i = 0$, made of a homogeneous material, $\rho = \text{const.}|_x$ and $z = \text{const.}|_x$, the electric charge will remain constant, when a charge enters the wire on one end, another charge exists the wire on the other end simultaneously. The energy that bounces the charged

[10] Charge per time, C/s, is called A(mpere) named for André Marie Ampère. The electric current (area) density, J_i, is in A/m^2.

particles is transported so quick that we observe it as instantaneous.[11] The charge remains constant in time, $z = \text{const.}|_t$. We can visualize this phenomenon as an "incompressible" flow of electric charge. In the coming sections we will amend the formulation for incorporation this effect, too. In this section, as a consequence of the incompressible charge flow, we obtain the balance of charge for a conducting, rigid, homogeneous wire

$$\frac{\partial \mathcal{J}_i}{\partial x_i} = 0 . \tag{3.7}$$

The simplest relation for the electric conduction current, $\mathcal{J}_i$, is to associate the current with a *force* or *moment* that accelerates the charged particles. The specific force, $z\mathcal{E}_i$, is given by the electromotive *force (charge) density*[12] or *intensity*, $\mathcal{E}_i$, measured in the co-moving frame. It is an objective quantity.[13] Again in the co-moving frame we can measure a moment due to the magnetic flux (area) density, $\mathcal{B}_i$.[14] The magnetic flux may be applied on the system or it is induced due to the conduction current. The total force density:

$$\mathcal{F}_i = \rho f_i^{\text{Lor.}} = \rho z \mathcal{E}_i + \epsilon_{ijk} \mathcal{J}_j \mathcal{B}_k , \tag{3.8}$$

containing the electromotive intensity and magnetic flux is referred to as the LORENTZ force density.[15] The first term can be comprehended as a result of a translational charge movement and the second term is because of a rotational motion. The translational motion of charged particles are measured as the conduction current such that the electric current depends on the first term of the LORENTZ force. In the next section we will see that the conduction current induces a magnetic force affecting conduction over the second term in an indirect way. The conduction current is, however, only due to the translational motion and a linear relation between them reads

$$\mathcal{J}_i = \varsigma \mathcal{E}_i , \tag{3.9}$$

---

[11] We know that the action fails to be instantaneous; however, our experimental machine is detecting slower than the information transported from one end to the other end of the wire.

[12] Since the continuum body is defined as a collection of charges, *density* here means a quantity per electric charge, hence we write a force charge density meaning a force per charge, N/C. Another adequate name is *intensity*, which we shall use further on.

[13] The electromotive intensity is measured in V(olt) per meter since 1 V is the energy per electric charge, 1 V $\widehat{=}$ 1 J/C such that 1 N/C $\widehat{=}$ 1 N m/(C m) $\widehat{=}$ 1 V/m. Moreover, a single V across a wire conducting a current of 1 A dissipates of 1 W $\widehat{=}$ 1 J/s of power, hence 1 W $\widehat{=}$ 1 A V.

[14] The unit of magnetic flux area density is in T or in Wb (Weber) per m^2 named after Wilhelm Eduard Weber. Since 1 Wb $\widehat{=}$ 1 kg m^2/(C s) $\widehat{=}$ 1 N s m/C $\widehat{=}$ 1 V s, magnetic flux *area* density is in V s/m^2.

[15] The force is named for Hendrik Antoon Lorentz. For a detailed physical interpretation of this force in the atomic scale, see [33, Sect. 4–5].

which is referred to as OHM's law[16] with the *electric conductivity*,[17] ς, the inverse of resistivity, $r = 1/\varsigma$. This constitutive equation will be derived in a thermodynamically consistent way in Sect. 3.3 on p. 213. In this section we take it for granted.

Measuring the LORENTZ force in a co-moving frame is difficult such that we want to transform it from the co-moving to the laboratory frame.[18] The magnetic flux remains simply the same:

$$B_i = \mathcal{B}_i \ , \tag{3.10}$$

in other words, we measure the same numerical values for the components of the magnetic flux (area) density in the co-moving as well as in the laboratory frame. From now on we will not distinguish between them and use B_i for the magnetic flux (area) density. The electric field measured on the laboratory frame, however, depends on the velocity and magnetic flux as follows

$$E_i = \mathcal{E}_i - (\boldsymbol{v} \times \boldsymbol{B})_i = \mathcal{E}_i - \epsilon_{ijk} v_j B_k \ . \tag{3.11}$$

Now by using Eq. (3.5) we obtain the LORENTZ force density measured in the laboratory frame:

$$\begin{aligned} \mathcal{F}_i = \rho f_i^{\text{Lor.}} = \rho z \mathcal{E}_i + \epsilon_{ijk} \mathcal{J}_j \mathcal{B}_k = \\ = \rho z (E_i + \epsilon_{ijk} v_j B_k) + \epsilon_{ijk} (J_j - \rho z v_j) B_k = \rho z E_i + \epsilon_{ijk} J_j B_k \ . \end{aligned} \tag{3.12}$$

The primitive variables are E_i and B_i fields. In this section we neglect the magnetic flux, $B_i = 0$, and set our goal to solve the electric field. Unfortunately, we cannot solve Eq. (3.7) and obtain three components in E_i, since Eq. (3.7) is a single differential equation. Therefore, we introduce another scalar quantity

$$E_i = -\frac{\partial \phi}{\partial x_i} \ , \tag{3.13}$$

where the *electric potential*, ϕ, is in V. In Sect. 3.2 on p. 178 we will generalize the latter formulation and introduce the definition of the electric potential in a more convenient way. By introducing the *scalar* or electric potential, ϕ, we find a primitive variable to be computed by satisfying Eq. (3.7). Hence, we multiply Eq. (3.7) with the test function, $\delta\phi$, within the continuum body of matter

---

[16] It is named after Georg Simon Ohm.

[17] Electrical conductivity is in S(iemens)/m where S $\hat{=}$ 1/Ω $\hat{=}$ A/V. Thus the conductivity is in A/(V m). The unit S is named for Werner von Siemens. The unit Ω is capital omega in Greek pronounced as Ohm if used as the unit named after Georg Simon Ohm.

[18] We refer to [24, Sect. 9] for the transformation of $\mathcal{E}_i$ and $\mathcal{B}_i$ fields from the co-moving to the laboratory frame.

$$\int_{\mathcal{B}} \mathcal{J}_{i,i}\, \delta\phi\, \mathrm{d}v = 0\,, \tag{3.14}$$

where again we have used a comma notation for partial derivative in space. In this section we assume that the body is rigid, $v_i = 0$, such that $\mathcal{E}_i = E_i$ and $\mathcal{J}_i = J_i$. Since the conduction current depends on the electric field that depends on the derivative of ϕ we have a twice differentiability condition of the electric potential.[19] We weaken this condition by integrating by parts

$$\mathrm{F}_\phi = -\int_{\mathcal{B}} J_i\, \delta\phi_{,i} \mathrm{d}v + \int_{\partial\mathcal{B}} J_i\, \delta\phi n_i \mathrm{d}a\,. \tag{3.15}$$

This weak form is in the unit of power. We want to model a conducting wire grounded on its left side whereas the electric potential is altered on wire's right side. Hence, we use DIRICHLET conditions on the left and right boundaries such that $\delta\phi = 0$ on these sides. For the other boundaries we assume an electric insulation, $J_i n_i = 0$.

A conducting body heats up. Actually, we experience this liberation of heat in our daily lives. A smartphone *produces* heat energy during phoning or watching a video. This heat energy increases the temperature. Since the temperature distribution is computed by using the balance of internal energy, we need the balance equation of internal energy for a rigid conductor[20]

$$\rho c \frac{\partial T}{\partial t} + \frac{\partial q_i}{\partial x_i} - \rho r = J_i E_i\,. \tag{3.16}$$

The right hand side is called the JOULE heating, *resistive* heating, or OHMic loss. It produces heat leading to a temperature increase with a positive electrical conductivity, $\varsigma > 0$, since $J_i = \varsigma E_i$ in a rigid body. For the heat flux, q_i, we implement FOURIER's law:

$$q_i = -\kappa T_{,i}\,. \tag{3.17}$$

The weak form is then acquired after discretizing in time, multiplying with $\delta T/T$, applying integration by parts, and using a comma for partial derivative in space, as follows

$$\begin{aligned}\mathrm{F}_T = \int_{\mathcal{B}} \Big(\rho c \frac{T - T^0}{\Delta t} \frac{\delta T}{T} - q_i \Big(\frac{\delta T}{T}\Big)_{,i} - \rho r \frac{\delta T}{T} + J_i \phi_{,i} \frac{\delta T}{T} \Big) \mathrm{d}v + \\ + \int_{\partial\mathcal{B}} q_i \frac{\delta T}{T} n_i \mathrm{d}a\,.\end{aligned} \tag{3.18}$$

[19] Mathematically speaking the function has to belong to C^2 or higher in order to ensure the twice differentiability condition. Since ϕ and $\delta\phi$ are of the same space according to the GALERKIN type finite element method, both have to be twice differentiable.

[20] In this section we deal with unpolarized systems.

This weak form is in the unit of power. For boundaries we use the natural convection as introduced in Sect. 2.1 as the mixed or ROBIN conditions:

$$q_i n_i = h(T - T_{\text{amb}}) , \tag{3.19}$$

with the convective heat transfer coefficient, h, in W/(m^2 K). The form for computing electric potential, ϕ, and temperature, T, with the aforementioned boundary conditions reads

$$\begin{aligned}\text{Form} = \text{F}_\phi + \text{F}_T = \int_{\mathcal{B}} \bigg(& - J_i \,\delta\phi_{,i} + \rho c \frac{T - T^0}{\Delta t} \frac{\delta T}{T} - q_i \Big(\frac{\delta T}{T}\Big)_{,i} - \\ & -\rho r \frac{\delta T}{T} + J_i \phi_{,i} \frac{\delta T}{T} \bigg) \text{d}v + \int_{\partial \mathcal{B}} h(T - T_{\text{amb}}) \frac{\delta T}{T} \text{d}a ,\end{aligned} \tag{3.20}$$

where the electric current and heat flux are given by

$$J_i = -\varsigma \phi_{,i} \ , \quad q_i = -\kappa T_{,i} \ . \tag{3.21}$$

We compute a copper wire with a constant thermal conductivity:

$$\kappa = 400 \, \text{W/(m K)} , \tag{3.22}$$

and a temperature dependent electrical conductivity:

$$\varsigma = \frac{\varsigma_0}{(1 + \tilde{\alpha}(T - T_{\text{ref.}}))} \ , \quad \varsigma_0 = 5.8 \cdot 10^7 \, \text{S/m} \ , \quad \tilde{\alpha} = 3.9 \cdot 10^{-3} \, \text{K}^{-1} \ . \tag{3.23}$$

A wire of length 1 m and of square shaped cross section[21] has been modeled. The wire is initially at the reference temperature, $T_{\text{ref.}} = 300\,\text{K}$. One end of the wire is grounded, i.e., 0 V, and a potential difference of 0.1 V is applied on the other end. The electric potential changes immediately and distributes linearly as to be depicted in Fig. 3.1. The temperature increases over time and for the case of free convection in air, $h \approx 10\,\text{W/(m}^2\,\text{K)}$, after 10 and 20 min we observe a significant increase in temperature in Fig. 3.2.

Actually, this configuration is unrealistic. We have implemented a wire with $\Delta V = 0.1\,\text{V}$ difference per meter and a resistivity of $r = 1/\varsigma \approx 1.7 \cdot 10^{-8}\,\Omega\,\text{m}$ for 1 m wire. Since the conductivity of copper is huge, its resistivity is quite low. In order to

[21] We should model a rounded wire since in reality wires have round cross sections, however, it is simpler and shorter to code it this way. The presented solutions and discussions are the same for both cases.

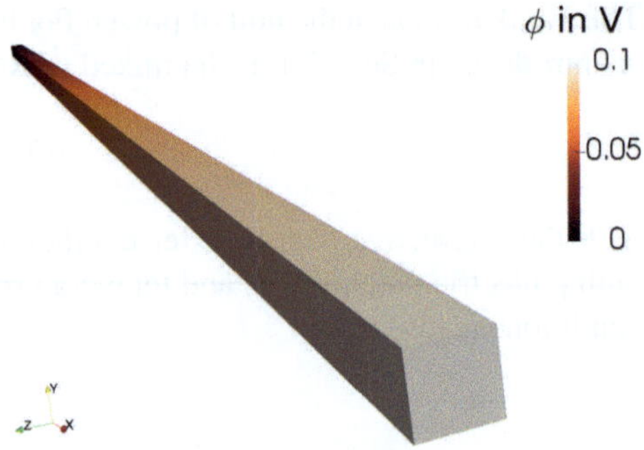

Fig. 3.1 Distribution of electric potential ϕ in V within a conducting wire out of copper

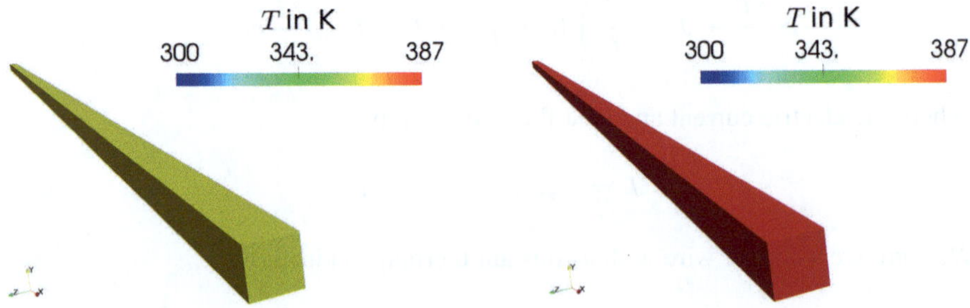

Fig. 3.2 Temperature distribution under 0.1 V difference in 1 m wire after 10 min (*on left*) and 20 min (*on right*) for the case of free convection in air

realize the consequences, we employ OHM's law as learned in school, $\Delta V = I\,R$, and calculate the current in ampere. We need to obtain the so-called *resistance*, R, from the *resistivity*, r, by the following relation:

$$R = \frac{r\ell}{a} , \tag{3.24}$$

where the length of the wire, $\ell = 1\,\text{m}$, and the cross-sectional area of the wire, $a = 0.01 \times 0.01\,\text{m}^2$, are known. Then the electric current reads $I = \Delta V / R = 580\,\text{A}$, which is too high. For example, at home we have an electrical fuse that breaks the circuit by a current higher than 15 A. This unrealistic simulation is owing to the implemented wire with a nearly zero resistance as a part of a circuit. In reality there is much greater resistance on the electric circuit since a light bulb or an LCD-monitor is connected to the wire, thus, much smaller currents occur in the circuit. Although the wire might transfer charges in such high current, as we observe from the simulation, its heat production is tremendous. A small current produces a small JOULE heating such that our cables at home remain nearly at the room temperature.

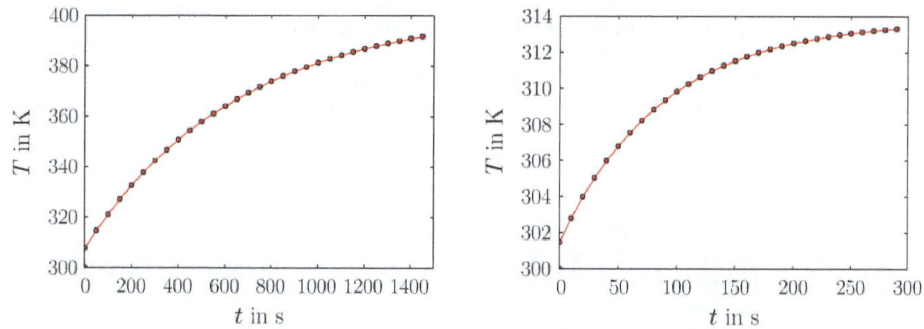

Fig. 3.3 Temperature increase under 0.1 V difference in the 1 m wire for free convection in air (*on left*) and for forced convection in air (*on right*)

The temperature increase is caused by the current conducting in the wire and heat convection across the boundary. We simulate two different cases in order to present the effect of the natural convection. First case is a free convection in air. Air remains still and the convection coefficient is set as $h = 10\,\text{W/(m}^2\,\text{K)}$. The value of this parameter is difficult to measure and it depends on the properties of air and copper. Usually, only the properties of air are accounted for estimating the convective heat transfer coefficient, h. For the second case we assume a forced convection, i.e., a fan is blowing air of 300 K to the wire. This time the heat transfer coefficient increases to $h = 100\,\text{W/(m}^2\,\text{K)}$ and the heat convection over the boundary gets more efficient. We present the results in Fig. 3.3.

Depending on the choice of the free or forced convection, the time until steady state is achieved and the value of temperature at steady state vary. Electric potential distribution remains constant, leading to a constant JOULE heating. As the temperature increases the heat exchange across boundaries grows, too. After a while the heat exchange catches the heat production such that the temperature converges to a steady state value. As in reality, the heat transferred over the boundary increases for greater differences between the boundary and ambient temperature depending on the parameter h. We all know that a fan in our laptop is necessary, since the convective heat transfer coefficient for a free convection in air is much lower than in case of a forced convection. For special computers where the power (and thus, current) is higher we even need an active cooling system working with water such that the convection parameter is increased further. We recall that the resistance in the wire is set to a unrealistically low value, we simulate a short circuit. The code below is used for all simulations in this section.

```
"""Computational reality 15, conducting wire"""
__author__ = "B. Emek Abali"
__license__ = "GNU GPL Version 3.0 or later"
#This code underlies the GNU General Public License, http://www.gnu.org/licenses/gpl-3.0.en.html

from fenics import *
import numpy
set_log_level(ERROR)
#units: m, kg, s, A, V, K
delta = Identity(3)
Tref = 300. #in K
t_alpha = 3.9E-3 #in 1/K
varsigma0 = 5.8E7 #in S/m or in 1/(Ohm m)
Tamb = Tref
rho = 8960. #in kg / m^3
kappa = 400.0 #in W/(m K)
capacity = 390. #in J / (kg K)
#approximate values: for convection over boundaries
#free convection in air h=10
#forced convection in air h=100
#forced convection in water h=1000
h = 10. #in W / (m^2 K)

tMax = 1500.0
Dt = 50.0
t = 0.0

xMin, xMax, xElements = 0.0, 1.0, 200
yMin, yMax, yElements = -0.005, +0.005, 2
zMin, zMax, zElements = -0.005, +0.005, 2
mesh = BoxMesh(Point(xMin,yMin,zMin), Point(xMax,yMax,zMax), xElements,yElements,zElements)
N = FacetNormal(mesh)

Scalar = FunctionSpace(mesh, 'P', 1)
Vector = VectorFunctionSpace(mesh, 'P', 1)
Tensor = TensorFunctionSpace(mesh, 'P', 1)
# phi, T
Space = MixedFunctionSpace([Scalar, Scalar])

cells = CellFunction('size_t', mesh)
facets = FacetFunction('size_t', mesh)
da = Measure('ds', domain=mesh, subdomain_data=facets)
dv = Measure('dx', domain=mesh, subdomain_data=cells)

left = CompiledSubDomain('near(x[0],l) && on_boundary',l=xMin)
right = CompiledSubDomain('near(x[0],l) && on_boundary',l=xMax)
facets.set_all(0)
bc = [DirichletBC(Space.sub(0), 0.0, left),\
DirichletBC(Space.sub(0), 0.1, right),]

```

```
dunkn = TrialFunction(Space)
test = TestFunction(Space)
del_phi, del_T = split(test)

unkn = Function(Space)
unkn0 = Function(Space)

unkn_init = Expression(('0.0','T_ini'),T_ini=Tref)
unkn0 = interpolate(unkn_init,Space)
unkn = interpolate(unkn_init,Space)

phi, T = split(unkn)
phi0, T0 = split(unkn0)

i, j, k, l = indices(4)
delta = Identity(3)
varsigma = varsigma0/(1.0 + t_alpha*(T-Tref))
J = -varsigma*grad(phi)
q = -kappa*grad(T)
r = Constant(0.0)

Form = (-J[i]*del_phi.dx(i) + rho*capacity*(T-T0)/Dt*del_T/T
    ↪ - q[i]*(del_T/T).dx(i) - rho*r*del_T/T + J[i]*phi.dx(i
    ↪ )*del_T/T )*dv + h*(T-Tamb)*del_T/T*da
Gain = derivative(Form, unkn, dunkn)

pwd='/calcul/CR15/'
file_phi = File(pwd+'phi.pvd')
file_T = File(pwd+'temp.pvd')

import matplotlib as mpl
mpl.use('Agg')
import matplotlib.pyplot as pylab
pylab.rc('text', usetex=True )
pylab.rc('font', family='serif', serif='cm', size=30 )
pylab.rc('legend', fontsize=30)
pylab.rc(('xtick.major','ytick.major'), pad=15)
pylab.subplots_adjust(top=0.90)
pylab.subplots_adjust(bottom=0.17)
pylab.subplots_adjust(left=0.20)
pylab.subplots_adjust(right=0.95)
fig = pylab.figure(1, figsize=(14,10))
ax1 = fig.add_subplot(111)
ax1.set_xlabel('$t$ in s')
ax1.set_ylabel('$T$ in K')
ax1.set_xlim([0.0,tMax])
t_plot, T_plot = [0], [Tref]

while t < tMax:
    print 'time: ',t,' T ',T_plot[-1]
    solve(Form==0, unkn, bc, J=Gain, \
        solver_parameters={"newton_solver":{"linear_solver":
            ↪ "mumps", "relative_tolerance": 1e-5} }, \
        form_compiler_parameters={"cpp_optimize": True, "
```

```
    representation": "quadrature", "
    quadrature_degree": 2} )

    #file_T << (unkn.split()[1], t)
    #file_phi << (unkn.split()[0], t)

    t_plot.append(t)
    T_plot.append(unkn.split()[1](xMax/2.,0.,0.))
    ax1.plot(t_plot,T_plot, 'o-', color='r')
    fig.savefig(pwd+'CompReal15_temp.pdf')
    unkn0.assign(unkn)
    t += Dt
```

To-do

An electrothermal coupling has been modeled. In many commercial codes this type of a simulation is achieved over a staggered scheme. We have implemented here in a monolithic way such that in each time step the electric potential and temperature are solved at once. The temperature rise occurs due to the current being conducted in the wire.

- Implement the code for a gold wire. Of course it costs more than copper but what can be achieved by using a gold wire instead of copper?
- Find out a forced convection parameter changing with the speed of air and compare the results for different fan speeds. During computation the fan of the laptop gets louder by rotating in an increased speed. Is the convection parameter, h, higher for an increased speed?

3.2 Polarized Materials

As a material specific property, the continuum body can be electrically or magnetically *polarized* subject to electromagnetic fields. The electric or magnetic polarization indicates a change in the electric charge distribution in the body. Suppose that the electric charge is homogeneously distributed. Under the influence of electromagnetic fields, charged particles deviate from their homogeneous distribution and the continuum body becomes polarized. Before we discuss this phenomenon deeply, we reinvent the MAXWELL equations and then bring in the ideas of polarization.

We start with a balance equation on a material surface,[22] $\mathcal{S}$, known as FARADAY's law:[23]

$$\Big(\int_{\mathcal{S}} B_i \mathrm{d}a_i\Big)^{\boldsymbol{\cdot}} = -\int_{\partial\mathcal{S}} \mathcal{E}_i \mathrm{d}\ell_i \,, \tag{3.25}$$

[22] A material surface is a material system without convection terms where the domain is a surface instead of a volume leading to an area density instead of a volume density.

[23] It is named after Michael Faraday.

where the magnetic flux area density, B_i, in a material surface, $\mathcal{S}$, is balanced with the electromotive intensity, $\mathcal{E}_i$, acting on the boundary of the surface, $\partial\mathcal{S}$. We recall that $\mathcal{B}_i = B_i$. Hence, FARADAY's law is defined on a material surface co-moving with the continuum body. This equation has to hold for any surface, for example, in the case of a closed surface without boundaries, $\partial\bar{\mathcal{S}} = \{\}$, it still holds

$$\left(\oint_{\bar{\mathcal{S}}} B_i \mathrm{d}a_i\right)^{\boldsymbol{\cdot}} = 0\,. \tag{3.26}$$

This closed surface can be visualized like a closed hull over a body, $\bar{\mathcal{S}} = \partial\mathcal{B}$, such that $\partial\partial\mathcal{B} = \{\}$. By integrating in time we obtain an integration constant to be determined by the known initial condition. We may set the initial magnetic flux as zero. By starting with zero magnetic flux area density, $B_i(t=0, x_j) = 0$, the integration constant vanishes

$$\int_{\partial\mathcal{B}} B_i \mathrm{d}a_i = 0\,. \tag{3.27}$$

Since the surface is closed we can apply GAUSS's law and obtain

$$\frac{\partial B_i}{\partial x_i} = 0\,. \tag{3.28}$$

The latter equation is one of MAXWELL's equations and it holds universally.[24] Now we want to obtain a local form from Eq. (3.25). First we use product rule and STOKES's law on the line integral

$$\int_{\mathcal{S}} B_i^{\boldsymbol{\cdot}} \mathrm{d}a_i + \int_{\mathcal{S}} B_i (\mathrm{d}a_i)^{\boldsymbol{\cdot}} = -\int_{\mathcal{S}} \mathrm{curl}(\boldsymbol{\mathcal{E}})_i \mathrm{d}a_i\,. \tag{3.29}$$

Since the domain is on a material surface, x_i denotes the current position of massive particles such that $x_i^{\boldsymbol{\cdot}} = v_i$ of matter. Thus, we can use the identities in Eqs. (1.120), (1.121) as follows

$$(\mathrm{d}v)^{\boldsymbol{\cdot}} = (J\mathrm{d}V)^{\boldsymbol{\cdot}} = J^{\boldsymbol{\cdot}}\mathrm{d}V = \frac{J^{\boldsymbol{\cdot}}}{J}\mathrm{d}v\,,\quad (\mathrm{d}v)^{\boldsymbol{\cdot}} = \frac{\partial v_i}{\partial x_i}\mathrm{d}v \Rightarrow J^{\boldsymbol{\cdot}} = J\frac{\partial v_i}{\partial x_i}\,, \tag{3.30}$$

and

$$(\mathrm{d}a_i)^{\boldsymbol{\cdot}} = \left(J(\boldsymbol{F}^{-1})_{ji}\mathrm{d}A_j\right)^{\boldsymbol{\cdot}} = \left(J\frac{\partial v_k}{\partial x_k}(\boldsymbol{F}^{-1})_{ji} + J(\boldsymbol{F}^{-1})^{\boldsymbol{\cdot}}_{ji}\right)\mathrm{d}A_j\,. \tag{3.31}$$

[24] A relation holds universally, if it is free of any dependence on the underlying material. In other words, a universal relation holds for all materials and even in the case of no material—vacuum.

Moreover, we know $F_{ij}(\boldsymbol{F}^{-1})_{jk} = \delta_{ik}$ such that the following identity:

$$
\begin{aligned}
\Big(F_{ij}(\boldsymbol{F}^{-1})_{jk}\Big)^{\bullet} &= 0\,,\\
F_{ij}(\boldsymbol{F}^{-1})^{\bullet}_{jk} &= -F^{\bullet}_{ij}(\boldsymbol{F}^{-1})_{jk}\,,\\
(\boldsymbol{F}^{-1})^{\bullet}_{jk} &= -F^{\bullet}_{il}(\boldsymbol{F}^{-1})_{lk}(\boldsymbol{F}^{-1})_{ji}\,,
\end{aligned}
\tag{3.32}
$$

can be used in order to acquire

$$
\begin{aligned}
(\mathrm{d}a_i)^{\bullet} &= J\frac{\partial v_k}{\partial x_k}(\boldsymbol{F}^{-1})_{ji}\mathrm{d}A_j - JF^{\bullet}_{nl}(\boldsymbol{F}^{-1})_{li}(\boldsymbol{F}^{-1})_{jn}\mathrm{d}A_j =\\
&= \frac{\partial v_k}{\partial x_k}\mathrm{d}a_i - F^{\bullet}_{nl}(\boldsymbol{F}^{-1})_{li}\mathrm{d}a_n = \Big(\frac{\partial v_k}{\partial x_k}\delta_{ji} - \frac{\partial v_j}{\partial x_i}\Big)\mathrm{d}a_j\,,
\end{aligned}
\tag{3.33}
$$

since

$$
F^{\bullet}_{nl}(\boldsymbol{F}^{-1})_{li} = \Big(\frac{\partial x_n}{\partial X_l}\Big)^{\bullet}\frac{\partial X_l}{\partial x_i} = \frac{\partial v_n}{\partial X_l}\frac{\partial X_l}{\partial x_i} = \frac{\partial v_n}{\partial x_i}\,.
\tag{3.34}
$$

Therefore, we obtain the following local form for FARADAY's law:

$$
\begin{aligned}
B^{\bullet}_j + B_i\Big(\frac{\partial v_k}{\partial x_k}\delta_{ji} - \frac{\partial v_j}{\partial x_i}\Big) &= -\,\mathrm{curl}(\boldsymbol{\mathfrak{E}})_j\,,\\
\frac{\partial B_j}{\partial t} + \frac{\partial B_j}{\partial x_k}v_k + B_j\frac{\partial v_k}{\partial x_k} - B_i\frac{\partial v_j}{\partial x_i} + \mathrm{curl}(\boldsymbol{\mathfrak{E}})_j &= 0\,,\\
\frac{\partial B_j}{\partial t} + \frac{\partial B_j v_k}{\partial x_k} - B_i\frac{\partial v_j}{\partial x_i} + \mathrm{curl}(\boldsymbol{\mathfrak{E}})_j &= 0\,.
\end{aligned}
\tag{3.35}
$$

Since $\epsilon_{ijk}\epsilon_{klm} = \delta_{il}\delta_{jm} - \delta_{im}\delta_{jl}$, we can insert the following relation:

$$
\begin{aligned}
\mathrm{curl}(\boldsymbol{v}\times\boldsymbol{B})_i = \epsilon_{ijk}\frac{\partial(\boldsymbol{v}\times\boldsymbol{B})_k}{\partial x_j} &= \epsilon_{ijk}\epsilon_{klm}\frac{\partial v_l B_m}{\partial x_j} = \frac{\partial v_i B_j}{\partial x_j} - \frac{\partial v_j B_i}{\partial x_j}\,,\\
\frac{\partial v_j B_i}{\partial x_j} &= \frac{\partial v_i B_j}{\partial x_j} - \mathrm{curl}(\boldsymbol{v}\times\boldsymbol{B})_i\,,
\end{aligned}
\tag{3.36}
$$

into the latter to obtain

$$
\begin{aligned}
\frac{\partial B_j}{\partial t} + \frac{\partial v_j B_i}{\partial x_i} - \mathrm{curl}(\boldsymbol{v}\times\boldsymbol{B})_j - B_i\frac{\partial v_j}{\partial x_i} + \mathrm{curl}(\boldsymbol{\mathfrak{E}})_j &= 0\,,\\
\frac{\partial B_j}{\partial t} + v_j\frac{\partial B_i}{\partial x_i} + \mathrm{curl}(\boldsymbol{\mathfrak{E}} - \boldsymbol{v}\times\boldsymbol{B})_j &= 0\,.
\end{aligned}
\tag{3.37}
$$

By using Eq. (3.11) and inserting one of MAXWELL's equations in Eq. (3.28) we express the latter in the laboratory frame

$$\begin{aligned} &\frac{\partial B_j}{\partial t} + \mathrm{curl}(\boldsymbol{E})_j = 0 \, , \\ &\frac{\partial B_i}{\partial t} + \epsilon_{ijk} \frac{\partial E_k}{\partial x_j} = 0 \, . \end{aligned} \tag{3.38}$$

This equation is another one of MAXWELL's equations holding universally. We have declared E_i and B_i as the primitive variables. Two of MAXWELL's equations, namely Eqs. (3.28) and (3.38), can be solved by using the following trial functions:

$$E_i = -\frac{\partial \phi}{\partial x_i} - \frac{\partial A_i}{\partial t} \, , \quad B_i = \epsilon_{ijk} \frac{\partial A_k}{\partial x_j} \, , \tag{3.39}$$

where we introduce the so-called *electric* and *magnetic potentials*, ϕ and A_i, respectively, as functions in space and time:

$$\phi = \phi(x_i, t) \, , \quad A_i = A_i(x_j, t) \, . \tag{3.40}$$

The electric potentials are the new primitive variables instead of E_i and B_i. It is important to recall that we introduce the electric and magnetic potentials as one possible solution of Eqs. (3.28) and (3.38). We just propose these *ansatz* functions[25] and insert them into Eqs. (3.28) and (3.38) in order to ensure that they satisfy the aforementioned MAXWELL's equations, see Appendix A.6 on p. 305.

There is one drawback in the proposed solution of Eqs. (3.28) and (3.38). Instead of E_i and B_i, i.e., six components in 3D space, we search now for ϕ, A_i, i.e., only fours components in 3D space. Hence we loose information given by two scalar functions. Concretely, we lack information of $\partial\phi/\partial t$ and $\partial A_i/\partial x_i$, which is called the *gauge freedom*.[26] We can choose $\partial\phi/\partial t$ and $\partial A_i/\partial x_i$ arbitrarily and Eqs. (3.39) still satisfy Eqs. (3.28) and (3.38). We have already seen one of the consequences of this free choice of $\partial\phi/\partial t$ in the last section. The electric potential has been set up instantaneously and remained the same. We need to deliver this missing information for an accurate consistent formulation. The most common choice is GAUSS's gauge:[27]

$$\frac{\partial \phi}{\partial t} = 0 \, , \quad \frac{\partial A_i}{\partial x_i} = 0 \, . \tag{3.41}$$

[25] The German word *ansatz* has the equal meaning of a trial function. We simply find out by trial the functions satisfying differential equations.

[26] For the motivation of the gauge freedom see Appendix A.6 on p. 305.

[27] This gauge is named after Carl Friedrich Gauß.

Since the choice is free, the latter is definitely admissible and the simplest choice at all. Another choice is called LORENZ's gauge:[28]

$$\frac{\partial \phi}{\partial t} = -c^2 \frac{\partial A_i}{\partial x_i} , \tag{3.42}$$

which will lead to some useful simplifications in the formulation and will be used herein. For the moment it is hard to see, how this choice shall simplify the formulation.

We need governing equations for solving electric potentials, ϕ, A_i, in space and time. These equations follow from the balance of electric charge:

$$\begin{aligned}
&\frac{\partial \rho z}{\partial t} + \frac{\partial J_i}{\partial x_i} = 0 , \quad J_i = \mathcal{J}_i + \rho z v_i , \\
&z\Big(\frac{\partial \rho}{\partial t} + \rho \frac{\partial v_i}{\partial x_i} + v_i \frac{\partial \rho}{\partial x_i}\Big) + \rho \frac{\partial z}{\partial t} + \rho v_i \frac{\partial z}{\partial x_i} + \frac{\partial \mathcal{J}_i}{\partial x_i} = 0 , \\
&z\Big(\rho^{\bullet} + \rho \frac{\partial v_i}{\partial x_i}\Big) + \rho z^{\bullet} + \frac{\partial \mathcal{J}_i}{\partial x_i} = 0 .
\end{aligned} \tag{3.43}$$

By employing the balance of mass we obtain the balance of electric charge:

$$\rho z^{\bullet} + \frac{\partial \mathcal{J}_i}{\partial x_i} = 0 . \tag{3.44}$$

This local form of a balance equation can be written in a global form for an arbitrary (fixed) domain, Ω, of an open system, which is unbounded to the matter

$$\begin{aligned}
&\int_\Omega \Big(\rho z^{\bullet} + \frac{\partial \mathcal{J}_i}{\partial x_i}\Big) \mathrm{d}v = 0 , \\
&\int_\Omega \rho z^{\bullet} \mathrm{d}v = -\int_{\partial\Omega} \mathcal{J}_i \mathrm{d}a_i , \\
&\Big(\int_\Omega \rho z \mathrm{d}v\Big)^{\bullet} = -\int_{\partial\Omega} \rho z v_i \mathrm{d}a_i - \int_{\partial\Omega} \mathcal{J}_i \mathrm{d}a_i ,
\end{aligned} \tag{3.45}$$

where we have used GAUSS's law and then the balance of mass for an open system as introduced in Eq. (1.261) on p. 86. By taking Eq. (3.5) on p. 169 into account, the balance of electric charge in a fixed domain (control volume) reads

$$\Big(\int_\Omega \rho z \mathrm{d}v\Big)^{\bullet} = -\int_{\partial\Omega} J_i \mathrm{d}a_i . \tag{3.46}$$

In an arbitrarily chosen control volume in space, we can compute the electric charge by using a so-called *charge potential*, D_i, representing the amount of charge escaping from the domain across its boundaries as follows

[28] The gauge is named for Ludvig Valentin Lorenz.

$$\int_{\Omega} \rho z \mathrm{d}v = \int_{\partial\Omega} D_i \mathrm{d}a_i \ . \tag{3.47}$$

Hence D_i describes in a way the displacement of electric charges. By using GAUSS's law we obtain another MAXWELL equation:

$$\rho z = \frac{\partial D_i}{\partial x_i} \ . \tag{3.48}$$

Moreover, we can now rewrite Eq. (3.46) and obtain a balance on an arbitrary surface, S, instead of a volume,

$$\begin{aligned} \Big(\int_{\partial\Omega} D_i \mathrm{d}a_i \Big)^{\bullet} &= - \int_{\partial\Omega} J_i \mathrm{d}a_i \ , \\ \Big(\int_{S} D_i \mathrm{d}a_i \Big)^{\bullet} &= \int_{\partial S} H_i \mathrm{d}\ell_i - \int_{S} J_i \mathrm{d}a_i \ , \end{aligned} \tag{3.49}$$

where H_i is called the *current potential*. It is of importance to clarify that a volume has an enclosed surface or hull. Hence the hull, $\partial\Omega$, has no boundaries. If we exchange the enclosed surface $\partial\Omega$ with an arbitrary surface S, we have to add a term on the boundary of the surface, ∂S, with its line element, $\mathrm{d}\ell_i$.[29] All quantities, D_i, H_i, and J_i, are measured in the laboratory frame. The arbitrary surface, S, may possess a velocity, w_i, then the local form becomes

$$\frac{\partial D_i}{\partial t} + w_i \frac{\partial D_j}{\partial x_j} - \mathrm{curl}(\boldsymbol{w} \times \boldsymbol{D})_i = \mathrm{curl}(\boldsymbol{H})_i - J_i \ , \tag{3.50}$$

by using the aforementioned transformation and identities between Eq. (3.30) and Eq. (3.37). As we want to use a fixed domain, $w_i = 0$, it reads

$$\begin{aligned} \frac{\partial D_i}{\partial t} &= \epsilon_{ijk} \frac{\partial H_k}{\partial x_j} - J_i \ , \\ -\frac{\partial D_i}{\partial t} + \epsilon_{ijk} \frac{\partial H_k}{\partial x_j} &= J_i \ , \end{aligned} \tag{3.51}$$

which is the final one of MAXWELL's equations. From Eqs. (3.48) and (3.51) we can compute the primitive variables, ϕ, A_i, after closing up the governing equations by defining constitutive equations for D_i and H_i. The MAXWELL–LORENTZ aether relations[30] define the necessary constitutive equations in free space[31]

[29] The line element is directed along the positive surface boundary. The positive direction is such that we "walk along" the surface boundary and the surface is on our left-hand side.

[30] They are named after James Clerk Maxwell and Hendrik Antoon Lorentz.

[31] Free space is a technical definition used as a reference for electromagnetic fields, E_i, B_i. It can be visualized as a perfect vacuum without any medium such as massive particles that may transport the electromagnetic fields. Even in this free space the fields do propagate (with the speed of light, c).

$$D_i = \varepsilon_0 E_i \ , \quad H_i = \frac{1}{\mu_0} B_i \ , \tag{3.52}$$

where $\varepsilon_0 = 8.85 \cdot 10^{-12}$ A s/(V m) and $\mu_0 = 12.6 \cdot 10^{-7}$ V s/(A m) are universal constants.[32] Additionally, there is a paramount relation:

$$\varepsilon_0 \mu_0 = \frac{1}{c^2} \ , \tag{3.53}$$

where the speed of light in the free space, c, is also a universal constant. We can rewrite the MAXWELL–LORENTZ aether relations as follows

$$\frac{\partial D_i}{\partial E_j} = \delta_{ij} \varepsilon_0 \ , \quad \frac{\partial H_i}{\partial B_j} = \delta_{ij} \mu_0^{-1} \ . \tag{3.54}$$

They hold in free space. If we want to amend the formulation such that it holds in matter then we separate the electric charges, z, in two parts: free and bound charges.

Basically the atomic structure is such that charged particles within core are bound and outer charged particles—valence electrons—may move between atoms and molecules. Therefore, there are charged particles that move *freely* in body and thus conduct an electric current. The displacement of free charges per mass, $z^{\text{fr.}}$, is given by the free charge potential: $\mathfrak{D}_i$. The atomic position (energy level) of valence electrons determines how much energy is necessary to conduct electric current. At most there are 8 valence electrons: the first 2 are in s-band and the rest 6 are in p-band. The energy levels of s and p bands varies with the occupancy. Monovalent metals such as copper and silver have only one valence electron in the s-band with high energy such that only a small portion of energy succeeds to move them to neighboring atoms. Copper, silver, and gold are the best conductors.[33] In case of aluminum, s-band is full with two electrons and there is 1 electron in p-band. The energy level is lower than in monovalent metals. Therefore, aluminum is a good conductor, however, not as good as the monovalent metals.[34] If the s-band is full without any p-band electrons, then metal is divalent and the energy level is even lower. Hence we have to supply more energy to move the valence electrons. Iron (steel) and titanium have more resistivity than aluminum.[35] Valence electrons are moving freely and enable a *free conduction current* in matter.

There are also charges per mass, $z - z^{\text{fr.}}$, which are *bound*. This quantity is rather difficult to visualize. Consider a massive particle consisting of many molecules. The molecules consisting of atoms possess many positively and negatively charged particles distributed in space. The center of positive charges and the center of negative charges coincide. We call this state *unpolarized*. As a consequence of an electric field, these bound charges shift a bit (less than the atomic radius) and so-called *dipoles*

---

[32] Universal constants hold for every material, even without matter (in free space).

[33] Electronic configurations: Copper (Cu) $3d^{10}4s^1$, Silver (Ag) $4d^{10}5s^1$, Gold (Au) $4f^{14}5d^{10}6s^1$.

[34] Electronic configuration: Aluminum (Al) $3s^2 3p^1$.

[35] Electronic configurations: Iron (Fe) $3d^6 4s^2$, Titanium (Ti) $3d^2 4s^2$.

appear—the material is now *electrically polarized.* This atomic displacement creates a *polarization current.*

By convention the direction of the electric polarization is given from the center of negative charges to the center of positive charges, in the opposite direction of (positive) charge escape, D_i. Suppose the charge density is $q = \rho z$ in C/m^3. If positive and negative charges in an atom moved apart a distance of d_i (pointing from $-q$ to $+q$), then the dipole moment $m_i = q d_i$ in C/m^2 creates a polarized material. For a molecule with N atoms we can sum up all m_i and divide them by the number, N, in order to obtain an average value, $\langle m_i \rangle$. In the continuum scale the electric polarization, $P_i = \langle m_i \rangle N$, is a charge area density in C/m^2 directing toward positive charges. Now the bound (positive) charges diverging from the domain can be given

$$\int_\Omega \rho(z - z^{\text{fr.}}) \mathrm{d}v = -\int_{\partial\Omega} P_i \mathrm{d}a_i \ , \tag{3.55}$$

since positive charges move from positive to negative, i.e., toward the opposite direction of P_i. By using GAUSS's law we obtain

$$\rho(z - z^{\text{fr.}}) = -\frac{\partial P_i}{\partial x_i} \ . \tag{3.56}$$

By inserting the latter into Eq. (3.48) we acquire

$$\begin{aligned} \frac{\partial D_i}{\partial x_i} - \rho z^{\text{fr.}} &= -\frac{\partial P_i}{\partial x_i} \ , \\ \frac{\partial D_i}{\partial x_l} + \frac{\partial P_i}{\partial x_i} &= \rho z^{\text{fr.}} \ , \end{aligned} \tag{3.57}$$

which is equal to

$$\frac{\partial \mathfrak{D}_i}{\partial x_i} = \rho z^{\text{fr.}} \ , \tag{3.58}$$

with the free charge potential, $\mathfrak{D}_i = D_i + P_i$. We can now obtain a total charge potential:

$$D_i = \mathfrak{D}_i - P_i \ , \tag{3.59}$$

where $\mathfrak{D}_i$ denotes a charge potential due to free charges and P_i due to bound charges. The minus sign is because of the convention that the direction of the electric polarization is against the direction of the total charge potential. A moving electric charge creates an electric current. The freely moving electric current is much greater than the displacement current occurring due to the electric polarization. For a conductor, the electric polarization fails to be significant. Practically, an electric polarization occurs in an insulator.

Additional to the electric polarization the material possesses a *magnetic polarization*. Consider again in the atomic scale the dipoles. According to RUTHERFORD–BOHR's atomic model[36] the electrons moving around the nucleus creates a current, j_i, in A/m^2 in the atomic scale. These dipole loops induce a moment, $m_i = \epsilon_{ijk} d_j j_k$ in A/m, due to the atomic current. This current is a *monopole* without positive and negative sides. The average value, $\langle m_i \rangle$, is measured as a magnetic polarization (or simply a magnetization) $M_i = \langle m_i \rangle N$. Magnetization in A/m is a current line density. In the macroscopic scale we comprehend the magnetization, M_i, as a property of bound charges. Unfortunately, if the bound charges creating an electric polarization have a circular motion they create $\boldsymbol{P} \times \boldsymbol{v}$ that we cannot distinguish from the magnetization, M_i, experimentally. Therefore, the sum:

$$\mathcal{M}_i = M_i + \epsilon_{ijk} P_j v_k \ , \tag{3.60}$$

is observed in an experiment and also used in modeling the magnetization.[37] Analogous to polarization we introduce the so-called total (free and bound) current potential:

$$H_i = \mathfrak{H}_i + \mathcal{M}_i \ , \tag{3.61}$$

with a plus sign since this time we have introduced the magnetic polarization in the direction of current, thus, it has the same sign as the current potential. Now by inserting Eqs. (3.59) and (3.61) into Eq. (3.51) we obtain

$$\begin{aligned} -\frac{\partial \mathfrak{D}_i}{\partial t} + \frac{\partial P_i}{\partial t} + \epsilon_{ijk} \frac{\partial \mathfrak{H}_k}{\partial x_j} + \epsilon_{ijk} \frac{\partial \mathcal{M}_k}{\partial x_j} &= J_i \ , \\ -\frac{\partial \mathfrak{D}_i}{\partial t} + \epsilon_{ijk} \frac{\partial \mathfrak{H}_k}{\partial x_j} &= J_i^{\text{fr.}} \ , \end{aligned} \tag{3.62}$$

with

$$J_i^{\text{fr.}} = J_i - \frac{\partial P_i}{\partial t} - \epsilon_{ijk} \frac{\partial \mathcal{M}_k}{\partial x_j} \ . \tag{3.63}$$

We can rewrite the latter for an interpretation of the total current, J_i , as a sum of free current, $J_i^{\text{fr.}}$, and polarization currents

$$J_i = J_i^{\text{fr.}} + \frac{\partial P_i}{\partial t} + \epsilon_{ijk} \frac{\partial \mathcal{M}_k}{\partial x_j} \ . \tag{3.64}$$

[36]This model fails to be correct since if electrons would rotate they would radiate electromagnetic waves. Since experimentally we cannot detect any radiation from atoms this visualization is false. Better models are proposed by using *quantum mechanics*. However, we keep up with *continuum mechanics*; for introducing magnetic polarization we use the nice visualization of RUTHERFORD–BOHR's model named for Ernest Rutherford and Niels Henrik David Bohr.

[37]The magnetization used for the modeling, $\mathcal{M}_i$, is an objective quantity.

All currents are measured in the laboratory frame. We can introduce objective electric current for the total current, J_i, as well as for the free current, $J_i^{\text{fr.}}$, as follows

$$J_i = \mathcal{J}_i + \rho z v_i \ , \quad J_i^{\text{fr.}} = \mathcal{J}_i^{\text{fr.}} + \rho z^{\text{fr.}} v_i \ . \tag{3.65}$$

For polarized materials the objective free current is given by OHM's law:

$$\mathcal{J}_i^{\text{fr.}} = \varsigma \mathcal{E}_i \ , \tag{3.66}$$

we postpone its derivation to Sect. 3.5 on p. 243. We shall define charge and current potentials, $\mathfrak{D}_i$ and $\mathfrak{H}_i$, respectively, as well as electric and magnetic polarizations, P_i and $\mathcal{M}_i$, respectively, in order to close Eqs. (3.58), (3.62)$_2$.

There are two similar methods used in the literature for defining the necessary constitutive equations. The first method is based on defining the charge and current potentials from the electric and magnetic polarizations. For a material with bound charges we need to define material equations for charge and current potentials, $\mathfrak{D}_i$ and $\mathfrak{H}_i$, respectively. Based on the MAXWELL–LORENTZ aether relations in Eqs. (3.54) we can motivate

$$\frac{\partial \mathfrak{D}_i}{\partial E_j} = \varepsilon_{ij}^{\text{el.}} \ , \quad \frac{\partial \mathfrak{H}_i}{\partial B_j} = (\mu_{\text{mag.}}^{-1})_{ij} \ , \tag{3.67}$$

hence we obtain relations for $\mathfrak{D}_i$ and $\mathfrak{H}_i$ by measuring the *permittivity* tensor, $\varepsilon_{ij}^{\text{el.}}$, and the *permeability* tensor, $\mu_{ij}^{\text{mag.}}$. For so-called *simple* materials the charge potential depends only on the electric field and the current potential depends only on the magnetic flux:

$$\mathfrak{D}_i = \varepsilon_{ij}^{\text{el.}} E_j \ , \quad \mathfrak{H}_i = (\mu_{\text{mag.}}^{-1})_{ij} B_j \ , \tag{3.68}$$

where the *dielectric permittivity*, $\varepsilon_{ij}^{\text{el.}}$, and the *magnetic permeability*, $\mu_{ij}^{\text{mag.}}$, consist of constant coefficients[38] (constant in E_i and B_i) for linear materials. For isotropic materials they are reduced to $\varepsilon_{ij}^{\text{el.}} = \varepsilon^{\text{el.}} \delta_{ij}$ and $\mu_{ij}^{\text{mag.}} = \mu^{\text{mag.}} \delta_{ij}$ such that we can now write

$$\begin{aligned} \mathfrak{D}_i &= \varepsilon^{\text{el.}} E_i = \varepsilon_0 \bar{\varepsilon}^{\text{el.}} E_i \ , \\ \mathfrak{H}_i &= \mu_{\text{mag.}}^{-1} B_i = (\mu_0 \bar{\mu}^{\text{mag.}})^{-1} B_i = \frac{1}{\mu_0 \bar{\mu}^{\text{mag.}}} B_i \ , \end{aligned} \tag{3.69}$$

by introducing the relative permittivity $\bar{\varepsilon}^{\text{el.}} = \varepsilon^{\text{el.}} / \varepsilon_0$ and the relative permeability $\bar{\mu}^{\text{mag.}} = \mu^{\text{mag.}} / \mu_0$ without unit. By measuring the permittivity and permeability we have defined the charge and current potentials. We deduce from them the electric and magnetic polarizations:

[38] The permittivity is measured in F(arad)/m $\hat{=}$ C/(V m) $\hat{=}$ A s/(V m) where F is named after Michael Faraday. The permeability is measured in H(enry)/m $\hat{=}$ Wb/(A m) $\hat{=}$ V s/(A m) where H is named for Joseph Henry.

$$
\begin{aligned}
P_i &= \mathfrak{D}_i - D_i = \varepsilon_0 \bar{\varepsilon}^{\text{el.}} E_i - \varepsilon_0 E_i = \varepsilon_0 (\bar{\varepsilon}^{\text{el.}} - 1) E_i = \varepsilon_0 \chi^{\text{el.}} E_i \,, \\
\mathcal{M}_i &= -\mathfrak{H}_i + H_i = -\frac{1}{\mu_0 \bar{\mu}^{\text{mag.}}} B_i + \frac{1}{\mu_0} B_i = \frac{\bar{\mu}^{\text{mag.}} - 1}{\mu_0 \bar{\mu}^{\text{mag.}}} B_i = \frac{\chi^{\text{mag.}}}{\mu_0 \bar{\mu}^{\text{mag.}}} B_i \,,
\end{aligned} \tag{3.70}
$$

where $\chi^{\text{el.}} = \bar{\varepsilon}^{\text{el.}} - 1$ and $\chi^{\text{mag.}} = \bar{\mu}^{\text{mag.}} - 1$ are the electric and magnetic *susceptibilities*, respectively. Often they are found in the literature in the following form:

$$
P_i = \varepsilon_0 \chi^{\text{el.}} E_i = \chi^{\text{el.}} D_i \,, \quad \mathcal{M}_i = \frac{\chi^{\text{mag.}}}{\mu_0 \bar{\mu}^{\text{mag.}}} B_i = \chi^{\text{mag.}} \mathfrak{H}_i \,. \tag{3.71}
$$

We have two different options for describing the constitutive equations for polarized matter. The first option is to measure dielectric permittivity and magnetic permeability such that Eqs. (3.69) define the constitutive equations for the electric and current potentials. The second possibility relies upon measurements of electric and magnetic susceptibilities[39] and using Eqs. (3.70) as constitutive equations for the electric and magnetic polarizations. Both are correct since we have started with their relation as in Eqs. (3.59), (3.61). In both ways we have related them to the primitive variables, ϕ, A_i, since E_i, B_i are given in terms of the electric and magnetic potentials in Eqs. (3.39). By having defined the polarization we have arrived at a constitutive equation for the free current, $J^{\text{fr.}}$, in Eq. (3.63). Hence the governing Eqs. (3.58), $(3.62)_2$ are now closed and can be solved.

Our goal is to compute ϕ and A_i, thus, we need two weak forms. Although we skip a thorough discussion of the balance equations on singular surfaces, we will make much use of them especially in the weak forms of the electromagnetic potentials. A singular surface denotes an area over which a function undergoes a discontinuity. This singularity is simply a jump in the value of the function by crossing the singular surface. Consider two different materials attached together; their interface is a singular surface. A material specific quantity like a free charge potential, $\mathfrak{D}_i$, or free current potential, $\mathfrak{H}_i$, have jumps over the interface, since the permittivities and permeabilities are different for the two adjacent materials. Technically, interface is a singular surface without its own mass density. It is a fictitious surface, not a material surface.[40] Moreover, we neglect any effect of the surface charges on the interface.[41]

[39]There are various methods for measuring the susceptibilities, see for example [26, 37].

[40]The interface is a fictitious surface without mass. If we have a thin layer between two different materials, we may declare it as a singular surface (surface has zero thickness) by neglecting the layers thickness. However, the singular would have then a mass. We consider herein singular surfaces without mass.

[41]In many applications the surface charges have no effect at all. In Sect. 3.5 on p. 243 we will simulate the piezoelectric effect under 100 V and have a small error less than 1 V by neglecting the surface charges, see for a detailed computation of surface charges in piezoelectric ceramics in [21]. For some applications concerning mass diffusion (electromigration) in mixtures, the surface charges may have a significant effect. In this book mixtures are out of scope.

Under these assumptions the balance equations on singular surfaces take the simple form:

$$\boldsymbol{n}\cdot\big[\mathfrak{D}\big]=0\,,\quad \boldsymbol{n}\times\big[\mathfrak{H}\big]=0\,, \tag{3.72}$$

where we have introduced squared brackets indicating a jump. Suppose the interface has material 1 and material 2 at both sides. The value of $\mathfrak{D}_i$ on the interface as a boundary of material 1 is different than the value on the interface as a boundary of material 2. In other words, $\mathfrak{D}_i^+$ indicates the value on the boundary of material 1, i.e., the interface adjacent to material 1. Analogously, $\mathfrak{D}_i^-$ is the value on the interface adjacent to material 2. At the same point on interface, the plane normal of the boundary belonging to material 1, n_i^+, is directed against the plane normal of the boundary of material 2, n_i^-, such that $n_i^+ = -n_i^-$. Then the balance equations read

$$\begin{aligned} n_i\big[\mathfrak{D}_i\big]&=n_i(\mathfrak{D}_i^+-\mathfrak{D}_i^-)=n_i^+\mathfrak{D}_i^+ + n_i^-\mathfrak{D}_i^-=0\,,\\ \epsilon_{ijk}n_j\big[\mathfrak{H}_k\big]&=\epsilon_{ijk}n_j(\mathfrak{H}_k^+-\mathfrak{H}_k^-)=\epsilon_{ijk}(n_j^+\mathfrak{H}_k^+ + n_j^-\mathfrak{H}_k^-)=0\,. \end{aligned} \tag{3.73}$$

In order to obtain a weak form for computing the electric potential, ϕ, we use the balance of electric charge:

$$\frac{\partial \rho z}{\partial t}+J_{i,i}=0\,, \tag{3.74}$$

again by starting to use the comma notation for partial derivatives in space. By inserting one of MAXWELL's equations in Eq. (3.48) and the total current as in Eq. (3.63) into the balance equation, we acquire the field equation for electric potential:

$$\frac{\partial D_{i,i}}{\partial t}+\Big(J_i^{\text{fr.}}+\frac{\partial P_i}{\partial t}+\epsilon_{ijk}\mathcal{M}_{k,j}\Big)_{,i}=0\,. \tag{3.75}$$

Within a domain where P_i is continuous, we can interchange the order of space and time derivative such that the field equation reads

$$\frac{\partial \mathfrak{D}_{i,i}}{\partial t}+\Big(J_i^{\text{fr.}}+\epsilon_{ijk}\mathcal{M}_{k,j}\Big)_{,i}=0\,. \tag{3.76}$$

First we utilize the time discretization. Secondly, by multiplying with the test function, $\delta\phi$, and integrating over the domain where P_i and $\mathfrak{D}_i$ are continuous, we obtain a variational form. In order to have it in the unit of energy we multiply the form with Δt, which is constant in space. Thirdly, by integrating by parts we lower the continuity condition and generate the weak form:

$$\begin{aligned} \mathrm{F}_\phi=&\int_{\Omega^*}\big(-(\mathfrak{D}_i-\mathfrak{D}_i^0)\,\delta\phi_{,i}-\Delta t\,J_i^{\text{fr.}}\,\delta\phi_{,i}-\Delta t\,\epsilon_{ijk}\mathcal{M}_{k,j}\,\delta\phi_{,i}\big)\mathrm{d}v+\\ &+\int_{\partial\Omega^*}n_i(\mathfrak{D}_i-\mathfrak{D}_i^0+\Delta t\,J_i^{\text{fr.}}+\Delta t\,\epsilon_{ijk}\mathcal{M}_{k,j})\,\delta\phi\,\mathrm{d}a\,. \end{aligned} \tag{3.77}$$

Consider a domain, Ω, consisting of two materials, Ω_1 and Ω_2. A polarized material surrounded by air is an adequate example. We simply state that $\Omega = \Omega_1 \cup \Omega_2$ and $\partial\Omega^I = \partial\Omega_1 \cap \partial\Omega_2$, where the interaction boundary between the different materials, $\partial\Omega^I$, is a fictitious, singular surface. The primitive variables are continuous within the whole domain: the electric and magnetic potentials, ϕ, A_i, are continuous in Ω. Hence, the electric field, E_i, as well as the magnetic flux density, B_i, are continuous in Ω. However, this case fails to be true for constitutive equations. For example, $\mathcal{J}_i^{\text{fr.}}$ has a jump on the interface since the electrical conductivities within Ω_1 and Ω_2 differ. Analogously P_i and $\mathfrak{D}_i$ have discontinuities on the interface.

By discretizing in space we solve the integral form in each finite element and sum it up over the elements. If we observe two elements on both sides of the interface, i.e., one element is in Ω_1 and the other one is in Ω_2, then the summation over elements lead to two boundary integrals coming from each element with a plane normal pointing outward the domain Ω_1 or Ω_2. Hence we obtain a jump on the interface and attain the following weak form:

$$\begin{aligned} \mathrm{F}_\phi = \int_\Omega \Big(-(\mathfrak{D}_i - \mathfrak{D}_i^0)\,\delta\phi_{,i} - \Delta t\, J_i^{\text{fr.}}\,\delta\phi_{,i} - \Delta t\,\epsilon_{ijk}\mathcal{M}_{k,j}\,\delta\phi_{,i}\Big)\mathrm{d}v + \\ + \int_{\partial\Omega^I} n_i \big[\mathfrak{D}_i - \mathfrak{D}_i^0 + \Delta t\, J_i^{\text{fr.}} + \Delta t\,\epsilon_{ijk}\mathcal{M}_{k,j}\big]\,\delta\phi\,\mathrm{d}a + \\ + \int_{\partial\Omega} n_i (\mathfrak{D}_i - \mathfrak{D}_i^0 + \Delta t\, J_i^{\text{fr.}} + \Delta t\,\epsilon_{ijk}\mathcal{M}_{k,j})\,\delta\phi\,\mathrm{d}a\,. \end{aligned} \tag{3.78}$$

From the balance equation on singular surfaces we know that $n_i\big[\mathfrak{D}_i\big] = 0$. Therefore, the weak form for computing the electric potential reads

$$\begin{aligned} \mathrm{F}_\phi = \int_\Omega \Big(-(\mathfrak{D}_i - \mathfrak{D}_i^0)\,\delta\phi_{,i} - \Delta t\, J_i^{\text{fr.}}\,\delta\phi_{,i} - \Delta t\,\epsilon_{ijk}\mathcal{M}_{k,j}\,\delta\phi_{,i}\Big)\mathrm{d}v + \\ + \int_{\partial\Omega^I} \Big(n_i \Delta t\big[J_i^{\text{fr.}}\big]\,\delta\phi + n_i \Delta t\,\epsilon_{ijk}\big[\mathcal{M}_{k,j}\big]\,\delta\phi\Big)\mathrm{d}a + \\ + \int_{\partial\Omega} n_i (\mathfrak{D}_i - \mathfrak{D}_i^0 + \Delta t\, J_i^{\text{fr.}} + \Delta t\,\epsilon_{ijk}\mathcal{M}_{k,j})\,\delta\phi\,\mathrm{d}a\,. \end{aligned} \tag{3.79}$$

We will employ DIRICHLET boundary conditions on $\partial\Omega$ such that $\delta\phi|_{\partial\Omega} = 0$ leads to the following weak form:

$$\begin{aligned} \mathrm{F}_\phi = \int_\Omega \Big(-(\mathfrak{D}_i - \mathfrak{D}_i^0)\,\delta\phi_{,i} - \Delta t\, J_i^{\text{fr.}}\,\delta\phi_{,i} - \Delta t\,\epsilon_{ijk}\mathcal{M}_{k,j}\,\delta\phi_{,i}\Big)\mathrm{d}v + \\ + \int_{\partial\Omega^I} \Big(n_i \Delta t\big[J_i^{\text{fr.}}\big]\,\delta\phi + n_i \Delta t\,\epsilon_{ijk}\big[\mathcal{M}_{k,j}\big]\,\delta\phi\Big)\mathrm{d}a\,. \end{aligned} \tag{3.80}$$

For the magnetic potential, A_i, we will use MAXWELL's Eq. (3.51), namely,

$$-\frac{\partial D_i}{\partial t} + \epsilon_{ijk} H_{k,j} = J_i \ , \tag{3.81}$$

after implementing LORENZ's gauge. This choice of the gauge is because of numerical reasons—LORENZ's gauge enables a simplification in the field equation. In order to see this simplification we employ the MAXWELL–LORENTZ aether relations

$$-\frac{\partial}{\partial t}(\varepsilon_0 E_i) + \epsilon_{ijk} \frac{\partial}{\partial x_j}\Big(\frac{1}{\mu_0} B_k\Big) = J_i \ . \tag{3.82}$$

After utilizing Eqs. (3.39) we obtain

$$\varepsilon_0 \frac{\partial}{\partial t}\Big(\frac{\partial \phi}{\partial x_i} + \frac{\partial A_i}{\partial t}\Big) + \frac{1}{\mu_0} \epsilon_{ijk}\epsilon_{kmn} \frac{\partial^2 A_n}{\partial x_j \partial x_m} = J_i \ . \tag{3.83}$$

Since $\epsilon_{ijk} = \epsilon_{kij}$ and additionally with the identity, $\epsilon_{kij}\epsilon_{kmn} = \delta_{im}\delta_{jn} - \delta_{in}\delta_{jm}$, holding in Cartesian coordinates, we acquire the following equation:

$$\begin{aligned} &\varepsilon_0 \frac{\partial^2 \phi}{\partial t \partial x_i} + \varepsilon_0 \frac{\partial^2 A_i}{\partial t^2} + \frac{1}{\mu_0}\Big(\frac{\partial^2 A_j}{\partial x_j \partial x_i} - \frac{\partial^2 A_i}{\partial x_j \partial x_j}\Big) = J_i \ , \\ &\frac{\partial}{\partial x_i}\Big(\varepsilon_0 \frac{\partial \phi}{\partial t} + \frac{1}{\mu_0}\frac{\partial A_j}{\partial x_j}\Big) + \varepsilon_0 \frac{\partial^2 A_i}{\partial t^2} - \frac{1}{\mu_0}\frac{\partial^2 A_i}{\partial x_j \partial x_j} = J_i \ , \end{aligned} \tag{3.84}$$

where SCHWARZ's theorem[42] has been used. The first term vanishes by applying LORENZ's gauge in Eq. (3.42). After inserting the total current from Eq. (3.63), the field equation for magnetic potential reads

$$\varepsilon_0 \frac{\partial^2 A_i}{\partial t^2} - \frac{1}{\mu_0}\frac{\partial^2 A_i}{\partial x_j \partial x_j} = J_i^{\text{fr.}} + \frac{\partial P_i}{\partial t} + \epsilon_{ijk} \mathcal{M}_{k,j} \ . \tag{3.85}$$

As usual, after discretizing in time we generate the weak form by multiplying with the test function and lower the differentiability by integrating by parts and acquire

$$\begin{aligned} \mathrm{F}_A = &\int_\Omega \Big(\varepsilon_0 \frac{A_i - 2A_i^0 + A_i^{00}}{\Delta t \Delta t} \,\delta A_i + \frac{1}{\mu_0} A_{i,j}\,\delta A_{i,j} - J_i^{\text{fr.}}\,\delta A_i - \\ &-\frac{P_i - P_i^0}{\Delta t}\,\delta A_i + \epsilon_{ijk}\mathcal{M}_k\,\delta A_{i,j}\Big) \mathrm{d}v - \int_{\partial\Omega} \Big(\frac{1}{\mu_0} A_{i,j} + \epsilon_{ijk}\mathcal{M}_k\Big)\,\delta A_i n_j \mathrm{d}a \ . \end{aligned} \tag{3.86}$$

Again by summing up over the finite elements we find out (by recalling that μ_0 is a universal constant)

[42]The interchangeability of the order of variables in a differentiation is named after Hermann Amandus Schwarz.

$$
\begin{aligned}
\mathrm{F}_A = \int_\Omega \Big(\varepsilon_0 \frac{A_i - 2A_i^0 + A_i^{00}}{\Delta t \, \Delta t} \, \delta A_i + \frac{1}{\mu_0} A_{i,j} \, \delta A_{i,j} - J_i^{\text{fr.}} \, \delta A_i - \\
- \frac{P_i - P_i^0}{\Delta t} \, \delta A_i + \epsilon_{ijk} \mathcal{M}_k \, \delta A_{i,j} \Big) \mathrm{d}v - \int_{\partial\Omega^I} \Big(\frac{1}{\mu_0} [\![A_{i,j}]\!] + \epsilon_{ijk} [\![\mathcal{M}_k]\!] \Big) \, \delta A_i n_j \mathrm{d}a - \\
- \int_{\partial\Omega} \Big(\frac{1}{\mu_0} A_{i,j} + \epsilon_{ijk} \mathcal{M}_k \Big) \, \delta A_i n_j \mathrm{d}a \, .
\end{aligned}
\tag{3.87}
$$

For the jump condition we use the following balance equation on singular surfaces:

$$
\epsilon_{ijk} n_j [\![\mathfrak{H}_k]\!] = 0 \, . \tag{3.88}
$$

By rewriting the latter we acquire

$$
\begin{aligned}
\epsilon_{ijk} n_j [\![H_k - \mathcal{M}_k]\!] = 0 \, , \\
\epsilon_{ijk} n_j [\![H_k]\!] = \epsilon_{ijk} n_j [\![\mathcal{M}_k]\!] \, , \\
\epsilon_{ijk} n_j \frac{1}{\mu_0} \epsilon_{klm} [\![A_{m,l}]\!] = \epsilon_{ijk} n_j [\![\mathcal{M}_k]\!] \, .
\end{aligned}
\tag{3.89}
$$

Now by using the identity $\epsilon_{ijk}\epsilon_{klm} = \delta_{il}\delta_{jm} - \delta_{im}\delta_{jl}$ we obtain

$$
\begin{aligned}
n_j \frac{1}{\mu_0} [\![A_{j,i} - A_{i,j}]\!] = \epsilon_{ijk} n_j [\![\mathcal{M}_k]\!] \, , \\
n_j \frac{1}{\mu_0} [\![A_{j,i}]\!] = n_j \frac{1}{\mu_0} [\![A_{i,j}]\!] + \epsilon_{ijk} n_j [\![\mathcal{M}_k]\!] \, .
\end{aligned}
\tag{3.90}
$$

Therefore, the weak form for computing the magnetic potential reads

$$
\begin{aligned}
\mathrm{Form}_A = \int_\Omega \Big(\varepsilon_0 \frac{A_i - 2A_i^0 + A_i^{00}}{\Delta t \, \Delta t} \, \delta A_i + \frac{1}{\mu_0} A_{i,j} \, \delta A_{i,j} - J_i^{\text{fr.}} \, \delta A_i - \\
- \frac{P_i - P_i^0}{\Delta t} \, \delta A_i + \epsilon_{ijk} \mathcal{M}_k \, \delta A_{i,j} \Big) \mathrm{d}v - \\
- \int_{\partial\Omega^I} \frac{1}{\mu_0} [\![A_{j,i}]\!] \, \delta A_i n_j \mathrm{d}a - \int_{\partial\Omega} \Big(\frac{1}{\mu_0} A_{i,j} + \epsilon_{ijk} \mathcal{M}_k \Big) \, \delta A_i n_j \mathrm{d}a \, .
\end{aligned}
\tag{3.91}
$$

Since the primitive variables are continuous across the interface, the integral on $\partial\Omega^I$ vanishes. Moreover, we will use DIRICHLET boundary conditions on $\partial\Omega$ such that $\delta A_i|_{\partial\Omega} = 0$ leads to the weak form:

$$
\begin{aligned}
\mathrm{F}_A = \int_\Omega \Big(\varepsilon_0 \frac{A_i - 2A_i^0 + A_i^{00}}{\Delta t \, \Delta t} \, \delta A_i + \frac{1}{\mu_0} A_{i,j} \, \delta A_{i,j} - J_i^{\text{fr.}} \, \delta A_i - \\
- \frac{P_i - P_i^0}{\Delta t} \, \delta A_i + \epsilon_{ijk} \mathcal{M}_k \, \delta A_{i,j} \Big) \mathrm{d}v \, .
\end{aligned}
\tag{3.92}
$$

By using the weak form: Form $= \mathrm{F}_\phi + \mathrm{F}_A$ with F_ϕ as in Eq. (3.80) and F_A as in Eq. (3.92), we compute three engineering examples: capacitor, transformer, skin and proximity effects in a conductor.

3.2.1 Capacitor Simulation

An electric insulator between two metal conductors is a capacitor. Since the insulator does not permit an electric current, any positive or negative charges brought on the conductors are held. In other words, the capacitor stores electric energy. The stored charges on the conductors can be used as a power supply by connecting them in a circuit. This method is used in camera flashes where the capacitor is first charged by the battery and then the flash gets its power from the capacitor. If the current (power consumption) is high the battery starts to sag (shows a high latency and provides a lower energy output due to the resistance). Capacitors are much more accurate and reliable especially for high power consumptions, like a bright light for a short period of time as being the case in a camera flash.

Consider an insulator between two metal plates. As insulator we will use PTFE,[43] metal plates are made of copper. The capacitor is surrounded by air. The geometry consists of three different parts, namely PTFE, copper, and air. For generating the geometry and meshing, i.e., for preprocessing we use Salome[44] and obtain the model in Fig. 3.4. Air is an insulator, $\varsigma_{\text{air}} = 3 \cdot 10^{-15}$, and its susceptibilities are zero. In other words, air can be undertaken as a free space (vacuum) by means of electromagnetic interaction with the following permittivity and permeability:

$$\varepsilon_0 = 8.85 \cdot 10^{-12}\,\text{A s/(V m)}\,, \quad \mu_0 = 12.6 \cdot 10^{-7}\,\text{V s/(A m)}\,. \tag{3.93}$$

PTFE is an insulator and shows a strong electric polarization and a weak magnetic polarization:

$$\varsigma = 10^{-25}\,\text{S/m}\,, \quad \chi^{\text{el.}}_{\text{PTFE}} = 1\,, \quad \chi^{\text{mag.}}_{\text{PTFE}} = 10^{-6} \tag{3.94}$$

Copper (Cu) is a conductor, thus, it shows no electric polarization, $\chi^{\text{el.}}_{\text{Cu}} = 0$, but a weak magnetization:

$$\varsigma = 58.5^6\,\text{S/m}\,, \quad \chi^{\text{mag.}}_{\text{Cu}} = -10^{-5}\,. \tag{3.95}$$

Since $\chi^{\text{mag.}}_{\text{Cu}} < 0$, copper is a diamagnetic material.

---

[43] PTFE stands for PolyTetraFluoroEthylene—its prominent brand-name is Teflon from DuPont in France.

[44] See Appendix A.3 on p. 297 for instructions how to mark the surfaces for applying the boundary conditions and to mark the volumes for different parts.

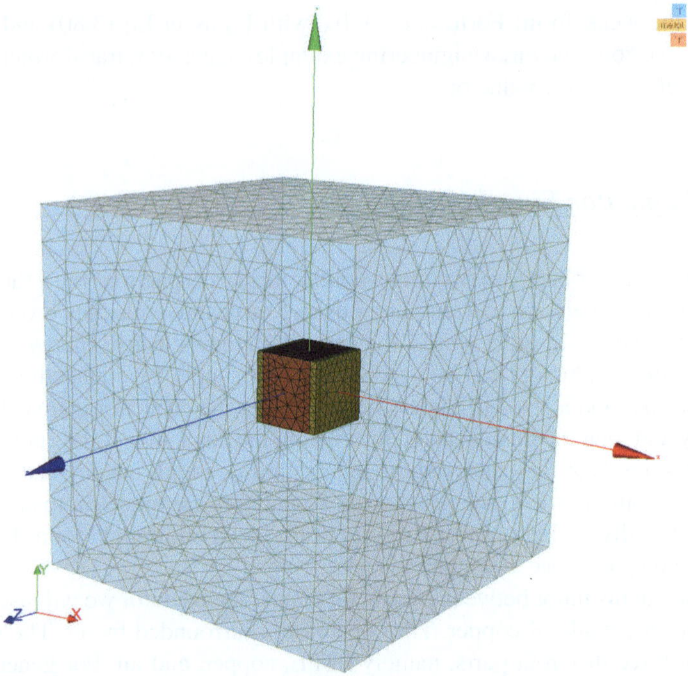

Fig. 3.4 The geometry is a capacitor in air with an insulator out of PTFE with two copper plates on both sides

For computing the primitive variables, viz., electric potential ϕ and magnetic potential A_i, we exploit the weak form as given in Eqs. (3.80), (3.92). At both sides of the capacitor we set the electric potential such that the difference increases (linearly) over time. After 1 s there is 0.2 V difference between the plates. Since PTFE is an insulator no current flows, however, we can measure an electric field due to the electric polarization. In other words, bound charges shift and this displacement of charges known as *dielectric displacement* creates an electric field within the capacitor as well as in the surrounding air. Moreover, this field varies in time leading to a magnetic field. Both effects can be seen in Fig. 3.5.

Electric field, E_i, and magnetic flux (area density), B_i, are orthogonal to each other. We show in Fig. 3.5 the electric field on z-plane and the magnetic flux on y-plane. The magnetic flux is small, however, it exists. Its magnitude depends on the rate of voltage on the plates, we generate 0.2 V difference in 1 s. In other words, the polarization current during charging is very low. Since PTFE is an insulator, there occurs no conduction current and the polarization current gains importance. If the capacitor is fully charged and the circuit is cut off, then the electric field becomes stationary, polarization current and thus the magnetic flux vanish completely.

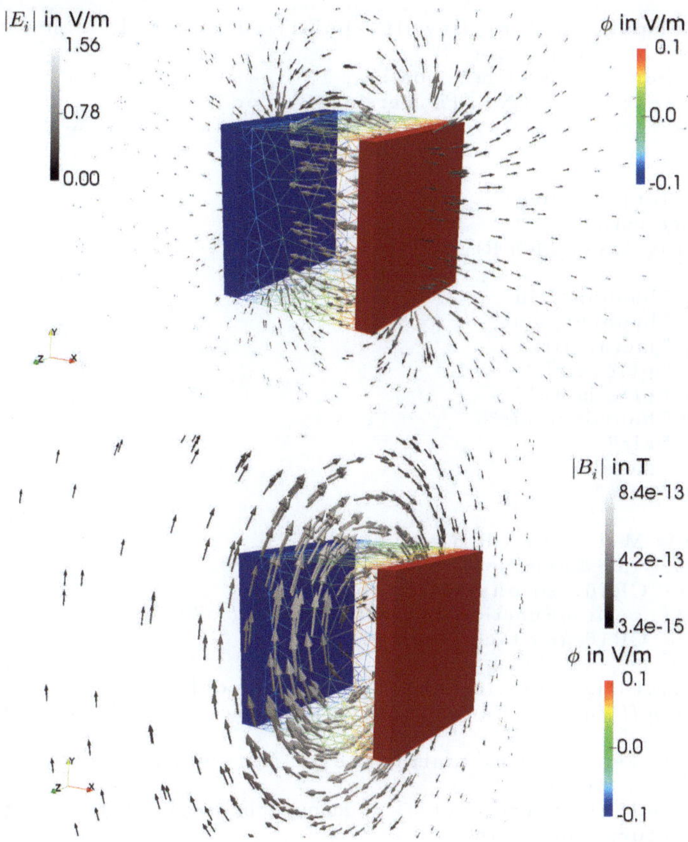

Fig. 3.5 PTFE insulator is visualized as wireframe and the copper plates as surfaces. The capacitor is colored by ϕ. The polarization of the capacitor creates an electric field shown on the *upper figure*. It creates also a magnetic flux during charging presented on the *lower figure*. The E_i and B_i fields are visualized by *arrows* only on cut planes for the sake of a better visualization. The transient solution is presented at 1 s

Far away from the capacitor—on the domain boundaries—electromagnetic fields vanish throughout the simulation, $\phi\big|_{\partial\Omega} = 0$, $A_i\big|_{\partial\Omega} = 0$, which is implemented as DIRICHLET conditions. On interfaces between air and copper, PTFE and copper, air and PTFE; the conditions from the balance equations on singular surfaces are implemented as aforementioned by deducing Eqs. (3.80), (3.92). The geometry for the computation can be found in [1]. Below, the code is given for the capacitor where standard finite element form functions are used and all primitive variables are computed at once in each iteration.

```
"""Computational reality 16, polarized material, simulation of a capacitor"""
__author__ = "B. Emek Abali"
__license__ = "GNU GPL Version 3.0 or later"
#This code underlies the GNU General Public License, http://www.gnu.org/licenses/gpl-3.0.en.html

from fenics import *
import numpy
set_log_level(ERROR)
'''
2D 1 "boundary_in"
2D 2 "boundary_out"
2D 3 "metal_air"
2D 4 "ptfe_air"
2D 5 "ptfe_metal"
2D 6 "boundary_air"
3D 1 "air"
3D 2 "metal"
3D 3 "ptfe"
'''
mesh = Mesh('geo/CR16_geo.xml')
cells = MeshFunction('size_t',mesh,'geo/CR16_geo_physical_region.xml')
facets = MeshFunction('size_t',mesh,'geo/CR16_geo_facet_region.xml')

def material_coefficient(target_mesh, cells_list, coeffs):
    coeff_func = Function(FunctionSpace(target_mesh, 'DG', 0))
    markers = numpy.asarray(cells_list.array(), dtype=numpy.int32)
    coeff_func.vector()[:] = numpy.choose(markers-1, coeffs)
    return coeff_func

n = FacetNormal(mesh)
#interface, area, volume elements
di = Measure('dS', domain=mesh, subdomain_data=facets)
da = Measure('ds', domain=mesh, subdomain_data=facets)
dv = Measure('dx', domain=mesh, subdomain_data=cells)

Scalar = FunctionSpace(mesh, 'P', 1)
Vector = VectorFunctionSpace(mesh, 'P', 1)
Tensor = TensorFunctionSpace(mesh, 'P', 1)
Space = MixedFunctionSpace([Scalar, Vector]) #phi, A

#units: m, kg, s, A, V, K
delta = Identity(3)
levicivita2 = as_matrix([ (0,1,-1) , (-1,0,1) , (1,-1,0) ])
levicivita3 = as_tensor([ ( (0,0,0),(0,0,1),(0,-1,0) ) , ( (0,0,-1),(0,0,0),(1,0,0) ) , ( (0,1,0),(-1,0,0),(0,0,0) ) ])
epsilon = levicivita3

```

```
eps_0 = 8.85E-12 #in A s/(V m)
mu_0 = 12.6E-7 #in V s/(A m)

null=1E-20 #for numerical reasons it is not zero

#air
varsigma_air = 3E-15
chi_el_air = null
chi_ma_air = null
mu_r_ma_air = chi_ma_air + 1.

#metal (copper)
varsigma_cu = 58.5E+6 #in S/m or in 1/(Ohm m)
chi_el_cu = null
chi_ma_cu = -1E-5
mu_r_ma_cu = chi_ma_cu + 1.

#Teflon (ptfe) is an insulator
varsigma_ptfe = 1E-25 #in S/m or in 1/(Ohm m)
chi_el_ptfe = 1.0
chi_ma_ptfe = 1E-6
mu_r_ma_ptfe = chi_ma_ptfe + 1.

chi_el = material_coefficient(mesh, cells, [chi_el_air,
    ↪ chi_el_cu, chi_el_ptfe])
chi_ma = material_coefficient(mesh, cells, [chi_ma_air,
    ↪ chi_ma_cu, chi_ma_ptfe])
mu_r_ma = material_coefficient(mesh, cells, [mu_r_ma_air,
    ↪ mu_r_ma_cu, mu_r_ma_ptfe])
varsigma = material_coefficient(mesh, cells, [varsigma_air,
    ↪ varsigma_cu, varsigma_ptfe])

tMax = 1.0
Dt = 0.1
t = 0.0

capacitor_1 = Expression('0.1*time', time=0)
capacitor_2 = Expression('-0.1*time', time=0)
bc01=DirichletBC(Space.sub(0), capacitor_1, facets, 1)
bc02=DirichletBC(Space.sub(0), capacitor_2, facets, 2)
bc03=DirichletBC(Space.sub(0), Constant(0.), facets, 6)
bc04=DirichletBC(Space.sub(1), Constant((0.,0.,0.)), facets,
    ↪ 6)

bc = [bc01,bc02,bc03,bc04]

dunkn = TrialFunction(Space)
test = TestFunction(Space)
del_phi, del_A = split(test)

unkn = Function(Space)
unkn0 = Function(Space)
unkn00 = Function(Space)
```

```

unkn_init = Expression(('0.0', '0.0','0.0','0.0'))
unkn00 = interpolate(unkn_init,Space)
unkn0.assign(unkn00)
unkn.assign(unkn0)

phi, A = split(unkn)
phi0, A0 = split(unkn0)
phi00, A00 = split(unkn00)

i, j, k, l = indices(4)
delta = Identity(3)
E = as_tensor(-phi.dx(i)-(A-A0)[i]/Dt, (i,))
E0 = as_tensor(-phi0.dx(i)-(A0-A00)[i]/Dt, (i,))
B = as_tensor(epsilon[i,j,k]*A[k].dx(j), (i,))

D = eps_0*E
D0 = eps_0*E0
H = 1./mu_0*B
P = eps_0*chi_el*E
P0 = eps_0*chi_el*E0
mD = D + P
mD0 = D0 + P0
MM =1./mu_0/mu_r_ma*chi_ma*B
J_fr = varsigma*E

F_phi = ( -(mD-mD0)[i]*del_phi.dx(i) - Dt*J_fr[i]*del_phi.dx(
    ↪ i) - Dt*epsilon[i,j,k]*MM[k].dx(j)*del_phi.dx(i) )*(dv
    ↪ (1)+dv(2)+dv(3)) + ( n('+')[i]*Dt*(J_fr('+') - J_fr('-
    ↪ '))[i]*del_phi('+') + n('+')[i]*Dt*epsilon[i,j,k]*(MM(
    ↪ '+')[k].dx(j) - MM('-')[k].dx(j))*del_phi('+') )*(di
    ↪ (3)+di(4)+di(5))

F_A = (eps_0*(A-2.*A0+A00)[i]/Dt/Dt*del_A[i] + 1./mu_0*A[i].
    ↪ dx(j)*del_A[i].dx(j) -J_fr[i]*del_A[i] - (P-P0)[i]/Dt*
    ↪ del_A[i] + epsilon[i,j,k]*MM[k]*del_A[i].dx(j) )*(dv
    ↪ (1)+dv(2)+dv(3))

Form = F_phi + F_A
Gain = derivative(Form, unkn, dunkn)

pwd='/calcul/CR16_capacitor/'
file_phi_metal = File(pwd+'phi_metal.pvd')
file_phi_ptfe = File(pwd+'phi_ptfe.pvd')
file_E = File(pwd+'E.pvd')
file_B = File(pwd+'B.pvd')

mesh_metal = SubMesh(mesh,cells,2)
mesh_ptfe = SubMesh(mesh,cells,3)

VectorSpace_metal = FunctionSpace(mesh_metal, 'P', 1)
VectorSpace_ptfe = FunctionSpace(mesh_ptfe, 'P', 1)
phi_metal = Function(VectorSpace_metal, name='$\phi$')
phi_ptfe = Function(VectorSpace_ptfe, name='$\phi$')
```

```


while t < tMax:
    t += Dt
    print 'time: ',t
    capacitor_1.time = t
    capacitor_2.time = t
    solve(Form==0, unkn, bc, J=Gain, \
        solver_parameters={"newton_solver":{"linear_solver":
            ↪ "mumps", "relative_tolerance": 1e-5} }, \
        form_compiler_parameters={"cpp_optimize": True, "
            ↪ representation": "quadrature", "
            ↪ quadrature_degree": 2} )

    phi_metal.assign(project(unkn.split(deepcopy=True)[0],
        ↪ VectorSpace_metal))
    file_phi_metal << (phi_metal, t)
    phi_ptfe.assign(project(unkn.split(deepcopy=True)[0],
        ↪ VectorSpace_ptfe))
    file_phi_ptfe << (phi_ptfe, t)
    file_B << (project(B,Vector),t)
    file_E << (project(E,Vector),t)

    unkn00.assign(unkn0)
    unkn0.assign(unkn)
```

3.2.2 *Transformer Simulation*

Every electronic device uses electricity. For example in a laptop the motherboard needs 12 V, however, the plug on the wall supplies 110–240 V depending on the country. For decreasing the voltage from the plug to the necessary voltage for the laptop, we need a transformer. The transformer consists of a core and two windings. The primary winding is connected to the input (to the plug on the wall) and the secondary winding is connected to the output (to the laptop). We model a simple transformer with a primary winding of 3 turns and a secondary winding of 2 turns, see Fig. 3.6.

The windings and core are good conductors; however, they are not in contact. Since both windings are connected to different circuits, we have an input and an output voltages. The input voltage due to the alternating current (A.C.) varies harmonically in $\nu = 50\,\text{Hz}$, and due to the coiled geometry of the winding, this induces a magnetic flux along the core. The core is chosen out of a ferromagnetic material with high permeability such that the magnetic flux is increased within the core. It generates a strong magnetic polarization directed along the core. Thus, the magnetic flux created

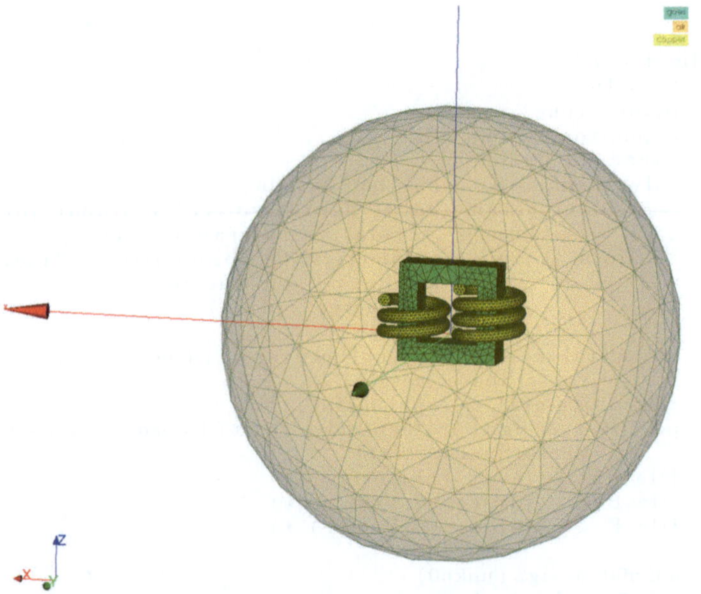

Fig. 3.6 The geometry is a transformer in air. A ferromagnetic electric steel is used as the transformer core and it transports the magnetic flux from the primary to the secondary winding, without contacting the windings

by the primary winding is increased and transported to the secondary winding. This flux induces a change in the electric potential on the secondary winding. As the input current in the primary winding is alternating, an induced A.C. is generated as the output.

If the magnetic core has no losses, then 3 turns input and 2 turns output would decrease the input voltage to 2/3. Of course the core material has some losses. Two different physical phenomena cause losses in the core. The first one is due to the magnetostriction, i.e., a deformation owing to the magnetic polarization. This effect causes a vibration in A.C. We assume rigid bodies in this section such that we ignore this effect in the simulation. The second aspect is mainly a characteristic of the chosen material. Modern transformers use a material called an *electric steel* with negligibly small losses. A small amount of silicon mixed into the steel is named as an electric steel. Moreover, the orientation of grains in the electric steel are directed along the core geometry. This choice reduces the loss further such that we assume that the material shows no hysteresis in A.C. In order to justify this assumption we consider a Grain Oriented Electric Steel (GOES) alloy with the experimental data as seen in

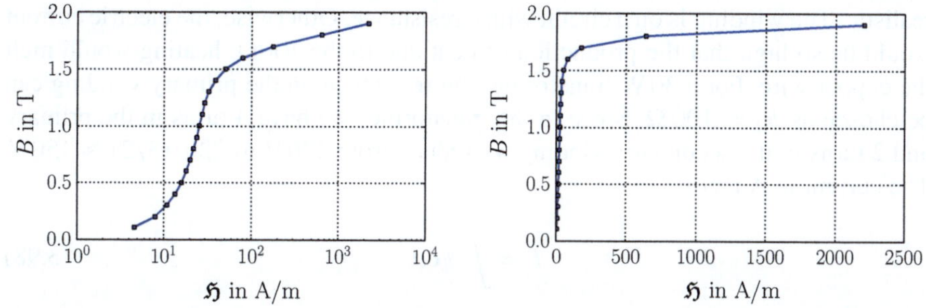

Fig. 3.7 Permeability A.C. measurement of a Grain Oriented Electric Steel (GOES) alloy, in logarithmic base (*left*) and in non-logarithmic base (*right*). The experimental data is taken from Allegheny Technologies Incorporated, www.ATImetals.com

Fig. 3.7. The experimental data is provided in logarithmic base and shows that no hysteresis occurs in A.C. In order to comprehend the data better, we plot it in normal (non-logarithmic) base. Obviously, the relation between B_i and $\mathfrak{H}_i$ is not linear in the whole range, in other words, the permeability is not a constant. However, by restricting to magnetic fluxes lower than 1.5 T we may assume a constant (relative) permeability:

$$\bar{\mu}^{\text{mag.}} = 20\,000 \,. \tag{3.96}$$

Copper and GOES alloy are conductors such that we implement Eqs. (3.80), (3.92) for computing the electric potentials. On the ends of the winding with 3 turns the electric potential is prescribed by DIRICHLET conditions:

$$\phi = A \sin(\nu 2\pi t) \,, \tag{3.97}$$

where $A = \pm 110$ V at both ends and $\nu = 50$ Hz such that we have a 220 V difference alternating with 50 Hz as usual in the home electricity in Europe. The core out of GOES alloy and the copper windings are embedded in air.[45] In reality there is a thin layer on the winding, a coating, suppressing a current in the plane normal direction. Hence, for a precise modeling, the current toward the interface normal is set to zero on the interface between air and winding. We employ Eqs. (3.80), (3.92) for computing the electromagnetic potentials leading to the magnetic flux due to the electric current in the primary winding. The winding made of copper possesses an electric conductivity as high as $\varsigma = 1/r = 58.5 \cdot 10^6$ S/m. However, this is not

[45] In reality, the transformer is housed in a polymer like epoxy, which is an insulator alike air. We just neglect the electric polarization occurring in the polymer housing.

realistic. The winding is on a circuit with a resistance. Otherwise, the electric current would be so high that the production of heat due to the JOULE heating would melt the copper wire. For a 30 W transformer the resistance on the primary winding can be chosen as $R_1 = 100\,\Omega$. Since in the transformer we have 3 turns in the primary and 2 turns in the secondary winding we reduce from 220 V to $220/(3/2) \approx 150$ V. The current in A reads

$$I = \int \mathcal{J}_i \mathrm{d}a_i \ , \tag{3.98}$$

and the voltage is $V_1 = 220$ V and $V_2 = 150$ V in two windings. Since the power is the same in each winding:

$$P = I_1 V_1 = I_2 V_2 \ , \tag{3.99}$$

we can find out the adequate resistance on the secondary winding

$$\begin{aligned} V_1 = I_1 R_1 \ , \quad V_2 = I_2 R_2 \ , \\ \frac{R_1}{R_2} = \frac{V_1 I_2}{V_2 I_1} = \Big(\frac{V_1}{V_2}\Big)^2 = \Big(\frac{3}{2}\Big)^2 \ , \\ R_2 = \frac{R_1}{(3/2)^2} \approx 45\,\Omega \ . \end{aligned} \tag{3.100}$$

As the coils of radius, $r_c = 0.004$ m, has a surface of $a = \pi r_c^2/2$, the conductivity of windings, $\varsigma = 1/r$, read

$$\varsigma_1 = \frac{\ell_1}{R_1 a} \ , \quad \varsigma_2 = \frac{\ell_2}{R_2 a} \ , \tag{3.101}$$

where the length of each winding is $\ell_w = 2\pi r_w$ with the winding radius of $r_w = 0.02$ m. The primary winding is then $\ell_1 = 3\ell_w$ and the secondary winding is of length $\ell_2 = 2\ell_w$. Since the resistance in the second winding is lower, the possible electric current is higher. The electric current in the first winding flows in a helix such that a magnetic flux is induced inside the coil, i.e., in the core in $-z$ direction. This flux creates a magnetic polarization in the core. The polarization is transferred over the core to the second winding. There the magnetic flux is in $+z$ direction and creates an electric current in the second winding in the opposite direction. In a power supply with A.C. the direction of current has no importance. We visualize the electric potentials in the windings, the magnetic polarization within the core, and the magnetic flux in the whole space in Fig. 3.8. The geometry for the computation is in [1]. The code below is used for the transient simulation of electrodynamics in rigid bodies.

```python
"""Computational reality 16, polarized material, simulation of a transformer"""
__author__ = "B. Emek Abali"
__license__ = "GNU GPL Version 3.0 or later"
#This code underlies the GNU General Public License, http://www.gnu.org/licenses/gpl-3.0.en.html

from fenics import *
import numpy
set_log_level(ERROR)
'''
2D 1 "copper_2_out"
2D 2 "copper_1_in"
2D 3 "copper_1_out"
2D 4 "air_goes"
2D 5 "copper1_air"
2D 6 "copper2_air"
2D 7 "air_boundary"
2D 8 "copper_2_in"
3D 1 "copper_1_with_3_turns"
3D 2 "copper_2_with_2_turns"
3D 3 "air"
3D 4 "goes"
'''
mesh = Mesh('geo/CR16_geo_trafo.xml')
cells = MeshFunction('size_t',mesh,'geo/CR16_geo_trafo_physical_region.xml')
facets = MeshFunction('size_t',mesh,'geo/CR16_geo_trafo_facet_region.xml')

def material_coefficient(target_mesh, cells_list, coeffs):
    coeff_func = Function(FunctionSpace(target_mesh, 'DG', 0))
    markers = numpy.asarray(cells_list.array(), dtype=numpy.int32)
    coeff_func.vector()[:] = numpy.choose(markers-1, coeffs)
    return coeff_func

n = FacetNormal(mesh)
#interface, area, volume elements
di = Measure('dS', domain=mesh, subdomain_data=facets)
da = Measure('ds', domain=mesh, subdomain_data=facets)
dv = Measure('dx', domain=mesh, subdomain_data=cells)

Scalar = FunctionSpace(mesh, 'P', 1)
Vector = VectorFunctionSpace(mesh, 'P', 1)
Tensor = TensorFunctionSpace(mesh, 'P', 1)
Space = MixedFunctionSpace([Scalar, Vector]) #phi, A

#units: m, kg, s, A, V, K
delta = Identity(3)
epsilon = as_tensor([ ( (0,0,0),(0,0,1),(0,-1,0) ), ( (0,0,-1),(0,0,0),(1,0,0) ), ( (0,1,0),(-1,0,0)
```

```
,(0,0,0) ) ])

eps_0 = 8.85E-12 #in A s/(V m)
mu_0 = 12.6E-7 #in V s/(A m)

null=1E-20 #for numerical reasons it is not zero

#Grain Oriented Electrical Steel (GOES) is a ferromagnetic material
varsigma_goes = 2.1E+6 #in S/m or in 1/(Ohm m)
chi_el_goes = null
mu_r_ma_goes = 20000. #approximately
chi_ma_goes = mu_r_ma_goes - 1.

#air
varsigma_air = 3E-15
chi_el_air = null
chi_ma_air = null
mu_r_ma_air = chi_ma_air + 1.

#metal (copper)
a = pi*0.004**2/2.
l_w = 2.*pi*0.02
#winding 1 with 3 turns
l_1 = 3.*l_w
R_1 = 100. #in Ohm or 1/S
V_1 = 220. #in V
I_1 = V_1/R_1
varsigma_cu_1 = l_1/(R_1*a) #in S/m
#winding 2 with 2 turns
l_2 = 2.*l_w
R_2 = 45.
varsigma_cu_2 = l_2/(R_2*a) #in S/m or in 1/(Ohm m)

chi_el_cu = null
chi_ma_cu = -1E-5
mu_r_ma_cu = chi_ma_cu + 1.

chi_el = material_coefficient(mesh, cells, [chi_el_cu, chi_el_cu, chi_el_air, chi_el_goes])
chi_ma = material_coefficient(mesh, cells, [chi_ma_cu, chi_ma_cu, chi_ma_air, chi_ma_goes])
mu_r_ma = material_coefficient(mesh, cells, [mu_r_ma_cu, mu_r_ma_cu, mu_r_ma_air, mu_r_ma_goes])
varsigma = material_coefficient(mesh, cells, [varsigma_cu_1, varsigma_cu_2, varsigma_air, varsigma_goes])

tMax = 0.02
Dt = tMax/20.
t = 0.0

bc01=DirichletBC(Space.sub(0), Constant(0.), facets, 7)
bc02=DirichletBC(Space.sub(1), Constant((0.,0.,0.)), facets, 7)
```

```
bc_cu_in = Expression('A*sin(nu*2.0*pi*time)',A=V_1/2.,nu=50.,time=0.)
bc_cu_out = Expression('A*sin(nu*2.0*pi*time)',A=-V_1/2.,nu=50.,time=0.)
bc03=DirichletBC(Space.sub(0), bc_cu_in, facets, 2)
bc04=DirichletBC(Space.sub(0), bc_cu_out, facets, 3)
bc = [bc01,bc02,bc03,bc04]

dunkn = TrialFunction(Space)
test = TestFunction(Space)
del_phi, del_A = split(test)

unkn = Function(Space)
unkn0 = Function(Space)
unkn00 = Function(Space)

unkn_init = Expression(('0.0','0.0','0.0','0.0'))
unkn00 = interpolate(unkn_init,Space)
unkn0.assign(unkn00)
unkn.assign(unkn0)

phi, A = split(unkn)
phi0, A0 = split(unkn0)
phi00, A00 = split(unkn00)

i, j, k, l = indices(4)
delta = Identity(3)
E = as_tensor(-phi.dx(i)-(A-A0)[i]/Dt, (i,))
E0 = as_tensor(-phi0.dx(i)-(A0-A00)[i]/Dt, (i,))
B = as_tensor(epsilon[i,j,k]*A[k].dx(j), (i,))

D = eps_0*E
D0 = eps_0*E0
H = 1./mu_0*B
P = eps_0*chi_el*E
P0 = eps_0*chi_el*E0
mD = D + P
mD0 = D0 + P0
MM = 1./mu_0/mu_r_ma*chi_ma*B
J_fr = varsigma*E

F_phi = ( -(mD-mD0)[i]*del_phi.dx(i) - Dt*J_fr[i]*del_phi.dx(i) - Dt*epsilon[i,j,k]*MM[k].dx(j)*del_phi.dx(i) )*(dv(1)+dv(2)+dv(3)+dv(4)) + n('+')[i]*Dt*epsilon[i,j,k]*(MM('+')[k].dx(j) - MM('-')[k].dx(j))*del_phi('+')*(di(1)+di(4)+di(5)+di(6)+di(8))

F_A = (eps_0*(A-2.*A0+A00)[i]/Dt/Dt*del_A[i] + 1./mu_0*A[i].dx(j)*del_A[i].dx(j) -J_fr[i]*del_A[i] - (P-P0)[i]/Dt*del_A[i] + epsilon[i,j,k]*MM[k]*del_A[i].dx(j))*(dv(1)+dv(2)+dv(3)+dv(4))

Form = F_phi + F_A
Gain = derivative(Form, unkn, dunkn)
```

```

pwd='/calcul/CR16_transformer/'
file_phi = File(pwd+'phi.pvd')
file_M = File(pwd+'M.pvd')
file_B = File(pwd+'B.pvd')

mesh_1 = SubMesh(mesh,cells,1)
mesh_2 = SubMesh(mesh,cells,2)
mesh_3 = SubMesh(mesh,cells,3)
mesh_4 = SubMesh(mesh,cells,4)

phi_copper_1_ = Function(FunctionSpace(mesh_1, 'P', 1), name='$\phi$ in V')
phi_copper_2_ = Function(FunctionSpace(mesh_2, 'P', 1), name='$\phi$ in V')
MM_goes_ = Function(VectorFunctionSpace(mesh_4, 'P', 1), name='$|\mathcal{M}_i|$ in A/m')
B_ = Function(VectorFunctionSpace(mesh, 'P', 1), name='$|B_i|$ in T')


while t < tMax:
    t += Dt
    print 'time: ',t
    bc_cu_in.time = t
    bc_cu_out.time = t
    solve(Form==0, unkn, bc, J=Gain, \
        solver_parameters={"newton_solver":{"linear_solver": "mumps", "relative_tolerance": 1e-5} }, \
        form_compiler_parameters={"cpp_optimize": True, "representation": "quadrature", "quadrature_degree": 2} )

    phi_copper_1_.assign(project(unkn.split(deepcopy=True)[0], FunctionSpace(mesh_1, 'P', 1)))
    file_phi << (phi_copper_1_, t)
    phi_copper_2_.assign(project(unkn.split(deepcopy=True)[0], FunctionSpace(mesh_2, 'P', 1)))
    file_phi << (phi_copper_2_, t)
    MM_goes_.assign(project(MM, VectorFunctionSpace(mesh_4, 'P', 1)))
    file_M << (MM_goes_,t)
    B_.assign(project(B, VectorFunctionSpace(mesh, 'P', 1)))
    file_B << (B_,t)

    unkn00.assign(unkn0)
    unkn0.assign(unkn)
```

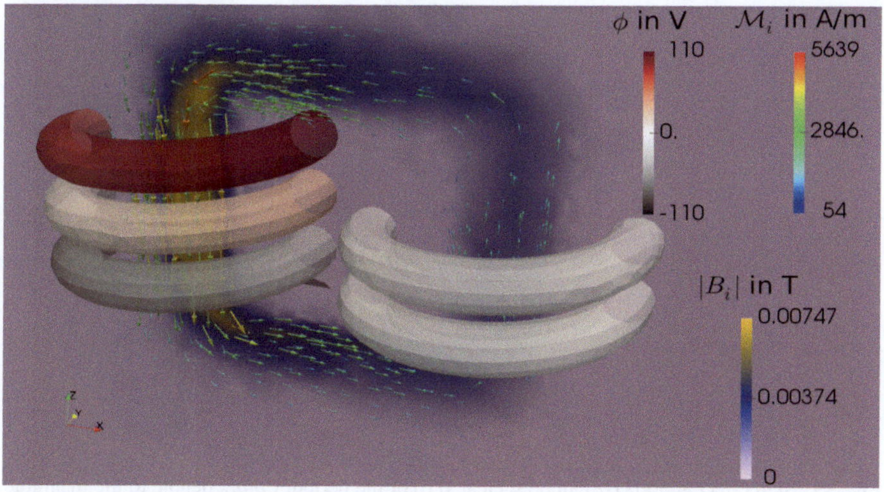

Fig. 3.8 Three fields are visualized at $t = 0.05$ s. The core out of GOES alloy is magnetized, the magnetic polarization is visualized as *arrows colored* by the values of $\mathcal{M}_i$. The primary winding with 3 turns and the secondary winding with 2 turns are colored by the electric potential, ϕ. On a slice in y-plane, the magnitude of the magnetic flux density, B_i, is shown as *colored*. The magnetic flux is immensely increased inside the core, so the leaded magnetic field from the primary to the secondary coil induces an electric potential in the secondary coil. Since the current is alternating on the primary coil, the induced current is alternating, too

3.2.3 Proximity and Skin Effects

In a conductor, for example in a copper wire, the charge carriers are valence electrons. They conduct the charge and this transport is called the free objective electric current, $\mathcal{J}_i^{\text{fr.}}$. It depends on the electromotive intensity, $\mathcal{E}_i$, which is the electric field measured on the co-moving frame. In this section we assume the copper wire as a rigid body; all objective variables, $\mathcal{J}_i^{\text{fr.}}$, $\mathcal{E}_i$, $\mathcal{M}_i$, equal to their corresponding variables measured in the laboratory frame, $J_i^{\text{fr.}} = \mathcal{J}_i^{\text{fr.}}$, $E_i = \mathcal{E}_i$, $M_i = \mathcal{M}_i$.

An alternating electric current induces a magnetic field, which again induces a current in the wire itself. This induced current is swirling within the wire and is called FOUCAULT or *eddy* current.[46] The eddy current is perpendicular to the cross-section of the wire and is directed along the current near the surface and against the current in the core of the wire. The net amount of current is greater on the outer shell than in the core of the wire. Even if we apply a constant electric potential over the cross-section, the current comes out as distributed. The effective conduction current (area density), $J_i^{\text{fr.}}$, is greater near surface than in core. In Fig. 3.9 the so-called *skin effect* is visualized at 500 kHz. Skin effect occurs in an alternating current, A.C., since a current is induced due to the varying charge potential. Impedance is an effective resistance of the wire against A.C., thus, the skin effect increases the impedance of the wire more in the core than on the surface. The deviation of the impedance between surface and core increases with increasing frequency. Especially for digital cables carrying signals in MHz, the skin effect results in an effective current transported

[46]Eddy current was discovered firstly by Jean Bernard Léon Foucault.

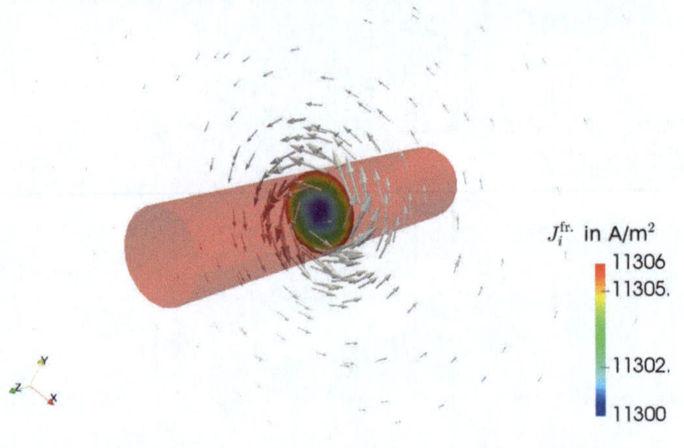

Fig. 3.9 Skin effect at 500 kHz can be seen at 1/10 of the period. *Colors* denote to the magnitude of $J_i^{\text{fr.}}$ and *arrows* denote the direction of the magnetic flux density, B_i

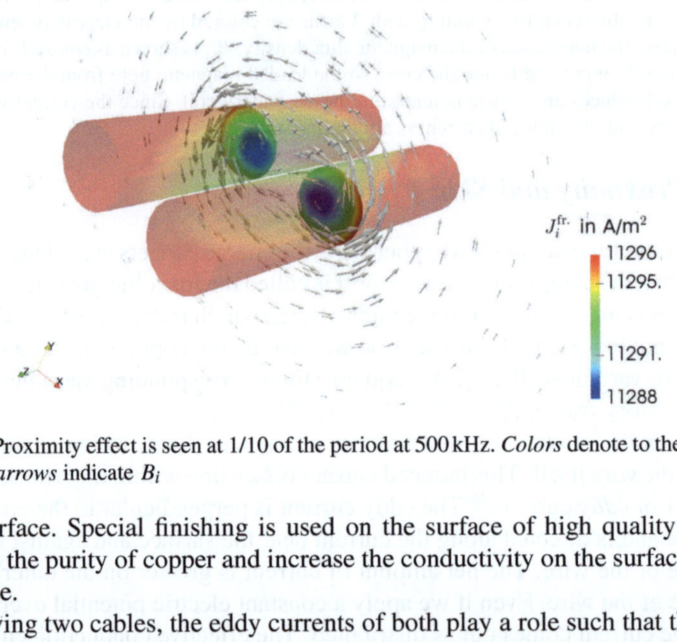

Fig. 3.10 Proximity effect is seen at 1/10 of the period at 500 kHz. *Colors* denote to the magnitude of $J_i^{\text{fr.}}$ and *arrows* indicate B_i

on the surface. Special finishing is used on the surface of high quality cables to maximize the purity of copper and increase the conductivity on the surface as much as possible.

By having two cables, the eddy currents of both play a role such that the current distribution on the cross-section changes, which is referred to as a *proximity effect*. The proximity effect is visualized in Fig. 3.10. Especially for cables in high frequencies there are many different designs reducing the skin and proximity effects. The general idea is to use bundles twisted around each other such that the proximity effect is eliminated by the neighboring cables in every direction. These cables are called *litz wires*.[47] The geometries for the computations can be found in [1] and the code used for computing the skin and proximity effects is given below.

[47] *Der Litzendraht* in German means *stranded wire*.

```
"""Computational reality 16, polarized material, skin and proximity effect"""
__author__ = "B. Emek Abali"
__license__ = "GNU GPL Version 3.0 or later"
#This code underlies the GNU General Public License, http://www.gnu.org/licenses/gpl-3.0.en.html

from fenics import *
import numpy
set_log_level(ERROR)

'''
2D 1 "in"
2D 2 "out"
2D 3 "air_boundary"
2D 4 "air_copper"
3D 1 "air"
3D 2 "copper"
'''
mesh = Mesh('geo/CR16_geo_proximity_two_wires.xml')
cells = MeshFunction('size_t',mesh,'geo/CR16_geo_proximity_two_wires_physical_region.xml')
facets = MeshFunction('size_t',mesh,'geo/CR16_geo_proximity_two_wires_facet_region.xml')

def material_coefficient(target_mesh, cells_list, coeffs):
    coeff_func = Function(FunctionSpace(target_mesh, 'DG', 0))
    markers = numpy.asarray(cells_list.array(), dtype=numpy.int32)
    coeff_func.vector()[:] = numpy.choose(markers-1, coeffs)
    return coeff_func

n = FacetNormal(mesh)
#interface, area, volume elements
di = Measure('dS', domain=mesh, subdomain_data=facets)
da = Measure('ds', domain=mesh, subdomain_data=facets)
dv = Measure('dx', domain=mesh, subdomain_data=cells)

Scalar = FunctionSpace(mesh, 'P', 1)
Vector = VectorFunctionSpace(mesh, 'P', 1)
Tensor = TensorFunctionSpace(mesh, 'P', 1)
Space = MixedFunctionSpace([Scalar, Vector]) #phi, A

#units: m, kg, s, A, V, K
delta = Identity(3)
epsilon = as_tensor([ ( (0,0,0),(0,0,1),(0,-1,0) ), ( (0,0,-1),(0,0,0),(1,0,0) ), ( (0,1,0),(-1,0,0),(0,0,0) ) ])

eps_0 = 8.85E-12 #in A s/(V m)
mu_0 = 12.6E-7 #in V s/(A m)
null=1E-20 #for numerical reasons it is not zero

```

```
#air
varsigma_air = 3E-15
chi_el_air = null
chi_ma_air = null
mu_r_ma_air = chi_ma_air + 1.

#copper
rho_cu = 8960. #in kg / m^3
a = pi*0.01**2/2.
l = 0.1
R = 100.
varsigma_cu = l/(R*a) #in S/m or in 1/(Ohm m)
V = 220. #in V, I = V/R in A
chi_el_cu = null
chi_ma_cu = -1E-5
mu_r_ma_cu = chi_ma_cu + 1.

chi_el = material_coefficient(mesh, cells, [chi_el_air, chi_el_cu])
chi_ma = material_coefficient(mesh, cells, [chi_ma_air, chi_ma_cu])
mu_r_ma = material_coefficient(mesh, cells, [mu_r_ma_air, mu_r_ma_cu])
varsigma = material_coefficient(mesh, cells, [varsigma_air, varsigma_cu])

freq = 500000. #in Hz
tMax = 1./freq
Dt = tMax/20.
t = 0.0

bc01=DirichletBC(Space.sub(0), Constant(0.), facets, 3)
bc02=DirichletBC(Space.sub(1), Constant((0.,0.,0.)), facets, 3)
bc_in = Expression('A*sin(nu*2.0*pi*time)',A=V/2.,nu=freq, time=0.)
bc_out = Expression('A*sin(nu*2.0*pi*time)',A=-V/2.,nu=freq, time=0.)
bc03=DirichletBC(Space.sub(0), bc_in, facets, 1)
bc04=DirichletBC(Space.sub(0), bc_out, facets, 2)
bc = [bc01,bc02,bc03,bc04]

dunkn = TrialFunction(Space)
test = TestFunction(Space)
del_phi, del_A = split(test)

unkn = Function(Space)
unkn0 = Function(Space)
unkn00 = Function(Space)

unkn_init = Expression(('0.0', '0.0','0.0','0.0'))
unkn00 = interpolate(unkn_init,Space)
unkn0.assign(unkn00)
unkn.assign(unkn0)
```

```

phi, A = split(unkn)
phi0, A0 = split(unkn0)
phi00, A00 = split(unkn00)

i, j, k, l = indices(4)
delta = Identity(3)
E = as_tensor(-phi.dx(i)-(A-A0)[i]/Dt, (i,))
E0 = as_tensor(-phi0.dx(i)-(A0-A00)[i]/Dt, (i,))
B = as_tensor(epsilon[i,j,k]*A[k].dx(j), (i,))

D = eps_0*E
D0 = eps_0*E0
H = 1./mu_0*B
P = eps_0*chi_el*E
P0 = eps_0*chi_el*E0
mD = D + P
mD0 = D0 + P0
MM = 1./mu_0/mu_r_ma*chi_ma*B
J_fr = varsigma*E

F_phi = ( -(mD-mD0)[i]*del_phi.dx(i) - Dt*J_fr[i]*del_phi.dx(i) - Dt*epsilon[i,j,k]*MM[k].dx(j)*del_phi.dx(i) )*(dv(1)+dv(2)) + ( n('+')[i]*Dt*(J_fr('+') - J_fr('-'))[i]*del_phi('+') + n('+')[i]*Dt*epsilon[i,j,k]*(MM('+')[k].dx(j) - MM('-')[k].dx(j))*del_phi('+') )*di(4)

F_A = (eps_0*(A-2.*A0+A00)[i]/Dt/Dt*del_A[i] + 1./mu_0*A[i].dx(j)*del_A[i].dx(j) -J_fr[i]*del_A[i] - (P-P0)[i]/Dt*del_A[i] + epsilon[i,j,k]*MM[k]*del_A[i].dx(j))*(dv(1)+dv(2))

Form = F_phi + F_A
Gain = derivative(Form, unkn, dunkn)

pwd='/calcul/CR16_proximity_two_wires/'
file_phi = File(pwd+'phi.pvd')
file_z = File(pwd+'z.pvd')
file_B = File(pwd+'B.pvd')
file_J_fr = File(pwd+'J_fr.pvd')

mesh_1 = SubMesh(mesh, cells, 1)
mesh_2 = SubMesh(mesh, cells, 2)

phi_copper_ = Function(FunctionSpace(mesh_2, 'P', 1), name='$\phi$ in V')
z_ = Function(FunctionSpace(mesh_2, 'P', 1), name='$z$ in C/kg')
B_ = Function(VectorFunctionSpace(mesh, 'P', 1), name='$|B_i|$ in T')
J_fr_ = Function(VectorFunctionSpace(mesh_2, 'P', 1), name='$J_i^\mathrm{fr.}$ in A/m$^2$')

```

```
while t < tMax:
    t += Dt
    print 'time: ',t
    bc_in.time = t
    bc_out.time = t
    tic()
    solve(Form==0, unkn, bc, J=Gain, \
        solver_parameters={"newton_solver":{"linear_solver": "mumps", "relative_tolerance": 1e-5} }, \
        form_compiler_parameters={"cpp_optimize": True, "representation": "quadrature", "quadrature_degree": 2} )

    print 'finished in ',toc(),' seconds'
    phi_copper_.assign(project(unkn.split(deepcopy=True)[0], FunctionSpace(mesh_2, 'P', 1)))
    file_phi << (phi_copper_, t)
    B_.assign(project(B, VectorFunctionSpace(mesh, 'P', 1)))
    file_B << (B_,t)
    z_.assign(project(D[i].dx(i)/rho_cu, FunctionSpace(mesh_2, 'P', 1)))
    file_z << (z_,t)
    J_fr_.assign(project(J_fr, VectorFunctionSpace(mesh_2, 'P', 1)))
    file_J_fr << (J_fr_,t)

    unkn00.assign(unkn0)
    unkn0.assign(unkn)
```

To-do

The electric field, E_i, and the magnetic flux (area density), B_i, exist in material and in free space. They are always orthogonal to each other.

- Implement the code for capacitor and plot on the same cut plane E_i as well as B_i in order to see that they are orthogonal.
- Simulate a transformer with a different core.
- Use the code for a conducting wire and plot $D_{i,i} = \rho z$ in order to test the simplification of incompressible flow of electric charges utilized in the last section.
- In the literature there are formulations attacking MAXWELL's equations in a way to solve directly the fields E_i and B_i without using the electromagnetic potentials, ϕ, A_i. In this configuration the numerical implementation of appropriate elements is quite difficult. There are different proposals. One of them is implementing special elements for E_i and B_i. Search for NEDELEC elements and solutions of electromagnetic problems by using NEDELEC elements in FEniCS.
- Make a web based search for the capacitors. Learn how the capacitive touchscreen of a smartphone works.

3.3 Thermoelectric Coupling

As we have seen in Sect. 3.1 the continuum body heats up due to the heat produced during conducting an electric current. Formally, this production is JOULE's heating and written on the right hand side in the balance of internal energy. Since the temperature changes, the material shrinks or expands. In order to incorporate this effect into the computational reality, we have to use the balance of linear momentum with the electromagnetic interactions (with matter). We will motivate the balance equation and then derive the constitutive equations in a thermodynamically consistent way. In this section we employ the formulation for an *unpolarized* material, $z - z^{\text{fr.}} = 0$. Thus, electric and magnetic polarizations vanish

$$P_i = 0 \,, \quad \mathcal{M}_i = 0 \,. \tag{3.102}$$

Total energy consists of the energy due to the *matter* and *field*. Matter denotes particles with mass and field means the electromagnetic fields due to particles with an electric charge. Of course materials like copper include molecules with mass and charge. However, mass and charge are treated separately, they are both assumed to exist independently. The thermodynamical formulation starts with the assertion that the total energy is conserved. In other words, a balance of total energy lacks a production term

$$\frac{\partial \rho e}{\partial t} - \frac{\partial}{\partial x_j}\big(- v_j \rho e + F_j \big) - \rho s = 0 \,, \tag{3.103}$$

where the specific total energy, e, its flux term, F_i, and its specific supply term, s, shall be defined. We postpone their derivation and proceed with the balance of momentum with the electromagnetic interactions:

$$\frac{\partial \rho v_i}{\partial t} - \frac{\partial}{\partial x_j}\big(- v_j \rho v_i + \sigma_{ji} \big) - \rho f_i = \mathcal{F}_i \,, \tag{3.104}$$

where the additional force density, $\mathcal{F}_i$, is caused by the electromagnetic fields. A moving particle "feels" this additional force density—it is the LORENTZ force density for unpolarized systems:

$$\mathcal{F}_i = \rho z E_i + \epsilon_{ijk} J_j B_k \,, \tag{3.105}$$

as given in Eq. (3.12) with the help of the specific electric charge, z, the mass density, ρ, and the electric current, J_i. In the case of electromagnetic interactions between matter and field, the momentum is not a conserved quantity and the LORENTZ force acts as a production. Now by using the balance of mass:

$$\frac{\partial \rho}{\partial t} + \frac{\partial \rho v_j}{\partial x_j} = 0 \,, \tag{3.106}$$

and the so-called total time rate:[48]

$$\frac{\mathrm{d}}{\mathrm{d}t} = \frac{\partial}{\partial t} + v_j \frac{\partial}{\partial x_j} , \tag{3.107}$$

we obtain from the balance of momentum

$$\rho \frac{\mathrm{d}v_i}{\mathrm{d}t} - \frac{\partial \sigma_{ji}}{\partial x_j} - \rho f_i = \mathcal{F}_i . \tag{3.108}$$

By multiplying the latter with the velocity, we acquire the balance of kinetic energy:

$$\begin{aligned} \rho \frac{\mathrm{d}v_i}{\mathrm{d}t} v_i - \frac{\partial \sigma_{ji}}{\partial x_j} v_i - \rho f_i v_i = \mathcal{F}_i v_i , \\ \rho \frac{\mathrm{d}}{\mathrm{d}t}\Big(\frac{1}{2} v_i v_i\Big) - \frac{\partial \sigma_{ji} v_i}{\partial x_j} - \rho f_i v_i = -\sigma_{ji} \frac{\partial v_i}{\partial x_j} + \mathcal{F}_i v_i . \end{aligned} \tag{3.109}$$

The balance of mass is used once more in order to bring the balance of kinetic energy in the following form:

$$\frac{\partial}{\partial t}\Big(\rho \frac{1}{2} v_i v_i\Big) - \frac{\partial}{\partial x_j}\Big(- \rho v_j \frac{1}{2} v_i v_i + \sigma_{ji} v_i \Big) - \rho f_i v_i = -\sigma_{ji} \frac{\partial v_i}{\partial x_j} + \mathcal{F}_i v_i , \tag{3.110}$$

where the right-hand side is the production term. Now by using

$$J_i = \mathcal{J}_i + \rho z v_i , \quad E_i = \mathcal{E}_i - \epsilon_{ijk} v_j B_k , \tag{3.111}$$

we can rewrite the final term in the production of kinetic energy,

$$\begin{aligned} \mathcal{F}_i v_i = \big(\rho z E_i + \epsilon_{ijk} J_j B_k\big) v_i = \\ = (J_i - \mathcal{J}_i) E_i + \epsilon_{ijk} v_i (\mathcal{J}_j + \rho z v_j) B_k = \\ = J_i E_i - \mathcal{J}_j (E_j - \epsilon_{ijk} v_i B_k) = J_i E_i - \mathcal{J}_j (E_j + \epsilon_{jik} v_i B_k) = \\ = J_i E_i - \mathcal{J}_j \mathcal{E}_j . \end{aligned} \tag{3.112}$$

By employing MAXWELL's Eq. $(3.51)_2$ and $\epsilon_{ijk} = -\epsilon_{ikj}$ we rewrite the latter,

[48]For a scalar and as a special case for the velocity the total time rate, $\frac{\mathrm{d}(\cdot)}{\mathrm{d}t}$, is equal to the objective time rate, $(\cdot)^{\bullet}$, for a fixed coordinate system, $w_i = 0$.

$$
\begin{aligned}
\mathcal{F}_i v_i &= \Big(- \frac{\partial D_i}{\partial t} + \epsilon_{ijk} \frac{\partial H_k}{\partial x_j} \Big) E_i - \mathcal{J}_i \mathcal{E}_i = \\
&= - \frac{\partial D_i}{\partial t} E_i + \epsilon_{ijk} \frac{\partial H_k E_i}{\partial x_j} - \epsilon_{ijk} H_k \frac{\partial E_i}{\partial x_j} - \mathcal{J}_i \mathcal{E}_i = \\
&= - \frac{\partial D_i}{\partial t} E_i - \epsilon_{ikj} \frac{\partial H_k E_i}{\partial x_j} + H_k \epsilon_{kji} \frac{\partial E_i}{\partial x_j} - \mathcal{J}_i \mathcal{E}_i \, .
\end{aligned} \tag{3.113}
$$

After inserting MAXWELL's Eq. (3.38) and employing the MAXWELL–LORENTZ aether relations in Eqs. (3.52) we acquire

$$
\begin{aligned}
\mathcal{F}_i v_i &= - \frac{\partial D_i}{\partial t} E_i - \frac{\partial (\boldsymbol{E} \times \boldsymbol{H})_j}{\partial x_j} - H_k \frac{\partial B_k}{\partial t} - \mathcal{J}_i \mathcal{E}_i = \\
&= - \frac{1}{2} \frac{\partial \varepsilon_0 E_i E_i}{\partial t} - \frac{\partial (\boldsymbol{E} \times \boldsymbol{H})_j}{\partial x_j} - \frac{1}{\mu_0} B_k \frac{\partial B_k}{\partial t} - \mathcal{J}_i \mathcal{E}_i = \\
&= - \frac{1}{2} \frac{\partial}{\partial t} \Big(\varepsilon_0 E_i E_i + \frac{1}{\mu_0} B_k B_k \Big) - \frac{\partial (\boldsymbol{E} \times \boldsymbol{H})_j}{\partial x_j} - \mathcal{J}_i \mathcal{E}_i = \\
&= - \frac{\partial}{\partial t} \Big(\frac{1}{2} (D_i E_i + H_i B_i) \Big) - \frac{\partial (\boldsymbol{E} \times \boldsymbol{H})_j}{\partial x_j} - \mathcal{J}_i \mathcal{E}_i \, .
\end{aligned} \tag{3.114}
$$

The latter is inserted into the balance of kinetic energy in Eq. (3.110) and we obtain

$$
\begin{aligned}
\frac{\partial}{\partial t} \Big(\rho \frac{1}{2} v_i v_i + \frac{1}{2} (D_i E_i + H_i B_i) \Big) - \frac{\partial}{\partial x_j} \Big(- \rho v_j \frac{1}{2} v_i v_i - (\boldsymbol{E} \times \boldsymbol{H})_j + \sigma_{ji} v_i \Big) - \\
- \rho f_i v_i = - \sigma_{ji} \frac{\partial v_i}{\partial x_j} - \mathcal{J}_i \mathcal{E}_i \, .
\end{aligned} \tag{3.115}
$$

The kinetic energy density has two components, $\rho e^{\text{kin.}} = \rho e^{\text{m.}} + e^{\text{f.}}$, one due to matter and one due to field:

$$
\rho e^{\text{m.}} = \rho \frac{1}{2} v_i v_i \, , \quad e^{\text{f.}} = \frac{1}{2} (D_i E_i + H_i B_i) \, . \tag{3.116}
$$

For simplicity we rewrite the balance of kinetic energy as follows

$$
\begin{aligned}
\frac{\partial}{\partial t} \Big(\rho e^{\text{kin.}} \Big) - \frac{\partial}{\partial x_j} \Big(- v_j (\rho e^{\text{kin.}} - e^{\text{f.}}) - (\boldsymbol{E} \times \boldsymbol{H})_j + \sigma_{ji} v_i \Big) - \rho f_i v_i = \\
= - \sigma_{ji} \frac{\partial v_i}{\partial x_j} - \mathcal{J}_i \mathcal{E}_i \, .
\end{aligned} \tag{3.117}
$$

Since $e^{\text{f.}}$ exists even in a vacuum (without massive particles) we refrain from introducing a specific energy (energy per mass). The total energy is composed of the kinetic energy (of matter and field) and of internal energy:

$$e = e^{\text{kin.}} + u \ . \tag{3.118}$$

Hence, we can subtract from the balance of total energy in Eq. (3.103) the balance of kinetic energy in Eq. (3.117),

$$\frac{\partial}{\partial t}\Big(\rho e - \rho e^{\text{kin.}}\Big) - \frac{\partial}{\partial x_j}\Big(- v_j(\rho e - \rho e^{\text{kin.}} + e^{\text{f.}}) + F_j + (\boldsymbol{E}\times\boldsymbol{H})_j - \sigma_{ji} v_i\Big) - \\ -\rho(s - f_i v_i) = \sigma_{ji}\frac{\partial v_i}{\partial x_j} + \mathcal{J}_i\mathcal{E}_i \ , \tag{3.119}$$

and obtain the balance of internal energy:

$$\frac{\partial \rho u}{\partial t} - \frac{\partial}{\partial x_j}\big(- v_j u - q_j\big) - \rho r = \Gamma \ , \tag{3.120}$$

with the so-called heat flux, q_i, supply term, r, and production term, Γ,

$$\begin{aligned} -q_j &= -v_j e^{\text{f.}} + F_j + (\boldsymbol{E}\times\boldsymbol{H})_j - \sigma_{ji} v_i \ , \\ r &= s - f_i v_i \ , \quad \Gamma = \sigma_{ji}\frac{\partial v_i}{\partial x_j} + \mathcal{J}_i\mathcal{E}_i \ . \end{aligned} \tag{3.121}$$

Especially the heat flux can be chosen differently than herein. Flux of field, $(\boldsymbol{E}\times\boldsymbol{H})_j$, is the radiation transporting heat. A typical example is the heat of the Sun reaching the Earth through the free space. Since the radiation is a heat flux the above definition is possible. However, we could leave out the radiation from the heat flux and continue with a balance of internal energy where its flux is $-q_j + (\boldsymbol{E}\times\boldsymbol{H})_j$. The difference is how we measure the heat flux. If the measurement is done by including the radiation term then the definition used herein is appropriate. After using the balance of mass we obtain

$$\rho\frac{\mathrm{d}u}{\mathrm{d}t} + \frac{\partial q_j}{\partial x_j} - \rho r = \Gamma \ , \tag{3.122}$$

as we can use the objective and total rates interchangeably for a scalar quantity,

$$\rho u^{\bullet} + \frac{\partial q_i}{\partial x_i} - \rho r = \Gamma = \sigma_{ji}\frac{\partial v_i}{\partial x_j} + \mathcal{J}_i\mathcal{E}_i \ . \tag{3.123}$$

The primitive variables are $\{\phi, A_i, u_i, T\}$. For the electromagnetic potentials, ϕ, A_i, we will use the same equations as in the last section, after utilizing the LORENZ gauge we obtain

$$\frac{\partial \rho z}{\partial t} + \frac{\partial J_i}{\partial x_i} = 0 \ , \quad \varepsilon_0\frac{\partial^2 A_i}{\partial t^2} - \frac{1}{\mu_0}\frac{\partial^2 A_i}{\partial x_j \partial x_j} = J_i \ , \tag{3.124}$$

where the total current is

$$J_i = J_i^{\text{fr.}} + \frac{\partial P_i}{\partial t} + \epsilon_{ijk}\frac{\partial \mathcal{M}_k}{\partial x_j} \ , \quad J_i^{\text{fr.}} = \mathcal{J}_i^{\text{fr.}} + \rho z^{\text{fr.}} v_i \ . \tag{3.125}$$

For the displacement, u_i, we will employ the balance of linear momentum:

$$\rho v_i^{\bullet} - \frac{\partial \sigma_{ji}}{\partial x_j} - \rho f_i = \mathcal{F}_i \ . \tag{3.126}$$

For the temperature, T, we may utilize the balance of internal energy:

$$\rho u^{\bullet} + \frac{\partial q_i}{\partial x_i} - \rho r = \sigma_{ji}\frac{\partial v_i}{\partial x_j} + \mathcal{J}_i \mathcal{E}_i \ . \tag{3.127}$$

In order to close these equations we need to determine the constitutive equations, $\mathcal{J}_i$, σ_{ij}, q_i. We start off with the balance of internal energy at the equilibrium state. We decompose the stress tensor into reversible and dissipative terms:

$$\sigma_{ij} = {}^{\text{r}}\sigma_{ij} + {}^{\text{d}}\sigma_{ij} \ , \quad {}^{\text{r}}\sigma_{ij} = -p\delta_{ij} + {}^{\text{e}}\sigma_{ij} \ , \tag{3.128}$$

where ${}^{\text{e}}\sigma_{ij}$ denotes the elastic stress. We ignore the pressure, p, since its effect is negligibly small for solid bodies. The stress tensor is symmetric for unpolarized materials

$$\sigma_{ji}\frac{\partial v_i}{\partial x_j} = \sigma_{ij} d_{ij} = \sigma_{ij}\varepsilon_{ij}^{\bullet} \ , \tag{3.129}$$

where we have used that the rate of strains is equal to the symmetric part of the velocity gradient, for the sake of brevity we prove this identity in Appendix A.4 on p. 301. We readily restricted the implementation to linearized strains, ε_{ij}, in other words, we assume that the displacements are so small that the deformation gradient is equal to the identity. In order to obtain the equilibrium state and the constitutive equations, we use the method introduced in Sect. 2.3 on p. 126. For the mechanical equilibrium we utilize the decomposition of stress and identify the dissipative term with the irreversible process such that it has to vanish at equilibrium, ${}^{\text{d}}\sigma_{ij} = 0$. For the thermal equilibrium we introduce the entropy rate density, $\rho\eta^{\bullet}$, as the minus divergence of heat flux per temperature. Moreover, $r = 0$ for the thermal equilibrium. For the electromagnetic equilibrium $\mathcal{J}_i = 0$ holds such that the production term vanishes. The balance of internal energy at equilibrium reads

$$\rho u^{\bullet} - \rho T \eta^{\bullet} = {}^{\text{e}}\sigma_{ij}\varepsilon_{ij}^{\bullet} \ . \tag{3.130}$$

By using the 1st law of thermodynamics we exchange the rates with differential forms and obtain GIBBS's equation:

$$
\begin{aligned}
\rho \, \mathrm{d}u &= \rho T \mathrm{d}\eta + {}^{\mathrm{e}}\sigma_{ij} \mathrm{d}\varepsilon_{ij} \,, \\
\mathrm{d}u &= T \mathrm{d}\eta + {}^{\mathrm{e}}\sigma_{ij} v \mathrm{d}\varepsilon_{ij} \,.
\end{aligned}
\tag{3.131}
$$

with the specific volume, $v = 1/\rho$. Obviously, the internal energy depends on the entropy and strain, $u = u(\eta, \varepsilon_{ij})$. In order to acquire a dependence on temperature instead on entropy, we introduce the specific free energy:

$$
\begin{aligned}
\psi = u - T\eta \,, \quad \mathrm{d}\psi &= \mathrm{d}u - \eta \, \mathrm{d}T - T \mathrm{d}\eta \,, \\
\mathrm{d}\psi &= -\eta \, \mathrm{d}T + {}^{\mathrm{e}}\sigma_{ij} v \mathrm{d}\varepsilon_{ij} \,.
\end{aligned}
\tag{3.132}
$$

The specific free energy depends on the temperature and strain, $\psi = \psi(T, \varepsilon_{ij})$, as follows

$$
\eta = -\frac{\partial \psi}{\partial T} \,, \quad {}^{\mathrm{e}}\sigma_{ij} v = \frac{\partial \psi}{\partial \varepsilon_{ij}} \,.
\tag{3.133}
$$

The dual variables, η and ${}^{\mathrm{e}}\sigma_{ij}$, depend on the same set of state variables such that we obtain

$$
\begin{aligned}
\mathrm{d}\eta &= A \mathrm{d}T + \bar{p}_{ij} \mathrm{d}\varepsilon_{ij} \,, \\
\mathrm{d}\,{}^{\mathrm{e}}\sigma_{ij} &= p_{ij} \mathrm{d}T + C_{ijkl} \mathrm{d}\varepsilon_{kl} \,,
\end{aligned}
\tag{3.134}
$$

where the coefficients depend on the state space, $A = A(T, \varepsilon_{ij})$, $\bar{p}_{ij} = \bar{p}_{ij}(T, \varepsilon_{ij})$, $p_{ij} = p_{ij}(T, \varepsilon_{ij})$, $C_{ijkl} = C_{ijkl}(T, \varepsilon_{ij})$. Instead of measuring entropy, the heat flux is measured, $\delta Q = T \mathrm{d}\eta = c \mathrm{d}T$, by a constant strain, $\mathrm{d}\varepsilon_{ij} = 0$. The specific heat capacity, $c = TA$, is determined by varying the temperature and measuring the heat flux. By holding the temperature constant, $\mathrm{d}T = 0$, at a specific temperature, the stiffness tensor, C_{ijkl}, is determined by varying strain and measuring stress. We employ the MAXWELL symmetry (reciprocal) relation:

$$
\bar{p}_{ij} = \frac{\partial \eta}{\partial \varepsilon_{ij}} = -\frac{\partial^2 \psi}{\partial \varepsilon_{ij} \partial T} = -\frac{\partial^2 \psi}{\partial T \partial \varepsilon_{ij}} = -v \frac{\partial\, {}^{\mathrm{e}}\sigma_{ij}}{\partial T} = -v p_{ij} \,,
\tag{3.135}
$$

since the mass density and thus the specific volume depends on space and time but not on temperature. We obtain

$$
\begin{aligned}
\mathrm{d}\eta &= \frac{c}{T} \mathrm{d}T - p_{ij} v \mathrm{d}\varepsilon_{ij} \,, \\
\mathrm{d}\,{}^{\mathrm{e}}\sigma_{ij} &= p_{ij} \mathrm{d}T + C_{ijkl} \mathrm{d}\varepsilon_{kl} \,.
\end{aligned}
\tag{3.136}
$$

The thermal pressure, p_{ij}, can be rewritten by introducing an experiment where temperature is varied and strain is measured, $\mathrm{d}\varepsilon_{ij} = \alpha_{ij} \mathrm{d}T$, in order to determine the thermal expansion coefficient, α_{ij}. Since this experiment is realized by holding stress constant, we have from Eq. (3.136)$_2$

$$
\begin{aligned}
0 = p_{ij} \mathrm{d}T + C_{ijkl} \mathrm{d}\varepsilon_{kl} &= p_{ij} \mathrm{d}T + C_{ijkl} \alpha_{kl} \mathrm{d}T \\
\Rightarrow p_{ij} &= -C_{ijkl} \alpha_{kl} \,.
\end{aligned}
\tag{3.137}
$$

We want to implement a linear material, i.e., all parameters, c, C_{ijkl}, and α_{ij} are constants, in other words, they do not depend on the state variables. In this case we simply integrate from a reference state, $T_{\text{ref.}}$, $\varepsilon_{ij} = 0$, without stress and entropy[49] to the actual state, T, ε_{ij}, and acquire the following constitutive equations:

$$\eta = c\big(\ln(T) - \ln(T_{\text{ref.}})\big) + C_{ijkl}\alpha_{kl}v\varepsilon_{ij} = c\ln\left(\frac{T}{T_{\text{ref.}}}\right) + C_{ijkl}\alpha_{kl}v\varepsilon_{ij} \,, \\ {}^{\text{e}}\sigma_{ij} = -C_{ijkl}\alpha_{kl}(T - T_{\text{ref.}}) + C_{ijkl}\varepsilon_{kl} = C_{ijkl}\big(\varepsilon_{kl} - \alpha_{kl}(T - T_{\text{ref.}})\big) \,. \tag{3.138}$$

Now the rate of internal energy density is determined completely

$$\rho u^{\bullet} = \rho T \eta^{\bullet} + {}^{\text{e}}\sigma_{ij}\varepsilon_{ij}^{\bullet} \,, \tag{3.139}$$

so we can insert the latter into the balance of internal energy in Eq. (3.127) and obtain

$$\rho T \eta^{\bullet} + \frac{\partial q_i}{\partial x_i} - \rho r = \mathcal{J}_i \mathcal{E}_i + {}^{\text{d}}\sigma_{ij} d_{ij} \,. \tag{3.140}$$

In this section we set the dissipative stress zero by assuming that the material is only elastic. After a reformulation we obtain the balance of entropy for an unpolarized elastic material:

$$\rho\eta^{\bullet} + \frac{\partial}{\partial x_i}\left(\frac{q_i}{T}\right) - \frac{1}{T}\rho r = q_i \frac{\partial}{\partial x_i}\left(\frac{1}{T}\right) + \frac{1}{T}\mathcal{J}_i \mathcal{E}_i \,, \tag{3.141}$$

with the entropy production:

$$\Sigma = q_i \frac{\partial}{\partial x_i}\left(\frac{1}{T}\right) + \frac{1}{T}\mathcal{J}_i \mathcal{E}_i = -\frac{q_i}{T^2}\frac{\partial T}{\partial x_i} + \frac{\mathcal{J}_i}{T}\mathcal{E}_i \,, \tag{3.142}$$

which has to be positive according to the 2nd law of thermodynamics, $\Sigma \geq 0$. For notational simplicity we again use

$$G_i = \frac{\partial T}{\partial x_i} \,. \tag{3.143}$$

By introducing the thermodynamical fluxes:

$$\mathfrak{F}^{\alpha} = \Big\{ -q_i \,,\ \mathcal{J}_i \Big\} \,, \tag{3.144}$$

---

[49]From a theoretical point of view this assumption is not satisfying. We shall consider $T = 0$ state as the zero state for entropy. At $T = T_{\text{ref.}}$ the entropy is then η_0 and it is unknown. Since we only employ the rate of entropy, the unknown value drops and in the end we reach the same formulation as presented herein. However, for strains the coefficient of thermal expansion, α_{ij}, has been measured by using a reference temperature, which is certainly not 0 K. In simulations we use $T_{\text{ref.}} = 300$ K.

and the thermodynamical forces:

$$\mathcal{K}^\alpha = \left\{ \frac{G_i}{T^2} \,, \; \frac{\mathcal{E}_i}{T} \right\} , \tag{3.145}$$

the 2nd law of thermodynamics reads

$$\Sigma = \mathcal{K}^\alpha \cdot \mathcal{F}^\alpha \geq 0 \,, \quad \alpha = 1, 2 \,, \tag{3.146}$$

where over α the summation convention is applied. Since both thermodynamical forces are of the same type (tensor of rank one) both thermodynamical fluxes depend on both thermodynamical forces

$$\mathcal{F}^1 = \mathcal{F}^1(\mathcal{K}^1, \mathcal{K}^2) \,, \quad \mathcal{F}^2 = \mathcal{F}^2(\mathcal{K}^1, \mathcal{K}^2) \,. \tag{3.147}$$

We propose the following relations:

$$-q_i = \lambda \frac{G_i}{T^2} + \gamma \frac{\mathcal{E}_i}{T} \,, \quad \mathcal{J}_i = \beta \frac{G_i}{T^2} + \theta \frac{\mathcal{E}_i}{T} \,. \tag{3.148}$$

Since the 2nd law has to hold for any process

$$\begin{aligned} -q_i \frac{G_i}{T^2} + \mathcal{J}_i \frac{\mathcal{E}_i}{T} \geq 0 \,, \\ \lambda \frac{G_i G_i}{T^4} + (\gamma + \beta) \frac{G_i \mathcal{E}_i}{T^3} + \theta \frac{\mathcal{E}_i \mathcal{E}_i}{T^2} \geq 0 \,, \end{aligned} \tag{3.149}$$

and since $T > 0$ we conclude

$$\lambda \geq 0 \,, \quad \gamma + \beta = 0 \,, \quad \theta \geq 0 \,. \tag{3.150}$$

The first and third relations are obvious, since $G_i G_i \geq 0$ and $\mathcal{E}_i \mathcal{E}_i \geq 0$ for any process. The second relation comes from the fact that $G_i \mathcal{E}_i$ can be positive or negative for different processes. In order to satisfy the 2nd law for any process, we have to restrict $\gamma = -\beta$ such that the second relation vanishes. This restriction is referred to as ONSAGER's reciprocal relation.[50] By renaming $\kappa = \lambda / T^2$, $\pi = \beta/(T^2 \varsigma)$, and $\varsigma = \theta / T$ we obtain

$$q_i = -\kappa \frac{\partial T}{\partial x_i} + \varsigma \pi T \mathcal{E}_i \,, \quad \mathcal{J}_i = \varsigma \pi \frac{\partial T}{\partial x_i} + \varsigma \mathcal{E}_i \,. \tag{3.151}$$

The simplest case occurs if the heat conduction parameter, κ, the electrical conductivity, ς, and the thermoelectric coupling, π, are all constant. The thermoelectric coupling is in V/K and measured by varying temperature and measuring electric

[50] In the literature the ONSAGER relation is motivated by microscopic calculations. Herein we reach the same conclusion by using thermodynamics. ONSAGER's relations are named after Lars Onsager.

field in a conductor. By having $\pi = 0$ we reach the usual FOURIER's and OHM's laws; and realize that these material models are thermodynamically sound relations for materials without thermoelectric coupling. In reality every conductor possesses a thermoelectric coupling. For every conductor even a small temperature gradient induces an electric current. This phenomenon in one conductor is called the THOMSON effect[51] and the same process between two different conductors is called the SEEBECK effect.[52] Basically this effect is used in thermocouples measuring the temperature. Moreover, we can have a heat conduction (thus entropy transport) without temperature difference but just due to an electric field. This process is called the PELTIER effect.[53]

Consider a conductor clamped on one side, which is held fixed at a reference temperature, $T_{\text{ref.}}$. The geometry is simply a beam surrounded by air, we only model the beam. In order to measure the temperature at the free end we connect the conductor to a circuit and measure the potential difference in both ends. This is basically how a thermocouple works and for such a simulation we need weak forms for computing the electric (scalar) potential, ϕ, the magnetic (vector) potential, A_i, the displacement, u_i, and the temperature, T. Since we want to compute the deformation, the LAGRANGEan frame is more appropriate. For the sake of simplicity we neglect the geometrical (and also material) nonlinearities such that the transformation of the balance equations from the current to the reference frame becomes an ease. As the reference frame we choose the initial frame, X_i. Since the geometric nonlinearities are ignored, the mass balance simplifies to $\rho = \rho_0$. Moreover, the volume element in the initial and current frame will be equal, $\mathrm{d}v = \mathrm{d}V$.

For the weak form of electric potential, ϕ, we use the following balance of electric charge in Eq. (3.124)$_1$ in the initial frame:

$$\frac{\partial \rho_0 z}{\partial t} + J_{i,i} = 0 \,, \quad J_{i,i} = \frac{\partial J_i}{\partial X_i} \,, \tag{3.152}$$

without geometric nonlinearities. The latter is very similar to Eq. (3.46) such that we follow the same steps and obtain Eq. (3.80). Since we have assumed that $P_i = 0$ and $\mathcal{M}_i = 0$, the weak form in the initial frame for small displacements reads

$$\mathrm{F}_\phi = \int_{\mathcal{B}_0} \big(-(D_i - D_i^0)\,\delta\phi_{,i} - \Delta t J_i\,\delta\phi_{,i} \big)\mathrm{d}V + \int_{\partial\mathcal{B}_0} N_i \Delta t J_i\,\delta\phi\,\mathrm{d}A \,, \tag{3.153}$$

in the unit of energy. The interface between beam and air is simply the boundary satisfying the balance laws on singular surfaces. Air is not modeled, its electric conduction is taken as zero. Zero polarization leads to $J_i^{\text{fr.}} \equiv J_i$ and the electric current is given by

$$J_i = \mathcal{J}_i + \rho z v_i \,, \quad \rho z = D_{i,i} \,, \tag{3.154}$$

[51]It is called for William Thomson (Lord Kelvin).

[52]It is named after Thomas Johann Seebeck.

[53]This effect is named after Jean Charles Athanase Peltier.

where $\mathcal{J}_i$ is defined in Eq. (3.151)$_2$. For computing the magnetic potential, A_i, we start with Eq. (3.124)$_2$ and transform it from the current to the initial frame by neglecting geometric nonlinearities and acquire

$$\varepsilon_0 \frac{\partial^2 A_i}{\partial t^2} - \frac{1}{\mu_0} A_{i,jj} = J_i \ , \quad A_{i,jj} = \frac{\partial^2 A_i}{\partial X_j \partial X_j} \ . \tag{3.155}$$

Its weak form is Eq. (3.92). Since polarization is omitted, there remains

$$\mathrm{F}_A = \int_{\mathcal{B}_0} \Big(\varepsilon_0 \frac{A_i - 2A_i^0 + A_i^{00}}{\Delta t \Delta t} \delta A_i + \frac{1}{\mu_0} A_{i,j} \, \delta A_{i,j} - J_i \, \delta A_i \Big) \mathrm{d}V \ , \tag{3.156}$$

in the unit of energy, where $J_i^{\text{fr.}} \equiv J_i$. In the initial frame the velocity equals to the partial derivative of displacement in time, thus,

$$v_i = \frac{\partial u_i}{\partial t} = \frac{u_i - u_i^0}{\Delta t} \ . \tag{3.157}$$

In order to compute the displacement we use the balance of linear momentum in the initial frame by neglecting geometric nonlinearities:

$$\rho_0 \frac{\partial^2 u_i}{\partial t^2} - \frac{\partial \sigma_{ji}}{\partial X_j} - \rho_0 f_i - \mathcal{F}_i = 0 \ , \quad \mathcal{F}_i = \frac{\partial D_j}{\partial X_j} E_i + \epsilon_{ijk} J_j B_k \ . \tag{3.158}$$

Hence, we obtain the following weak form in the unit of energy:

$$\begin{aligned} \mathrm{F}_u = \int_{\mathcal{B}_0} \Big(& \rho_0 \frac{u_i - 2u_i^0 + u_i^{00}}{\Delta t \Delta t} \delta u_i + \sigma_{ji} \, \delta u_{i,j} - \rho_0 f_i \, \delta u_i - \\ & - \mathcal{F}_i \, \delta u_i \Big) \mathrm{d}V - \int_{\partial \mathcal{B}_0} \hat{t}_i \, \delta u_i \mathrm{d}A \ . \end{aligned} \tag{3.159}$$

The boundaries will vanish for the clamped end by using DIRICHLET conditions and also for the other boundaries by assuming free boundaries, $\hat{t}_i = n_j \sigma_{ji} = 0$. For computing temperature we utilize the balance of entropy in Eqs. (3.141), (3.142) in the initial frame without geometric nonlinearities:

$$\begin{aligned} & \rho_0 \frac{\partial \eta}{\partial t} + \frac{\partial}{\partial X_i} \Big(\frac{q_i}{T} \Big) - \frac{1}{T} \rho_0 r = - \frac{q_i}{T^2} \frac{\partial T}{\partial X_i} + \frac{1}{T} \mathcal{J}_i \mathcal{E}_i + \frac{1}{T} {}^{\text{d}}\sigma_{ji} \frac{\partial v_i}{\partial X_j} \ , \\ & \rho_0 \frac{\partial \eta}{\partial t} + \frac{1}{T} \frac{\partial q_i}{\partial X_i} - \frac{1}{T} \rho_0 r = \frac{1}{T} \mathcal{J}_i \mathcal{E}_i + \frac{1}{T} {}^{\text{d}}\sigma_{ji} \frac{\partial v_i}{\partial X_j} \ . \end{aligned} \tag{3.160}$$

We assume that the deformation is purely elastic, ${}^{\text{d}}\sigma_{ji} = 0$. Hence, the weak form for temperature reads in the unit of energy

$$
\begin{aligned}
\mathrm{F}_T = \int_{\mathcal{B}_0} \Big(\rho_0(\eta - \eta^0)\,\delta T - \Delta t q_i \Big(\frac{\delta T}{T}\Big)_{,i} - \frac{\Delta t}{T} \rho_0 r\,\delta T - \\
- \frac{\Delta t}{T} \mathcal{J}_i \mathcal{E}_i\,\delta T \Big) \mathrm{d}V + \int_{\partial \mathcal{B}_0} \frac{\Delta t}{T} h(T - T_{\text{ref.}})\,\delta T \mathrm{d}A \,,
\end{aligned}
\tag{3.161}
$$

where for boundaries we readily applied the natural boundary condition, $q_i N_i = h(T - T_{\text{ref.}})$, with an ambient temperature as equal as the reference and initial temperature. The nonlinear weak form is the sum of all forms above:

$$
\text{Form} = \mathrm{F}_\phi + \mathrm{F}_A + \mathrm{F}_{\boldsymbol{u}} + \mathrm{F}_T \,, \tag{3.162}
$$

with the following constitutive equations:

$$
\begin{aligned}
\mathcal{J}_i = \varsigma \pi T_{,i} + \varsigma \mathcal{E}_i \,, \quad D_i = \varepsilon_0 E_i \,, \quad H_i = \frac{1}{\mu_0} B_i \,, \quad \sigma_{ij} = {}^{\mathrm{e}}\sigma_{ij} + {}^{\mathrm{d}}\sigma_{ij} \,, \\
{}^{\mathrm{e}}\sigma_{ij} = C_{ijkl}\big(\varepsilon_{kl} - \alpha_{kl}(T - T_{\text{ref.}})\big) \,, \quad {}^{\mathrm{d}}\sigma_{ij} = 0 \,, \\
\eta = c \ln\Big(\frac{T}{T_{\text{ref.}}}\Big) + C_{ijkl}\alpha_{kl} v \varepsilon_{ij} \,, \quad q_i = -\kappa T_{,i} + \varsigma \pi T \mathcal{E}_i \,,
\end{aligned}
\tag{3.163}
$$

and

$$
\mathcal{E}_i = E_i + \epsilon_{ijk} v_j B_k \,, \quad E_i = -\phi_{,i} - \frac{A_i - A_i^0}{\Delta t} \,, \quad B_i = \epsilon_{ijk} A_{k,j} \,. \tag{3.164}
$$

The material parameters, C_{ijkl}, α_{ij}, κ, c, π, and ς are constant. For an isotropic material the stiffness tensor and coefficients of thermal expansion read

$$
C_{ijkl} = \lambda \delta_{ij}\delta_{kl} + \mu \delta_{ik}\delta_{jl} + \mu \delta_{il}\delta_{jk} \,, \quad \alpha_{ij} = \alpha \delta_{ij} \,, \tag{3.165}
$$

by reducing to three materials parameters, viz., the LAME parameters, λ, μ, and the coefficient of thermal expansion, α.

The beam is made out of chromel, which is a nickel and chromium alloy with a relatively large thermoelectric coupling such that it is used as a thermocouple. The temperatures at both ends of the beam are given by DIRICHLET boundary conditions. The beam possesses $T_{\text{ref.}}$ on one end and a linearly increasing temperature on the other end. The temperature difference induces in addition to a thermal flux also an electric flux, i.e., the electric conduction current $\mathcal{J}_i$ since $\pi \neq 0$ for chromel material. Therefore, an electric potential difference occurs, see Fig. 3.11. After 30 s the temperature is almost in steady state and the potential difference is approximately 0.3 mV. The magnetic (vector) potential is directed in the same direction (no curl exists) such that magnetic flux vanishes. By measuring the potential difference we can estimate the temperature difference in a real application. Moreover, the body is a conductor such that in an electric circuit the potential difference implies an electric current. Hence, by using a PELTIER element we can light a bulb due to a temperature difference;

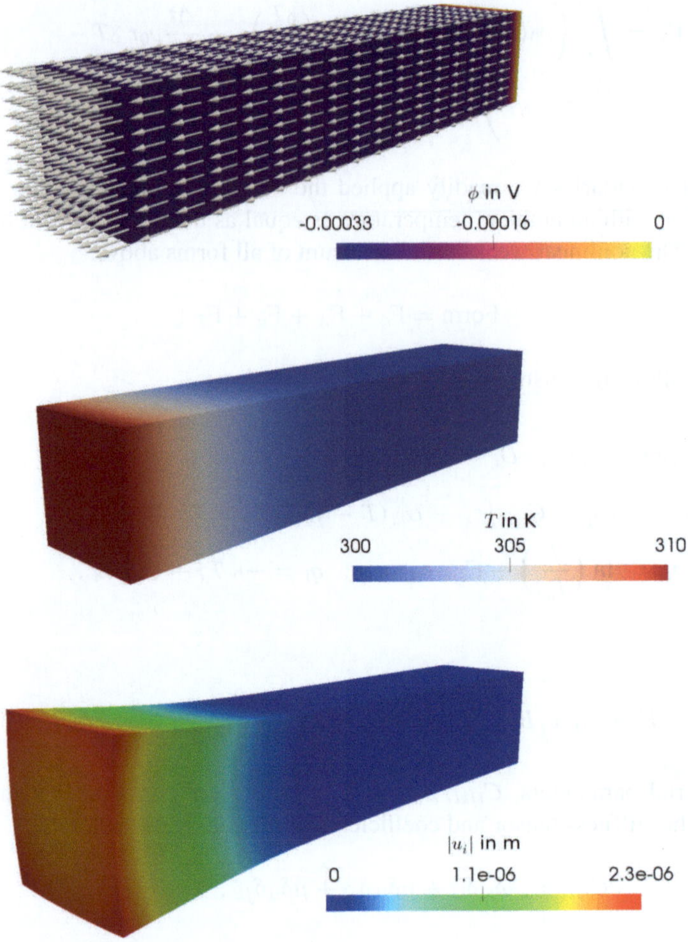

Fig. 3.11 Electric potential, temperature, and displacements after 30 s due to the temperature difference at both ends. Displacements are 2000 times enlarged for a better visualization. *Arrows* denote the magnetic potential

however, it is quite inefficient. A PELTIER element is mainly used for measuring the temperature accurately or for tuning the temperature precisely by pumping heat flux in or out of the system. In many devices performing material tests, the control of temperature is established by using PELTIER elements. The following code has been used for the simulation of nonlinear and coupled field equations, where all primitive variables, ϕ, A_i, u_i, T, are solved at once.

```
"""Computational reality 17, thermoelectric coupling"""
__author__ = "B. Emek Abali"
__license__ = "GNU GPL Version 3.0 or later"
#This code underlies the GNU General Public License, http://www.gnu.org/licenses/gpl-3.0.en.html

from fenics import *
import numpy
set_log_level(ERROR)
#units: m, kg, s, A, V, K
delta = Identity(3)
levicivita2 = as_matrix([ (0,1,-1) , (-1,0,1) , (1,-1,0) ])
levicivita3 = as_tensor([ ( (0,0,0),(0,0,1),(0,-1,0) ) , ( (0,0,-1),(0,0,0),(1,0,0) ) , ( (0,1,0),(-1,0,0),(0,0,0) ) ])
epsilon = levicivita3

#thermocouple of type E is made of
#chromel (nickel-chromium alloy) and
#is non-magnetic in reality, we also
#assume that it is non-polarizable
resistivity = 0.706E-6 #in Ohm m
varsigma = 1./resistivity #0.625 #in S/m
pi = 68E-6 #V/K
kappa = 19.0 #in W/(m K)
capacity = 390. #in J / (kg K)
alpha = 12.8E-6 #in 1/K
EModul = 186E+9 #in Pa
nu = 0.32
h = 10. #in J / (s m^2 K)

eps_0 = 8.85E-12 #in A s/(V m)
mu_0 = 12.6E-7 #in V s/(A m)

rho0 = 8500. #in kg / m^3
T_ref = 300.0 # K

tMax = 30.0
Dt = 1.0
t = 0.0

mesh = BoxMesh(Point(-0.05,-0.01,-0.01), Point(0.05,0.01,0.01), 50,10,10)
N = FacetNormal(mesh)

Scalar = FunctionSpace(mesh, 'P', 1)
Vector = VectorFunctionSpace(mesh, 'P', 1)
Tensor = TensorFunctionSpace(mesh, 'P', 1)
#phi, A, u, T
Space = MixedFunctionSpace([Scalar,Vector, Vector, Scalar])

cells = CellFunction('size_t', mesh)
facets = FacetFunction('size_t', mesh)
dV = Measure('dx', domain=mesh, subdomain data=cells)
```

```
dA = Measure('ds', domain=mesh, subdomain_data=facets)

cells.set_all(0)
facets.set_all(0)
left = CompiledSubDomain('near(x[0],-0.05) && on_boundary')
right = CompiledSubDomain('near(x[0],0.05) && on_boundary')
boundaries = CompiledSubDomain('on_boundary')

#phi, A, u, T
bc_T = Expression('T_r + 1.0*time', T_r=T_ref,time=0.)
bc01=DirichletBC(Space.sub(0), 0.0, right)
bc02=DirichletBC(Space.sub(2), Constant((0.,0.,0.)), right)
bc03=DirichletBC(Space.sub(3), T_ref, right)
bc04=DirichletBC(Space.sub(3), bc_T, left)

bc = [bc01,bc02,bc03,bc04]

dunkn = TrialFunction(Space)
test = TestFunction(Space)
unkn = Function(Space)
unkn0 = Function(Space)
unkn00 = Function(Space)

unkn_init = Expression(('0.','0.','0.','0.','0.','0.','0.','T_r'),T_r=T_ref)
unkn00 = interpolate(unkn_init,Space)
unkn0.assign(unkn00)
unkn.assign(unkn0)

del_phi,del_A,del_u,del_T = split(test)
phi,A,u,T = split(unkn)
phi0,A0,u0,T0 = split(unkn0)
phi00,A00,u00,T00 = split(unkn00)

i, j, k, l = indices(4)
delta = Identity(3)

lam = EModul * nu / (1.+nu) / (1.-2.*nu)
mu = 0.5 * EModul / (1.+nu)
C = as_tensor(lam*delta[i,j]*delta[k,l] + mu*delta[i,k]*delta[j,l] + mu*delta[i,l]*delta[j,k], (i,j,k,l))
alfa = alpha*delta

eps = sym(grad(u))
eps0 = sym(grad(u0))
v = as_tensor( (u-u0)[i]/Dt, (i,))
E = as_tensor(-phi.dx(i)-(A-A0)[i]/Dt, (i,))
E0 = as_tensor(-phi0.dx(i)-(A0-A00)[i]/Dt, (i,))
B = as_tensor(epsilon[i,j,k]*A[k].dx(j), (i,))
EE = as_tensor(E[i]+epsilon[i,j,k]*v[j]*B[k], (i,))

D = eps_0*E
D0 = eps_0*E0
H = 1./mu_0*B
```

```
JJ = as_tensor( varsigma*pi*T.dx(i) + varsigma*EE[i], (i,))
J = as_tensor( JJ[i] + D[j].dx(j)*v[i], (i,))
sigma = as_tensor( C[i,j,k,l]*(eps[k,l]-alfa[k,l]*(T-T_ref)), (i,j))
eta = as_tensor( capacity*ln(T/T_ref) + C[i,j,k,l]*alfa[k,l]/rho0*eps[i,j], ())
eta0 = as_tensor( capacity*ln(T0/T_ref) + C[i,j,k,l]*alfa[k,l]/rho0*eps0[i,j], ())
q = as_tensor(-kappa*T.dx(i)+varsigma*pi*T*EE[i], (i,))
FF = as_tensor(D[j].dx(j)*E[i] + epsilon[i,j,k]*J[j]*B[k], (i,))
f = Constant((0.,0.,0.))
r = Constant(0.0)

F_phi = (-(D-D0)[i]*del_phi.dx(i) - Dt*J[i]*del_phi.dx(i) )*dV + Dt*J[i]*del_phi*N[i]*dA
F_A = (eps_0*(A-2.*A0+A00)[i]/Dt/Dt*del_A[i] + 1./mu_0*A[i].dx(j)*del_A[i].dx(j) - J[i]*del_A[i] )*dV
F_u = (rho0*(u-2.*u0+u00)[i]/Dt/Dt*del_u[i] + sigma[j,i]*del_u[i].dx(j) - rho0*f[i]*del_u[i] - FF[i]*del_u[i])*dV
F_T = (rho0*(eta-eta0)*del_T - Dt*q[i]*(del_T/T).dx(i) - Dt/T*rho0*r*del_T - Dt/T*JJ[i]*EE[i]*del_T )*dV + Dt/T*h*(T-T_ref)*del_T*dA

Form = F_phi + F_A + F_u + F_T
Gain = derivative(Form, unkn, dunkn)

pwd='/calcul/CR17/'
file_phi = File(pwd+'phi.pvd')
file_A = File(pwd+'A.pvd')
file_E = File(pwd+'E.pvd')
file_B = File(pwd+'B.pvd')
file_u = File(pwd+'u.pvd')
file_T = File(pwd+'T.pvd')

phi_ = Function(Scalar, name='$\phi$')
A_ = Function(Vector, name='$A_i$')
u_ = Function(Vector, name='$u_i$')
T_ = Function(Scalar, name='$T$')
E_ = Function(Vector, name='$E_i$')
B_ = Function(Vector, name='$B_i$')

while t < tMax:
    print 'time: ',t
    if t <= 10.: bc_T.time = t
    solve(Form==0, unkn, bc, J=Gain, \
        solver_parameters={"newton_solver":{"linear_solver": "mumps", "relative_tolerance": 1e-5} }, \
        form_compiler_parameters={"cpp_optimize": True, "representation": "quadrature", "quadrature_degree": 2} )

    phi_.assign(unkn.split(deepcopy=True)[0])
```

```
        A_.assign(unkn.split(deepcopy=True)[1])
        u_.assign(unkn.split(deepcopy=True)[2])
        T_.assign(unkn.split(deepcopy=True)[3])
        E_.assign(project(E,Vector))
        B_.assign(project(B,Vector))

        print '    max. electric potential: ',max(phi_.vector()), ' V'
        print '    min. electric potential: ',min(phi_.vector()), ' V'

        file_phi << (phi_, t)
        file_A << (A_, t)
        file_E << (E_,t)
        file_B << (B_,t)
        file_u << (u_, t)
        file_T << (T_, t)

        unkn00.assign(unkn0)
        unkn0.assign(unkn)
        t += Dt
```

To-do

Thermoelectric coupling is discussed and implemented.

- Which equation needs to be changed, if we want to include the polarization?
- Inspect the boundary conditions. Repeat and list the physical meanings of the applied boundary conditions.
- Search for such computations in the literature. Try to find a coupled monolithic solution by using a staggered scheme or by solving all unknowns at once as above.
- Perform a web-based search in order to grasp the different types of PELTIER elements. Find out the crystallographic defects for an n-type and p-type PELTIER element.

3.4 Plastic Fatigue in a Circuit Board

In every electronic gadget, such as laptops, smartphones, or engine control systems in cars; there is a circuit board providing electrical connections between transistors. A circuit board consists of copper tracks embedded in an epoxy insulator. These copper tracks are called *vias*. An electric signal is propagated through a via and passes a semiconductor. The semiconductor allows the signal to pass by if its amplitude is above a threshold value.

An electric signal is a current in the via, which is a good conductor such as copper. As used in the last sections the electric current passing a conductor produces heat due to the JOULE heating. The temperature in the via increases. The via is embedded in the board such that the temperature in the board also increases. The via and the board expand due to the temperature increase. Unfortunately, the via and the board have different expansion coefficients. They try to expand differently and they are coherent; thermal stresses occur. In many applications the value of the thermal stress trespasses the yield stress and the via undergoes a plastic deformation. The board has a lower YOUNG's modulus than the via such that the via deforms plastically whereas the board remains elastic.

Consider a smartphone during phoning, some vias are in active use leading to thermal stresses and a plastic deformation. After finishing phoning the temperature decreases back to the ambient temperature (due to the heat exchange over the boundary). Hence, the via and board try to deform back to their initial geometries. Unfortunately, this trial results in stresses once more, since the geometry did deform plastically. Increasing and decreasing temperature generates a plastic deformation in the via. Repeated use of the smartphone adds a plastic deformation such that the plastic strain accumulates. By reaching a limit value the accumulated plastic strain initiates a crack in the via. By further plastic deformation the crack grows and cut off the circuit. This process is known as *fatigue*.[54]

For testing a new design there are standard experiments concerning fatigue failure. The board is tested under cyclic thermal or electric loading. Consider that the board is subject to a cyclic electric loading, which is indeed accelerated in order to achieve a fatigue failure in couple of days. In each cycle energy dissipates from the system, which is used for the plastic deformation. In other words, the plastic deformation accumulates in time and brings in new plastic deformation. The amount of dissipated energy (or accumulated plastic strain) in each cycle remains the same. This phenomenon is simply due to the reversible expansion of via and board. The mismatch of the expansion coefficients results in the same thermal stress and plastic strain in each cycle. After $N_{\text{f.}}$ cycles the accumulated plastic strain attains a value at which the via breaks or initiates a crack.

In order to simulate a fatigue experiment we need to involve plasticity.[55] We furthermore neglect the polarization in the material

$$P_i = 0 \,, \quad \mathcal{M}_i = 0 \,. \tag{3.166}$$

The primitive variables are the electric and magnetic potentials, the displacement, and the temperature, $\{\phi, A_i, u_i, T\}$. Since we compute a solid body, all primitive variables are functions in space, X_i, and time, t, in the LAGRANGEan frame where we choose the reference as initial frame with X_i denoting the positions of particles in the beginning, $t = 0$. However, we will neglect the geometric nonlinearities such that the deformation gradient equals the KRONECKER delta and partial derivatives with respect to X_i and x_i become identical. The effects of magnetic flux is negligible.

[54]The material never reaches an ultimate strain, where a failure is expected. The accumulated deformation causes such a failure over time. For an electronic device this time frame is more than couple of years.

[55]We have introduced associated plasticity in Sect. 1.6 and applied into the thermodynamical formulation in Sect. 3.3.

Hence, we neglect the magnetic potential and spare computational time by solving only 5 fields, $\{\phi, u_i, T\}$, instead of 8 fields, $\{\phi, A_i, u_i, T\}$. This estimation results in a failure by computing the electric field without the rate of magnetic potential. This error is small for electrical loadings at low frequencies, which is the case in the following application. By setting $A_i = 0$ we obtain the following electromagnetic fields:

$$
\begin{aligned}
E_i &= -\phi_{,i} - \frac{A_i - A_i^0}{\Delta t} = -\phi_{,i} \ , \quad B_i = \epsilon_{ijk} A_{k,j} = 0 \ , \\
\mathcal{E}_i &= E_i + \epsilon_{ijk} \frac{(u_j - u_j^0)}{\Delta t} B_k = E_i \ ,
\end{aligned}
\tag{3.167}
$$

of course the MAXWELL–LORENTZ aether relations hold as well:

$$
D_i = \varepsilon_0 E_i \ , \quad H_i = \frac{1}{\mu_0} B_i = 0 \ . \tag{3.168}
$$

The electric potential is then computed by the following weak from:

$$
\begin{aligned}
\mathrm{F}_\phi = &\int_{\mathcal{B}_0} \big(-(D_i - D_i^0)\,\delta\phi_{,i} - \Delta t\, J_i\,\delta\phi_{,i} \big) \mathrm{d}V + \\
&+ \int_{\partial\mathcal{B}_0^I} N_i \Delta t \big[J_i \big]\,\delta\phi \mathrm{d}A + \int_{\partial\mathcal{B}_0 \backslash \partial\mathcal{B}_0^I} N_i \Delta t\, J_i\,\delta\phi \mathrm{d}A \ .
\end{aligned}
\tag{3.169}
$$

The interface between via and board is denoted by $\partial\mathcal{B}_0^I$. For computing the displacement we utilize the balance of linear momentum and generate the weak form in the unit of energy:

$$
\begin{aligned}
\mathrm{F}_{\boldsymbol{u}} = &\int_{\mathcal{B}_0} \Big(\rho_0 \frac{u_i - 2u_i^0 + u_i^{00}}{\Delta t \Delta t}\,\delta u_i + \sigma_{ji}\,\delta u_{i,j} - \rho_0 f_i\,\delta u_i - \mathcal{F}_i\,\delta u_i \Big) \mathrm{d}V - \\
&- \int_{\partial\mathcal{B}_0} \hat{t}_i\,\delta u_i \mathrm{d}A \ ,
\end{aligned}
\tag{3.170}
$$

with the LORENTZ force density:

$$
\mathcal{F}_i = \rho z E_i + \epsilon_{ijk} J_j B_k = D_{j,j} E_i \ , \tag{3.171}
$$

since we have set $A_i = 0$. The computation of the temperature is realized by the weak form in the unit of energy generated from the balance of entropy

$$
\begin{aligned}
\mathrm{F}_T = &\int_{\mathcal{B}_0} \Big(\rho_0(\eta - \eta^0)\,\delta T - \Delta t \Phi_i\,\delta T_{,i} - \Delta t \rho_0 \frac{r}{T}\,\delta T - \Delta t \Sigma\,\delta T \Big) \mathrm{d}V + \\
&+ \int_{\partial\mathcal{B}_0} \frac{\Delta t}{T} h(T - T_{\text{amb}})\,\delta T \mathrm{d}A \ ,
\end{aligned}
\tag{3.172}
$$

with ROBIN boundary conditions. The entropy flux, Φ_i, and production, Σ, read

$$\Phi_i = \frac{q_i}{T}\,, \quad \Sigma = -\frac{q_i}{T^2}T_{,i} + \frac{1}{T}\mathcal{J}_i\mathcal{E}_i + \frac{1}{T}\sigma_{ij}\,{}^{\mathrm{p}}\dot{\varepsilon}_{ij}\,. \tag{3.173}$$

For all constitutive equations we basically repeat the procedure in Sect. 3.3 with the thermodynamical consideration discussed in Sect. 2.5. By using the additive decomposition in strains

$$\varepsilon_{ij} = {}^{\mathrm{r}}\varepsilon_{ij} + {}^{\mathrm{p}}\varepsilon_{ij}\,, \tag{3.174}$$

into a reversible and an irreversible (plastic) term and by introducing the GIBBS free energy:

$$g = u - T\eta - v\sigma_{ij}\,{}^{\mathrm{r}}\varepsilon_{ij}\,, \quad v = \frac{1}{\rho_0}\,, \tag{3.175}$$

we obtain the following linear constitutive equations with constant coefficients:

$$\begin{aligned} \eta &= c\ln\left(\frac{T}{T_{\text{ref.}}}\right) + v\alpha_{ij}\sigma_{ij}\,, \\ \sigma_{ij} &= C_{ijkl}\big(\varepsilon_{kl} - {}^{\mathrm{p}}\varepsilon_{kl} - {}^{\mathrm{th}}\varepsilon_{kl}\big)\,, \quad {}^{\mathrm{th}}\varepsilon_{ij} = \alpha_{ij}(T - T_{\text{ref.}})\,. \end{aligned} \tag{3.176}$$

In order to calculate the plastic strains incrementally,

$${}^{\mathrm{p}}\varepsilon_{ij} = {}^{\mathrm{p}}\varepsilon^0_{ij} + \Delta t\,{}^{\mathrm{p}}\dot{\varepsilon}_{ij}\,, \tag{3.177}$$

we apply the associated plasticity with the kinematic hardening

$${}^{\mathrm{p}}\dot{\varepsilon}_{mn} = \langle\gamma\rangle\frac{(\sigma^0_{|ij|} - \beta^0_{ij})C_{ijkl}(\dot{\varepsilon}_{kl} - {}^{\mathrm{p}}\dot{\varepsilon}_{kl} - {}^{\mathrm{th}}\dot{\varepsilon}_{kl})}{\frac{4}{9}h\sigma_{\mathrm{Y}}^2 + (\sigma^0_{|ij|} - \beta^0_{ij})C_{ijkl}(\sigma^0_{|kl|} - \beta^0_{kl})}(\sigma^0_{|mn|} - \beta^0_{mn})\,, \tag{3.178}$$

together with its corresponding back stress:

$$\beta_{ij} = \beta^0_{ij} + \Delta t\dot{\beta}_{ij}\,, \quad \dot{\beta}_{ij} = \frac{(\sigma^0_{|kl|} - \beta^0_{kl})\dot{\sigma}_{kl}}{\frac{2}{3}\sigma_{\mathrm{Y}}^2}(\sigma^0_{|ij|} - \beta^0_{ij})\,. \tag{3.179}$$

After using them in the balance of entropy and assuming that the choice of h and σ_{Y} such that $\sigma_{ij}\,{}^{\mathrm{p}}\dot{\varepsilon}_{ij}$ is positive for every processes, we can derive from the 2nd law

$$\mathcal{J}_i = \varsigma\pi T_{,i} + \varsigma\mathcal{E}_i\,, \quad q_i = -\kappa T_{,i} + \varsigma\pi T\mathcal{E}_i\,. \tag{3.180}$$

Since $A_i = 0$ we have $\mathcal{E}_i \equiv E_i$. The obtained weak form:

$$\text{Form} = \mathrm{F}_\phi + \mathrm{F}_{\boldsymbol{u}} + \mathrm{F}_T\,, \tag{3.181}$$

is coupled and nonlinear because of the entropy production.

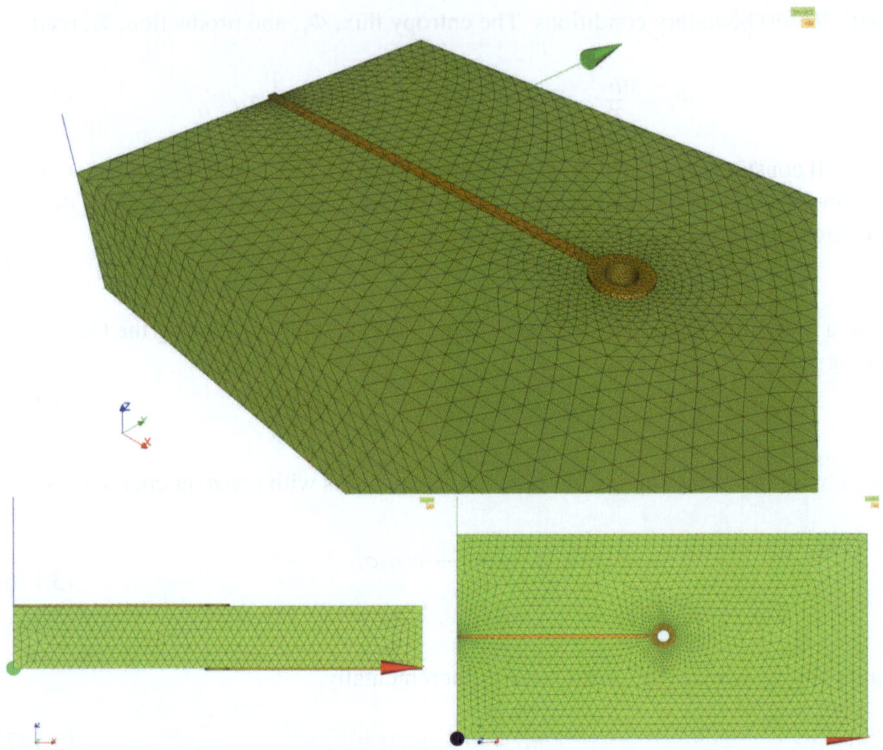

Fig. 3.12 Geometry of a simplified board with a single via in *upper figure*: perspective view; *lower left figure*: side view; *lower right figure*: top view. The board (*green*) is of fiber reinforced epoxy and the via (*yellow*) is of copper

Consider a simplified board with a single via as shown in Fig. 3.12. The board is a *laminate* composed of glass fibers and epoxy. Glass fibers are oriented along x and y axis such that the laminate is an orthotropic material. Fibers are embedded in a so-called *matrix* material. For circuit boards typically epoxy is used as the matrix material. The glass fiber reinforced epoxy used as a circuit board possesses the following materials properties at room temperature (293 K):[56]

$$C_{IJ}^{\text{lam.}} = \begin{pmatrix} 66242 & 41797 & 37814 & 0 & 0 & 0 \\ & 50460 & 32290 & 0 & 0 & 0 \\ & & 31591 & 0 & 0 & 0 \\ & & & 2250 & 0 & 0 \\ & \text{sym.} & & & 2250 & 0 \\ & & & & & 6630 \end{pmatrix} \cdot 10^6\,\text{Pa}\,, \tag{3.182}$$

[56]Measurement of the stiffness tensor is undertaken in Fraunhofer IZM in Berlin (Germany) for a typical laminate used in many circuit boards, the values at the room temperature are taken from [3], the values between 233–443 K can be found in [2, Table 1].

and

$$\alpha_{ij}^{\text{lam}} = \begin{pmatrix} 13.2 & 0 & 0 \\ 0 & 16.7 & 0 \\ 0 & 0 & 39 \end{pmatrix} \cdot 10^{-6}\, 1/\text{K}\,, \tag{3.183}$$

where the stiffness matrix is given in the VOIGT notation:

$$C_{IJ} = \begin{pmatrix} C_{1111} & C_{1122} & C_{1133} & C_{1123} & C_{1113} & C_{1112} \\ & C_{2222} & C_{2233} & C_{2223} & C_{2213} & C_{2212} \\ & & C_{3333} & C_{3323} & C_{3313} & C_{3312} \\ & & & C_{2323} & C_{2313} & C_{2312} \\ & \text{sym.} & & & C_{1313} & C_{1312} \\ & & & & & C_{1212} \end{pmatrix} . \tag{3.184}$$

Epoxy is a good insulator such that the laminate becomes an insulator. Indeed, we need an insulating board such that the signal is transferred along conducting vias. Being a cubic material, copper is frequently used as vias. Within each *grain*, the orientation of the cubic material varies randomly. For a geometry in a macroscopic length scale many grains with different orientations lead to an isotropic behavior. For electroplated copper used as vias, the grain size reaches approximately 50 μm. The geometric dimensions are in the same length scale such that we need to consider copper as a cubic material. Unfortunately, we lack the knowledge of the correct orientation in each grain. Thus, we model the copper as a single crystal with the axis along the coordinate system with the stiffness matrix in the VOIGT notation:

$$C_{IJ}^{\text{Cu}} = \begin{pmatrix} 169.1 & 122.2 & 122.2 & 0 & 0 & 0 \\ & 169.1 & 122.2 & 0 & 0 & 0 \\ & & 169.1 & 0 & 0 & 0 \\ & & & 75.42 & 0 & 0 \\ & \text{sym.} & & & 75.42 & 0 \\ & & & & & 75.42 \end{pmatrix} \cdot 10^{9}\,\text{Pa}\,, \tag{3.185}$$

and the thermal expansion coefficient tensor:

$$\alpha_{ij}^{\text{Cu}} = \begin{pmatrix} 17 & 0 & 0 \\ 0 & 17 & 0 \\ 0 & 0 & 17 \end{pmatrix} \cdot 10^{-6}\, 1/\text{K}\,. \tag{3.186}$$

In order to model a fatigue failure test, we perform a simulation of an electronic signal produced by an electric potential on one end of the via:

$$\hat{\phi} = A \sin(2\pi \nu t)\,, \tag{3.187}$$

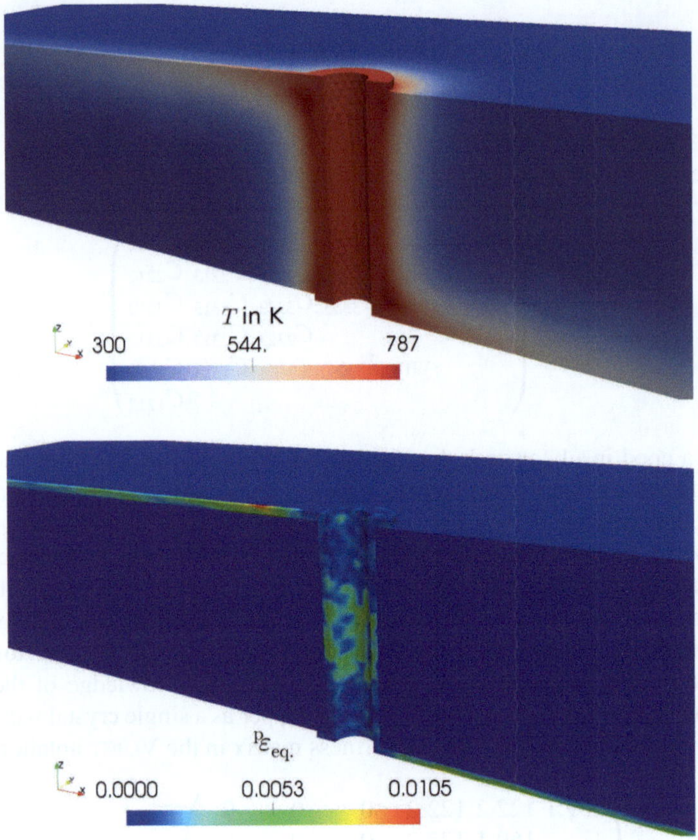

Fig. 3.13 Temperature distribution and accumulated plastic equivalent strain after one cycle

where the amplitude is $A = 12\,\text{V}$ and the frequency is $\nu = 10\,\text{Hz}$. The electrical loading is cyclic such that the direction of the electric field alternates. However, JOULE's heating, $\mathcal{J}_i \mathcal{E}_i$, becomes always positive since OHM's law asserts that the current and electric field have the same sign. Therefore, the internal energy and thus the temperature increase. We have applied ROBIN boundary conditions for a realistic modeling and after one cycle the temperature distribution is shown in Fig. 3.13. Temperature increase implies a deformation in the body consisting of different materials with different expansion coefficients. Different expansion behaviors of via and board generate thermal stresses in each cycle leading to a plastic deformation. The plastic strain rate is given in Eq. (3.178) such that the equivalent plastic strain rate according to VON MISES hypothesis reads

$$ {}^{\mathrm{p}}\dot{\varepsilon}_{\mathrm{eq.}} = \sqrt{\frac{2}{3}\, {}^{\mathrm{p}}\dot{\varepsilon}_{ij}\, {}^{\mathrm{p}}\dot{\varepsilon}_{ij}} \,, \tag{3.188} $$

where rate of the plastic strain is deviatoric since the plastic strain remains deviatoric.[57] The equivalent plastic strain reads

$$ {}^{\mathrm{p}}\varepsilon_{\mathrm{eq.}} = \int {}^{\mathrm{p}}\dot{\varepsilon}_{\mathrm{eq.}}\, \mathrm{d}t \,. \tag{3.189} $$

This equivalent plastic strain accumulates over time and the evolution of it is given by ${}^{\mathrm{p}}\dot{\varepsilon}_{\mathrm{eq.}}$. It is important to notice that we cannot formulate the latter as a potential, ${}^{\mathrm{p}}\dot{\varepsilon}_{\mathrm{eq.}}\mathrm{d}t \neq \mathrm{d}\varepsilon_{\mathrm{eq.}}$, since all values over time affect the process; not only the values in the beginning and at the end. We can demonstrate this phenomenon in a plastic sheet, which we bend back and forth. Although we just bend it back and forth, in other words, at the start and end the sheet has the same geometry, we know that after couple of cycles (bending back and forth) the accumulated plastic strain attains the ultimate strain limit and the sheet breaks. The accumulation process is only possible to calculate if we establish the evolution of the plastic strain in time with ${}^{\mathrm{p}}\dot{\varepsilon}_{\mathrm{eq.}}$.

The key parameter in the simulation is the electrical conductivity of copper, ς^{Cu}. This parameter varies depending on the circuit modeling, we have used an estimated value. As we know from the last sections, the real or effective conductivity depends on the resistance (caused by the semiconductors in a circuit board) installed on the wire. By having a lower conductivity on the wire, the electric current decreases, which leads to a smaller JOULE's heating such that the internal energy increase rate decreases. In reality the copper via undergoes a plastic deformation in each cycle, which accumulates and leads to a breakage at the end of a so-called *fatigue lifetime*. An accurate estimation of the lifetime due to fatigue is still an open question in the electronic industry. The geometry for the computation is in [1] and the code is given below.

57 The factor 2/3 is due to the tensile test. See Sect. 1.6 for a motivation of the deviatoric plastic strain.

```python
"""Computational reality 18, fatigue in a circuit board"""
__author__ = "B. Emek Abali"
__license__ = "GNU GPL Version 3.0 or later"
#This code underlies the GNU General Public License, http://www.gnu.org/licenses/gpl-3.0.en.html

from fenics import *
parameters["allow_extrapolation"]=True
import numpy
set_log_level(ERROR)
'''
2D 1 "voltage_in"
2D 2 "voltage_out"
2D 3 "clamped"
2D 4 "via_air"
2D 5 "board_air"
2D 6 "via_board"
3D 1 "board"
3D 2 "via"
'''
mesh = Mesh('geo/CR18_board.xml')
cells = MeshFunction('size_t',mesh,'geo/CR18_board_physical_region.xml')
facets = MeshFunction('size_t',mesh,'geo/CR18_board_facet_region.xml')

def material_coefficient(target_mesh, cells_list, coeffs):
    coeff_func = Function(FunctionSpace(target_mesh, 'DG', 0))
    markers = numpy.asarray(cells_list.array(), dtype=numpy.int32)
    coeff_func.vector()[:] = numpy.choose(markers-1, coeffs)
    return coeff_func

N = FacetNormal(mesh)
dI = Measure('dS', domain=mesh, subdomain_data=facets)
dA = Measure('ds', domain=mesh, subdomain_data=facets)
dV = Measure('dx', domain=mesh, subdomain_data=cells)

Scalar = FunctionSpace(mesh, 'P', 1)
Vector = VectorFunctionSpace(mesh, 'P', 1)
Tensor = TensorFunctionSpace(mesh, 'P', 1)
#phi, u, T
Space = MixedFunctionSpace([Scalar, Vector, Scalar])

mesh_board = SubMesh(mesh, cells, 1)
mesh_via = SubMesh(mesh, cells, 2)
Scalar_board = FunctionSpace(mesh_board, 'P', 1)
Scalar_via = FunctionSpace(mesh_via, 'P', 1)
Vector_via = VectorFunctionSpace(mesh_via, 'P', 1)
Tensor_via = TensorFunctionSpace(mesh_via, 'P', 1)

delta = Identity(3)
epsilon = as_tensor([ ( (0,0,0),(0,0,1),(0,-1,0) ), (
```

```
    ↪ (0,0,-1),(0,0,0),(1,0,0) ) , ( (0,1,0),(-1,0,0)
    ↪ ,(0,0,0) ) ])
T_ref = 300.0 # K
T_amb = T_ref
eps_0 = 8.85E-12 #in A s/(V m)
mu_0 = 12.6E-7 #in V s/(A m)
null = 1E-20

i,j,k,l,m,n,o,p = indices(8)

freq = 10.
cycle = 1.
tMax = cycle/freq
Dt = tMax/(cycle*10.)
t = 0.0

bc_in = Expression('amp*sin(2.0*pi*nu*time)', amp=12., nu=
    ↪ freq, time=0)
bc01=DirichletBC(Space.sub(0), bc_in, facets, 1)
bc02=DirichletBC(Space.sub(0), 0.0, facets, 2)
bc03=DirichletBC(Space.sub(1), Constant((0.,0.,0.)), facets,
    ↪ 3)

bc = [bc01,bc02,bc03]

dunkn = TrialFunction(Space)
test = TestFunction(Space)
unkn = Function(Space)
unkn0 = Function(Space)
unkn00 = Function(Space)

unkn_init = Expression(('0.', '0.','0.','0.', 'T_int'),T_int=
    ↪ T_amb)
unkn00 = interpolate(unkn_init,Space)
unkn0.assign(unkn00)
unkn.assign(unkn0)

del_phi,del_u,del_T = split(test)
phi,u,T = split(unkn)
phi0,u0,T0 = split(unkn0)
phi00,u00,T00 = split(unkn00)

def VoigtToTensor(A):
    A11, A12, A13, A14, A15, A16 = A[0,0], A[0,1], A[0,2], A
        ↪ [0,3], A[0,4], A[0,5]
    A22, A23, A24, A25, A26 = A[1,1], A[1,2], A[1,3], A[1,4],
        ↪ A[1,5]
    A33, A34, A35, A36 = A[2,2], A[2,3], A[2,4], A[2,5]
    A44, A45, A46 = A[3,3], A[3,4], A[3,5]
    A55, A56 = A[4,4], A[4,5]
    A66 = A[5,5]
    A21, A31, A41, A51, A61 = A12, A13, A14, A15, A16
    A32, A42, A52, A62 = A23, A24, A25, A26
    A43, A53, A63 = A34, A35, A36
```

```
    A54, A64 = A45, A46
    A65 = A56
    return as_tensor([ \
    [ \
    [ [A11,A16,A15], [A16,A12,A14], [A15,A14,A13]] , \
    [ [A61,A66,A65], [A66,A62,A64], [A65,A64,A63]] , \
    [ [A51,A56,A55], [A56,A52,A54], [A55,A54,A53]] \
    ] , [ \
    [ [A61,A66,A65], [A66,A62,A64], [A65,A64,A63]] , \
    [ [A21,A26,A25], [A26,A22,A24], [A25,A24,A23]] , \
    [ [A41,A46,A45], [A46,A42,A44], [A45,A44,A43]] \
    ] , [ \
    [ [A51,A56,A55], [A56,A52,A54], [A55,A54,A53]] , \
    [ [A41,A46,A45], [A46,A42,A44], [A45,A44,A43]] , \
    [ [A31,A36,A35], [A36,A32,A34], [A35,A34,A33]] ] \
    ])

#laminate materials data (for board)
varsigma_l = null #in S/m
pi_l = null #V/K
kappa_l =  1.3 #in W/(m K)
capacity_l = 800. #in J / (kg K)
alp11_l = 13.2E-6 #in 1/K
alp22_l = 16.7E-6 #in 1/K
alp33_l = 39E-6 #in 1/K
alpha_l = as_tensor([ [alp11_l ,0. ,0.] ,[0. , alp22_l ,0.] ,[0. ,0. ,
    ↪ alp33_l] ])
h_l = 10. #in J / (s m^2 K)
rho_l = 2.5E3 #in kg/m^3
C_voigt_l = numpy.array([ \
[66242.E6, 41797.E6, 37814.E6, 0, 0, 0],\
[41797.E6, 50460.E6, 32290.E6, 0, 0, 0],\
[37814.E6, 32290.E6, 31591.E6, 0, 0, 0],\
[0, 0, 0, 2250.E6, 0, 0],\
[0, 0, 0, 0, 2250.E6, 0],\
[0, 0, 0, 0, 0, 6630.E6]  ]) #in Pa
C_l = VoigtToTensor(C_voigt_l)

#copper data for a via in Printed Circuit Boards (PCBs)
varsigma_cu = 1.7E4 #in S/m
pi_cu = 1.8E-6 #V/K
kappa_cu = 385. #in W/(m K)
capacity_cu = 390. #in J / (kg K)
alp_cu = 17E-6 #in 1/K
alpha_cu = as_tensor([ [alp_cu ,0. ,0.] ,[0. , alp_cu ,0.] ,[0. ,0. ,
    ↪ alp_cu] ])
h_cu = 10. #in J / (s m^2 K)
rho_cu = 8.94E3 #kg/m^3
sigmaY_cu = 100.E6 #in Pa
h_pl_cu = 615.E6 #in Pa
p1,p2,p3=169100.E6,122200.E6,75420.E6 #in Pa
C_voigt_cu = numpy.array([ \
[p1, p2, p2, 0, 0, 0],\
[p2, p1, p2, 0, 0, 0],\
```

```
[p2, p2, p1, 0, 0, 0],\
[0, 0, 0, p3, 0, 0],\
[0, 0, 0, 0, p3, 0],\
[0, 0, 0, 0, 0, p3]  ])
C_cu = VoigtToTensor(C_voigt_cu)

rho = material_coefficient(mesh, cells, [rho_l, rho_cu])
varsigma = material_coefficient(mesh, cells, [varsigma_l, varsigma_cu])
pi_ = material_coefficient(mesh, cells, [pi_l, pi_cu])
kappa = material_coefficient(mesh, cells, [kappa_l, kappa_cu])
capacity = material_coefficient(mesh, cells, [capacity_l, capacity_cu])
h = material_coefficient(mesh, cells, [h_l, h_cu])

eps = sym(grad(u))
eps0 = sym(grad(u0))
epsDot = (eps-eps0)/Dt
v = as_tensor( (u-u0)[i]/Dt, (i,))
v0 = as_tensor( (u0-u00)[i]/Dt, (i,))
E = -grad(phi)
E0 = -grad(phi0)
EE = E
D = eps_0*E
D0 = eps_0*E0
JJ = as_tensor( varsigma*pi_*T.dx(i) + varsigma*EE[i], (i,))
J = as_tensor( D[j].dx(j)*v[i] + JJ[i], (i,))
FF = as_tensor(D[j].dx(j)*E[i], (i,))
q = as_tensor(-kappa*T.dx(i)+varsigma*pi_*T*EE[i], (i,))
Phi = q/T

f = Constant((0.,0.,0.))
r = Constant(0.0)

#laminate
peps_eq_l = Function(Scalar_board)
teps_l = as_tensor( alpha_l[i,j]*(T-T_ref), (i,j))
sigma_l = as_tensor( C_l[i,j,k,l]*(eps[k,l] - teps_l[k,l]), (i,j))
eta_l = as_tensor( capacity*ln(T/T_ref) + alpha_l[i,j]/rho*sigma_l[i,j], ())
eta0_l = as_tensor( capacity*ln(T0/T_ref) + C_l[i,j,k,l]*alpha_l[k,l]/rho*eps0[i,j], ())
Sigma_l = as_tensor(-q[i]/T/T*T.dx(i) + JJ[i]/T*EE[i] , ())

#copper
gamma_cu = Function(Scalar)
sigma0_cu = Function(Tensor)
peps0_cu = Function(Tensor)
beta0_cu = Function(Tensor)
peps_eq_cu = Function(Scalar_via)

dev_sigma0_cu = as_tensor(sigma0_cu[i,j]-1./3.*sigma0_cu[k,k
```

```
]*delta[i,j] , (i,j))
pepsDot_cu = as_tensor(gamma_cu*(dev_sigma0_cu[i,j]-beta0_cu[i,j])*C_cu[i,j,k,l]*epsDot[k,l] / (4./9.*sigmaY_cu**2*h_pl_cu +(dev_sigma0_cu[i,j]-beta0_cu[i,j])*C_cu[i,j,k,l]*(dev_sigma0_cu[k,l]-beta0_cu[k,l]) ) * (dev_sigma0_cu[m,n]-beta0_cu[m,n]) , (m,n))
teps_cu = as_tensor( alpha_cu[i,j]*(T-T_ref), (i,j))
tepsDot_cu = as_tensor( alpha_cu[i,j]*(T-T0)/Dt, (i,j))
sigmaDot_cu = as_tensor( C_cu[i,j,k,l]*( epsDot[k,l]-tepsDot_cu[k,l]-pepsDot_cu[k,l] ), (i,j))

betaDot_cu = as_tensor( gamma_cu*(dev_sigma0_cu[k,l]-beta0_cu[k,l])*sigmaDot_cu[k,l] / (2.0/3.0*sigmaY_cu**2)*(dev_sigma0_cu[i,j]-beta0_cu[i,j]) ,(i,j))

sigma_cu = sigma0_cu + Dt*sigmaDot_cu
beta_cu = beta0_cu + Dt*betaDot_cu
peps_cu = peps0_cu + Dt*pepsDot_cu

eta_cu = as_tensor( capacity*ln(T/T_ref) + alpha_cu[i,j]/rho*sigma_cu[i,j], ())
eta0_cu = as_tensor( capacity*ln(T0/T_ref) + alpha_cu[i,j]/rho*sigma0_cu[i,j], ())
Sigma_cu = as_tensor(-q[i]/T/T*T.dx(i) + JJ[i]/T*EE[i] + sigma_cu[i,j]/T*peps_cu[i,j] , ())


F_phi = (-(D-D0)[i]*del_phi.dx(i) - Dt*J[i]*del_phi.dx(i) )*(dV(1)+dV(2)) + N('+')[i]*Dt*(J('+')-J('-'))[i]*del_phi('+')*dI(6) + N[i]*Dt*J[i]*del_phi*(dA(1)+dA(2)+dA(3)+dA(4)+dA(5))

F_u = (rho*(v-v0)[i]/Dt*del_u[i] - rho*f[i]*del_u[i] - FF[i]*del_u[i] )*(dV(1)+dV(2)) + sigma_l[j,i]*del_u[i].dx(j)*dV(1) + sigma_cu[j,i]*del_u[i].dx(j)*dV(2)

F_T = ( - Dt*Phi[i]*del_T.dx(i) - Dt*rho*r/T*del_T )*(dV(1)+dV(2)) + (rho*(eta_l-eta0_l)*del_T - Dt*Sigma_l*del_T )*dV(1) + (rho*(eta_cu-eta0_cu)*del_T - Dt*Sigma_cu*del_T )*dV(2) + Dt*h*(T-T_ref)/T*del_T*(dA(1)+dA(2)+dA(3)+dA(4)+dA(5))


Form = F_phi + F_u + F_T
Gain = derivative(Form, unkn, dunkn)

pwd='/calcul/CR18/'
file_phi_via = File(pwd+'phi.pvd')
file_J_via = File(pwd+'J.pvd')
file_u = File(pwd+'u.pvd')
file_peps_via = File(pwd+'peps_via.pvd')
file_peps_board = File(pwd+'peps_board.pvd')
file_T = File(pwd+'T.pvd')
```

```
phi_via_ = Function(Scalar_via, name='$\phi$ in V')
J_via_ = Function(Vector_via, name='$|J_i|$ in A/m$^2$')
u_ = Function(Vector, name='$|u_i|$')
peps_via_ = Function(Scalar_via, name='$^\mathrm{p}\!\
    ↪ varepsilon_\mathrm{eq.}$')
T_ = Function(Scalar, name='$T$ in K')

while t < tMax:
    print 'time: ',t
    bc_in.time = t
    tic()
    solve(Form==0, unkn, bc, J=Gain, \
        solver_parameters={"newton_solver":{"linear_solver":
        ↪ "mumps", "relative_tolerance": 1e-5} }, \
        form_compiler_parameters={"cpp_optimize": True, "
            ↪ representation": "quadrature", "
            ↪ quadrature_degree": 2} )

    phi_via_.assign(project(unkn.split(deepcopy=True)[0],
        ↪ Scalar_via))
    J_via_.assign(project(J,Vector_via))
    u_.assign(unkn.split(deepcopy=True)[1])
    T_.assign(unkn.split(deepcopy=True)[2])
    file_phi_via << (phi_via_, t)
    file_J_via << (J_via_, t)
    file_u << (u_, t)
    file_T << (T_, t)
    print toc(),' seconds long computed, T max: ',max(unkn.
        ↪ split(deepcopy=True)[2].vector().array())

    sigma_cu_ = project(sigma_cu, Tensor, solver_type="mumps"
        ↪ , \
        form_compiler_parameters={"cpp_optimize": True, "
            ↪ representation": "quadrature", "
            ↪ quadrature_degree": 2})
    sigma0_cu.assign(sigma_cu_)

    beta_cu_ = project(beta_cu, Tensor, solver_type="mumps",
        ↪ \
        form_compiler_parameters={"cpp_optimize": True, "
            ↪ representation": "quadrature", "
            ↪ quadrature_degree": 2})
    beta0_cu.assign(beta_cu_)
    flow_ = project(1./2.*(dev_sigma0_cu[i,j]-beta0_cu[i,j])
        ↪ *(dev_sigma0_cu[i,j]-beta0_cu[i,j]) -1./3.*
        ↪ sigmaY_cu**2, \
        Scalar, solver_type="mumps", form_compiler_parameters
            ↪ ={"cpp_optimize": True, "representation": "
            ↪ quadrature", "quadrature_degree": 2})
    flow_bool = flow_.vector().array() >= 0.

    direction_ = project((dev_sigma0_cu[i,j]-beta0_cu[i,j])*
        ↪ epsDot[i,j], \
        Scalar, solver_type="mumps", form_compiler_parameters
```

```
        ↪ ={"cpp_optimize": True, "representation": "
        ↪ quadrature", "quadrature_degree": 2})

    direction_bool = direction_.vector().array() > 0
    gamma_cu.vector()[:] = numpy.array(flow_bool*
        ↪ direction_bool)

    peps_eq_ = project(peps_eq_cu + Dt*(2./3.*pepsDot_cu[i,j
        ↪ ]*pepsDot_cu[i,j])**0.5, \
        Scalar_via, solver_type="mumps",
        ↪ form_compiler_parameters={"cpp_optimize": True
        ↪ , "representation": "quadrature", "
        ↪ quadrature_degree": 2})
    print 'Plasticity: ',max(gamma_cu.vector().array())
    peps_eq_cu.assign(peps_eq_)
    file_peps_via << (peps_eq_cu,t)
    file_peps_board << (peps_eq_l,t)

    unkn00.assign(unkn0)
    unkn0.assign(unkn)
    t += Dt
```

To-do

The accumulated plastic strain is a good measure for estimating fatigue. There are many different proposals for this estimation. Find out some of them by searching for:

- Coffin–Manson model,
- Paris's law,
- Basquin equation.

Estimating fatigue is of paramount importance. Unfortunately, there have been fatigue failures in the history, where a fatigue was not foreseen but lead to a disaster: British Overseas Airways Corporation Flight 781, on January 10, 1954; Alexander Kielland Platform in North Sea, on March 27, 1980; Petrochemical Plant in Edmonton, Canada, on April 18, 1982; Japan Airlines Flight 123, on August 12, 1985; Aloha Airlines Flight 243, Boeing 737, on April 28, 1988; Deutsche Bahn Hanover-Hamburg train in Eschede/Germany on June 3, 1998; Crude Oil Pipeline in Winchester Kentucky, US, on January 27, 2000; China Airlines Flight CI611, B-18255, on May 25, 2002.

3.5 Piezoelectric Transducer

In 1880 Pierre and Paul-Jacques Curie brothers[58] had discovered an interesting phenomenon occurring on some non-conductive materials. An applied voltage results in stretching (or shrinking), thus, creating a mechanical stress in these materials. Nowadays, this property is used in *actuators*; an applied electric potential (leading to an electric field) generates a displacement (motion). The reverse relation holds, too. In *sensors* a force acts on this type of a material and generates an electric field measured as an electric potential difference. This coupling of electric field with mechanical stress is called *piezoelectricity*. Actually, we all experience this effect in tendons, which connect muscles to bones and shorten by an applied electric signal sent by nerves.

There are organic, natural, and synthetic piezoelectric materials. Starting from 1950s quartz (SiO_2), barium-titanate ($BaTiO_3$), and later on lead-zirconate-titanate (PZT[59]) are known as *piezoceramics* showing the phenomenon of piezoelectricity. There are even polymers having piezoelectric properties, for example, polyvinylidene-fluoride (PVDF). Another coupling between temperature and electric field is known as *pyroelectricity*. A temperature change generates an electric field as a material specific property. The inverse coupling also occurs. The piezo- and pyroelectricity are electromagnetic coupling effects used in pressure and thermal sensors, respectively. We want to include them into our computational reality.

Electromagnetic coupling effects such as piezo- and pyroelectricity occur in polarized materials. For polarized materials the GIBBS equation is given differently in the literature. Especially, the different formulations for the electromagnetic coupling is quite confusing. We want to derive the GIBBS equation and present a general methodology leading to the electromagnetic coupling such that the different formulations can be comprehended. No formulation is better than another, in principle, they are all admissible. Only by experimental validation we can verify their accuracy. The reality may be modeled in various ways. We need to understand the differences between the formulations such that we can estimate the most feasible selection in a specific system.

For the sake of presenting the differences between various formulations in the literature, it is beneficial to start with the balance of electromagnetic momentum:

$$\frac{\partial \mathcal{G}_i}{\partial t} = \frac{\partial m_{ji}}{\partial x_j} - \mathcal{F}_i \,, \tag{3.190}$$

where the relation is defined between an electromagnetic momentum density, $\mathcal{G}_i$, an electromagnetic flux term, m_{ji}, and an electromagnetic supply term, $\mathcal{F}_i$. The electromagnetic momentum density, $\mathcal{G}_i$, is in the same unit of (linear or angular) momentum density such as the flux term is a stress and the supply term is a force

---

[58] Pierre Curie had been the husband of Marie Skłodowska-Curie.

[59] PZT is the general name for lead-zirconate-titanate, $Pb_{1.1}Zr_{0.3}Ti_{0.7}O_3$ and $PbZr_{0.52}Ti_{0.48}O_3$ are the common used compositions of PZT.

density. The definition of the electromagnetic stress, m_{ji}, depends on the definition of the electromagnetic force density, $\mathcal{F}_i$. Their definitions are in some sense arbitrary; however, as given in Eq. (3.190), the divergence of the electromagnetic stress minus the electromagnetic force density is unique and equals the rate of the electromagnetic momentum density. This methodology leads to a consistent formulation.[60] We only need the definition of the electromagnetic momentum density, $\mathcal{G}_i$. If we have a definition for the electromagnetic momentum then we can propose an electromagnetic stress[61] as well as an electromagnetic force satisfying Eq. (3.190).

Unfortunately, the definition of the electromagnetic momentum density, $\mathcal{G}_i$, is discussed heavily in the literature without a general consensus.[62] There are mainly three different variants for the definition of the electromagnetic momentum, viz., the POYNTING vector, $\mathcal{G}_i^{\mathrm{P}}$, MINKOWSKI's definition, $\mathcal{G}_i^{\mathrm{M}}$, and ABRAHAM's choice, $\mathcal{G}_i^{\mathrm{A}}$, as follows[63]

$$\mathcal{G}_i^{\mathrm{P}} = (\boldsymbol{D} \times \boldsymbol{B})_i \ , \quad \mathcal{G}_i^{\mathrm{M}} = (\boldsymbol{\mathfrak{D}} \times \boldsymbol{B})_i \ , \quad \mathcal{G}_i^{\mathrm{A}} = \frac{1}{c^2}(\boldsymbol{E} \times \boldsymbol{\mathfrak{H}})_i \ . \tag{3.191}$$

Each definition has been proposed by using an evident motivation and convincing *gedankenexperiments*[64] such that each can be declared as correct.[65] In order to compare them with each other, we employ

$$c^2 = \frac{1}{\varepsilon_0 \mu_0} \ , \tag{3.192}$$

and the MAXWELL–LORENTZ aether relations:

$$D_i = \varepsilon_0 E_i \ , \quad H_i = \frac{1}{\mu_0} B_i \ . \tag{3.193}$$

Moreover, we need to employ the following relations between charge and current potentials:

---

[60] See [15, Chap. XIV].

[61] The same formulation in Eq. (3.190) can also be found by using an electromagnetic *energy-momentum* tensor, S_{ij}, where this tensor is just the negative of the electromagnetic stress, $S_{ij} = -m_{ij}$. Confusingly, in some books the divergence is taken regarding the second index. For the electromagnetic stress as well as for the energy-momentum tensor the choice matters, since they are not required to be symmetric.

[62] See for example [14, 25, 29].

[63] The different electromagnetic momenta are named after John Henry Poynting, Hermann Minkowski, Max Abraham.

[64] Thought experiments.

[65] See [8] for a review about experimental evidences for each choice. In [7, Sect. 4.4 and 8.7] there is some insight about different choices of electrodynamic forces and energies in the literature. See [20] for an experimental discussion about ABRAHAM and MINKOWSKI choices. See [6] for simulations of experiments with different choices and their comparison to experiments from the literature.

$$D_i = \mathfrak{D}_i - P_i \ , \quad H_i = \mathfrak{H}_i + M_i + (\boldsymbol{P} \times \boldsymbol{v})_i = \mathfrak{H}_i + \mathcal{M}_i \ , \tag{3.194}$$

which are introduced and discussed in Sect. 3.2. We use an *objective* magnetic polarization, $\mathcal{M}_i = M_i + (\boldsymbol{P} \times \boldsymbol{v})_i$, since an experiment fails to distinguish between a magnetic polarization, M_i, and a rotary motion of particles with an electric polarization, $\boldsymbol{P} \times \boldsymbol{v}$. Not only the formulation is simpler but the term $\mathcal{M}_i$ is an objective tensor: It transforms according to the tensor laws. Now by using the latter relations, we can rewrite ABRAHAM's electromagnetic momentum density:

$$\mathcal{G}_i^{\mathrm{A}} = \varepsilon_0 \mu_0 \epsilon_{ijk} E_j \mathfrak{H}_k = \epsilon_{ijk} D_j (B_k - \mu_0 \mathcal{M}_k) = \mathcal{G}_i^{\mathrm{P}} - \mu_0 (\boldsymbol{D} \times \boldsymbol{\mathcal{M}})_i \ . \tag{3.195}$$

Analogously for MINKOWKI's choice we obtain

$$\mathcal{G}_i^{\mathrm{M}} = \epsilon_{ijk} \mathfrak{D}_j B_k = \epsilon_{ijk} (D_j + P_j) B_k = \mathcal{G}_i^{\mathrm{P}} + (\boldsymbol{P} \times \boldsymbol{B})_i \ . \tag{3.196}$$

Obviously, for a material without polarization, all choices are identical to the POYNTING vector. The formulation by choosing $\mathcal{G}_i = \mathcal{G}_i^{\mathrm{P}}$ is often used in textbooks about electrodynamics. This choice is certainly correct for a material without polarization. Suppose that the system is unpolarized and we use POYNTING vector as the electromagnetic momentum. After defining the electromagnetic momentum, we can now select a possible set of electromagnetic stress and force. This derivation is well-known in the literature and it is often referred to as POYNTING's equation. We start off with the following MAXWELL equations expressed in (fixed) Cartesian coordinates

$$-\frac{\partial D_i}{\partial t} + \epsilon_{ijk} \frac{\partial H_k}{\partial x_j} = J_i \ , \quad \frac{\partial B_i}{\partial t} + \epsilon_{ijk} \frac{\partial E_k}{\partial x_j} = 0 \ . \tag{3.197}$$

Then we calculate the rate of the POYNTING vector by inserting the MAXWELL equations

$$\begin{aligned}
\frac{\partial}{\partial t} (\boldsymbol{D} \times \boldsymbol{B})_i &= \epsilon_{ijk} \frac{\partial D_j B_k}{\partial t} = \epsilon_{ijk} \frac{\partial D_j}{\partial t} B_k + \epsilon_{ijk} D_j \frac{\partial B_k}{\partial t} - \\
&= \epsilon_{ijk} \Big(\epsilon_{jlm} \frac{\partial H_m}{\partial x_l} - J_j \Big) B_k - \epsilon_{ijk} D_j \epsilon_{klm} \frac{\partial E_m}{\partial x_l} = \\
&= -\frac{\partial H_k}{\partial x_i} B_k + \frac{\partial H_i}{\partial x_k} B_k - (\boldsymbol{J} \times \boldsymbol{B})_i - D_j \frac{\partial E_j}{\partial x_i} + D_j \frac{\partial E_i}{\partial x_j} \ ,
\end{aligned} \tag{3.198}$$

where $\epsilon_{ijk} = -\epsilon_{ikj}$ and $\epsilon_{ijk}\epsilon_{klm} = \delta_{il}\delta_{jm} - \delta_{im}\delta_{jl}$ have been used. The first and forth terms of the latter can be rewritten by using the MAXWELL–LORENTZ aether relations:

$$-\frac{\partial H_k}{\partial x_i}B_k - D_j\frac{\partial E_j}{\partial x_i} = -\frac{1}{\mu_0}\frac{\partial B_k}{\partial x_i}B_k - \varepsilon_0 E_k\frac{\partial E_k}{\partial x_i} = -\frac{1}{2\mu_0}\frac{\partial}{\partial x_j}(B_k B_k \delta_{ij}) - \\ -\frac{\varepsilon_0}{2}\frac{\partial}{\partial x_j}(E_k E_k \delta_{ij}) = -\frac{\partial}{\partial x_j}\left(\frac{1}{2}\delta_{ji}(H_k B_k + D_k E_k)\right). \tag{3.199}$$

Next, we employ the following MAXWELL equations:

$$\frac{\partial B_i}{\partial x_i} = 0\,, \quad \frac{\partial D_i}{\partial x_i} = \rho z\,. \tag{3.200}$$

By using the latter we can rewrite the second and fifth terms in Eq. (3.198) as follows

$$\frac{\partial H_i}{\partial x_j}B_j + D_j\frac{\partial E_i}{\partial x_j} = \frac{\partial}{\partial x_j}(H_i B_j + D_j E_i) - E_i \rho z\,. \tag{3.201}$$

Hence we obtain the POYNTING equation:

$$\frac{\partial}{\partial t}(\boldsymbol{D}\times\boldsymbol{B})_i = \frac{\partial}{\partial x_j}\left(-\frac{1}{2}\delta_{ji}(H_k B_k + D_k E_k) + H_i B_j + D_j E_i\right) - \\ -\rho z E_i - (\boldsymbol{J}\times\boldsymbol{B})_i\,. \tag{3.202}$$

This equation equals the balance of electromagnetic momentum in Eq. (3.190) by associating

$$\mathcal{G}_i = \mathcal{G}_i^{\mathrm{P}} = (\boldsymbol{D}\times\boldsymbol{B})_i\,, \\ m_{ji} = -\frac{1}{2}\delta_{ji}(H_k B_k + D_k E_k) + H_i B_j + D_j E_i\,, \\ \mathcal{F}_i = \rho z E_i + (\boldsymbol{J}\times\boldsymbol{B})_i\,. \tag{3.203}$$

The electromagnetic stress, m_{ji}, is referred to as MAXWELL's stress[66] tensor and the supply term is called LORENTZ's force (density). An unpolarized matter "feels" the volumetric force density $\mathcal{F}_i = \rho z E_i + (\boldsymbol{J}\times\boldsymbol{B})_i$ in every particles.

For a polarized material we may propose $\mathcal{G}_i^{\mathrm{M}}$ or $\mathcal{G}_i^{\mathrm{A}}$ as the correct electromagnetic momentum density.[67] We choose $\mathcal{G}_i^{\mathrm{M}}$ in the following. With the analogous calculation we obtain

$$\mathcal{G}_i^{\mathrm{M}} = \mathcal{G}_i^{\mathrm{P}} + \epsilon_{ijk}P_j B_k\,, \quad m_{ji} = -\frac{1}{2}\delta_{ji}(H_k B_k + D_k E_k) + H_i B_j + D_j E_i\,, \\ \mathcal{F}_i = \rho z E_i + \epsilon_{ijk}J_j B_k - \epsilon_{ijk}\frac{\partial P_j}{\partial t}B_k - \epsilon_{ijk}P_j\frac{\partial B_k}{\partial t}\,. \tag{3.204}$$

---

[66] It is named for James Clerk Maxwell.

[67] A thermodynamical formulation for the choice of $\mathcal{G}_i^{\mathrm{A}}$ can be found in [15, Chap. XIV, Sect. 2].

The electromagnetic stress is chosen as MAXWELL's stress and the supply term is changed such that Eq. (3.190) holds. A polarized matter "feels" additional volumetric forces due to the electric polarization and its rate.

We develop a thermodynamical formulation of polarized materials by starting with the balance of mass and the balance of linear momentum. The balance of mass expressed in a fixed, Cartesian coordinate system reads

$$\frac{\partial \rho}{\partial t} + \frac{\partial \rho v_i}{\partial x_i} = 0 . \tag{3.205}$$

Mass is a conserved quantity since its balance equation lacks a production term. Total momentum has to be a conserved quantity, too. Since the polarized material generates an electromagnetic momentum density, $\mathcal{G}_i$, in addition to the mechanical momentum density, ρv_i, balance of the total momentum becomes

$$\frac{\partial \rho v_i + \mathcal{G}_i}{\partial t} - \frac{\partial}{\partial x_j}\big(- v_j \rho v_i + \sigma_{ji} + m_{ji}\big) - \rho f_i = 0 . \tag{3.206}$$

We can insert the balance of the electromagnetic momentum in Eq. (3.190) in order to obtain

$$\frac{\partial \rho v_i}{\partial t} - \frac{\partial}{\partial x_j}\big(- v_j \rho v_i + \sigma_{ji}\big) - \rho f_i = \mathcal{F}_i , \tag{3.207}$$

with the electromagnetic supply term as given in Eq. (3.204)$_3$. By inserting balance of mass into the balance of total momentum we acquire

$$\rho \frac{\mathrm{d}v_i}{\mathrm{d}t} - \frac{\partial \sigma_{ji}}{\partial x_j} - \rho f_i = \mathcal{F}_i . \tag{3.208}$$

We write the electromagnetic supply term on the right-hand side for denoting the fact that it tends to be a production term since it cannot be "switched off." A moving particle with an electric charge creates this supply term. We fail to establish a configuration where this supply term vanishes. The mechanical momentum possesses a production term, thus, it is not a conserved quantity in case of a polarized material. The total momentum is a conserved quantity.

By multiplying the latter form of the balance of momentum with v_i and using the mass balance once more, we obtain the balance of kinetic energy:

$$\begin{aligned} \rho \frac{\mathrm{d}}{\mathrm{d}t}\Big(\frac{1}{2} v_i v_i\Big) - \frac{\partial \sigma_{ji} v_i}{\partial x_j} - \rho f_i v_i &= -\sigma_{ji} \frac{\partial v_i}{\partial x_j} + \mathcal{F}_i v_i , \\ \frac{\partial}{\partial t}\Big(\rho \frac{1}{2} v_i v_i\Big) - \frac{\partial}{\partial x_j}\Big(- \rho v_j \frac{1}{2} v_i v_i + \sigma_{ji} v_i\Big) - \rho f_i v_i &= -\sigma_{ji} \frac{\partial v_i}{\partial x_j} + \mathcal{F}_i v_i . \end{aligned} \tag{3.209}$$

The final term is a production term due to the supply term of electromagnetic momentum given in Eq. (3.204)$_3$. We will rewrite this term in order to see a relation leading to the GIBBS equation in polarized materials. The formulation is lengthy and unfortunately not straight-forward. Therefore, we present it herein in a fully detailed manner. We start by writing the term explicitly,

$$\mathcal{F}_i v_i = \frac{\partial D_j}{\partial x_j} E_i v_i + v_i \epsilon_{ijk} J_j B_k - v_i \epsilon_{ijk} \frac{\partial P_j}{\partial t} B_k + v_i \epsilon_{ijk} P_j \epsilon_{klm} \frac{\partial E_m}{\partial x_l} , \tag{3.210}$$

after inserting MAXWELL's equations (3.197)$_2$ and (3.200)$_2$. We can also insert Eqs. (3.194) into MAXWELL's equations and obtain

$$\frac{\partial \mathfrak{D}_i}{\partial x_i} = \rho z^{\text{fr.}} , \quad -\frac{\partial \mathfrak{D}_i}{\partial t} + \epsilon_{ijk} \frac{\partial \mathfrak{H}_k}{\partial x_j} = J_i - \frac{\partial P_i}{\partial t} - \epsilon_{ijk} \frac{\partial \mathcal{M}_k}{\partial x_j} = J_i^{\text{fr.}} . \tag{3.211}$$

The latter equations are known as MAXWELL's equation in polarized matter, they have been introduced in Sect. 3.2. The definition of the free current, $J_i^{\text{fr.}}$, can be inserted into the production term of electromagnetic momentum as follows

$$\begin{aligned} \mathcal{F}_i v_i = & \frac{\partial D_j}{\partial x_j} E_i v_i + v_i \epsilon_{ijk} \Big(J_j^{\text{fr.}} + \frac{\partial P_j}{\partial t} + \epsilon_{jlm} \frac{\partial \mathcal{M}_m}{\partial x_l} \Big) B_k - \\ & - v_i \epsilon_{ijk} \frac{\partial P_j}{\partial t} B_k + v_i \epsilon_{ijk} P_j \epsilon_{klm} \frac{\partial E_m}{\partial x_l} . \end{aligned} \tag{3.212}$$

By using the tensorial identities:

$$\epsilon_{ijk} = -\epsilon_{jik} , \quad \epsilon_{jik} \epsilon_{jlm} = \delta_{il} \delta_{km} - \delta_{im} \delta_{kl} , \tag{3.213}$$

we acquire

$$\begin{aligned} \mathcal{F}_i v_i = & \frac{\partial D_j}{\partial x_j} E_i v_i + v_i \epsilon_{ijk} J_j^{\text{fr.}} B_k - v_i \left(\frac{\partial \mathcal{M}_k}{\partial x_i} - \frac{\partial \mathcal{M}_i}{\partial x_k} \right) B_k + \\ + v_i P_j \left(\frac{\partial E_j}{\partial x_i} - \frac{\partial E_i}{\partial x_j} \right) = & \frac{\partial D_j E_i}{\partial x_j} v_i - (D_j + P_j) \frac{\partial E_i}{\partial x_j} v_i + v_i \epsilon_{ijk} J_j^{\text{fr.}} B_k - \\ & - v_i \frac{\partial \mathcal{M}_k}{\partial x_i} B_k + v_i \frac{\partial \mathcal{M}_i}{\partial x_k} B_k + v_i P_j \frac{\partial E_j}{\partial x_i} . \end{aligned} \tag{3.214}$$

After inserting Eqs. (3.200)$_1$, (3.194) into the latter, we obtain

$$\mathcal{F}_i v_i = \frac{\partial D_j E_i}{\partial x_j} v_i - \mathfrak{D}_j \frac{\partial E_i}{\partial x_j} v_i + v_i \epsilon_{ijk} J_j^{\text{fr.}} B_k - \left(\frac{\mathrm{d}\mathcal{M}_k}{\mathrm{d}t} - \frac{\partial \mathcal{M}_k}{\partial t} \right) B_k +$$
$$+ v_i \frac{\partial \mathcal{M}_i B_k}{\partial x_k} + P_j \left(\frac{\mathrm{d}E_j}{\mathrm{d}t} - \frac{\partial E_j}{\partial t} \right) = \frac{\partial D_j E_i}{\partial x_j} v_i - \mathfrak{D}_j \frac{\partial E_i}{\partial x_j} v_i + v_i \epsilon_{ijk} J_j^{\text{fr.}} B_k -$$
$$- \left(\frac{\mathrm{d}\mathcal{M}_k}{\mathrm{d}t} - \frac{\partial \mathcal{M}_k}{\partial t} \right) B_k + v_i \frac{\partial \mathcal{M}_i B_k}{\partial x_k} + P_j \frac{\mathrm{d}E_j}{\mathrm{d}t} - \frac{\partial P_j E_j}{\partial t} + E_j \frac{\partial P_j}{\partial t} . \tag{3.215}$$

By using the material conduction current:

$$\mathcal{J}_i^{\text{fr.}} = J_i^{\text{fr.}} - \rho z^{\text{fr.}} v_i , \tag{3.216}$$

as well as Eq. (3.211)$_2$, we acquire

$$\mathcal{F}_i v_i = \frac{\partial D_j E_i}{\partial x_j} v_i - \mathfrak{D}_j \frac{\partial E_i}{\partial x_j} v_i + v_i \epsilon_{ijk} (\mathcal{J}_j^{\text{fr.}} + \rho z^{\text{fr.}} v_j) B_k - B_i \frac{\mathrm{d}\mathcal{M}_i}{\mathrm{d}t} +$$
$$+ \frac{\partial B_i \mathcal{M}_i}{\partial t} - \mathcal{M}_i \frac{\partial B_i}{\partial t} + v_i \frac{\partial \mathcal{M}_i B_k}{\partial x_k} + P_j \frac{\mathrm{d}E_j}{\mathrm{d}t} - \frac{\partial P_j E_j}{\partial t} - E_j \frac{\partial D_j}{\partial t} -$$
$$+ E_j \left(\epsilon_{jkl} \frac{\partial \mathfrak{H}_l}{\partial x_k} - J_j^{\text{fr.}} \right) = v_i \frac{\partial}{\partial x_j} (D_j E_i - \mathfrak{D}_j E_i + \mathcal{M}_i B_j) + v_i \frac{\partial \mathfrak{D}_j}{\partial x_j} E_i + \tag{3.217}$$
$$+ \frac{\partial}{\partial t} (B_i \mathcal{M}_i - P_j E_j) - E_j \frac{\partial D_j}{\partial t} - v_i \epsilon_{jik} \mathcal{J}_j^{\text{fr.}} B_k - \mathcal{M}_i \frac{\partial B_i}{\partial t} +$$
$$+ P_i \frac{\mathrm{d}E_i}{\mathrm{d}t} - B_i \frac{\mathrm{d}\mathcal{M}_i}{\mathrm{d}t} + E_j \epsilon_{jkl} \frac{\partial \mathfrak{H}_l}{\partial x_k} - E_j J_j^{\text{fr.}} ,$$

since $\epsilon_{ijk} v_i v_j = 0$ owing to Eq. (3.213)$_1$. After inserting Eqs. (3.197)$_2$, (3.211)$_1$ into the latter, the production term becomes

$$\mathcal{F}_i v_i = v_i \frac{\partial}{\partial x_j} \big(- P_j E_i + \mathcal{M}_i B_j \big) + (v_i \rho z^{\text{fr.}} - J_i^{\text{fr.}}) E_i + \frac{\partial}{\partial t} \Big(B_i \mathcal{M}_i - P_j E_j \Big) -$$
$$- E_j \frac{\partial D_j}{\partial t} - (\mathcal{E}_j - E_j) \mathcal{J}_j^{\text{fr.}} + \mathcal{M}_i \epsilon_{ijk} \frac{\partial E_k}{\partial x_j} + P_i \frac{\mathrm{d}E_i}{\mathrm{d}t} - B_i \frac{\mathrm{d}\mathcal{M}_i}{\mathrm{d}t} + E_j \epsilon_{jkl} \frac{\partial \mathfrak{H}_l}{\partial x_k} =$$
$$= v_i \frac{\partial}{\partial x_j} \big(- P_j E_i + \mathcal{M}_i B_j \big) + \frac{\partial}{\partial t} \Big(B_i \mathcal{M}_i - P_j E_j \Big) - E_j \frac{\partial D_j}{\partial t} -$$
$$- \mathcal{E}_i \mathcal{J}_i^{\text{fr.}} + \mathcal{M}_i \epsilon_{ijk} \frac{\partial E_k}{\partial x_j} + P_i \frac{\mathrm{d}E_i}{\mathrm{d}t} - B_i \frac{\mathrm{d}\mathcal{M}_i}{\mathrm{d}t} + E_j \epsilon_{jkl} \frac{\partial \mathfrak{H}_l}{\partial x_k} . \tag{3.218}$$

The fifth and eight terms can be rewritten as follows

$$
\begin{aligned}
\mathcal{M}_i\epsilon_{ijk}\frac{\partial E_k}{\partial x_j}+E_j\epsilon_{jkl}\frac{\partial \mathfrak{H}_l}{\partial x_k}&=-\mathcal{M}_i\epsilon_{jik}\frac{\partial E_k}{\partial x_j}+E_j\epsilon_{jkl}\frac{\partial \mathfrak{H}_l}{\partial x_k}=\\
&=-\frac{\partial}{\partial x_j}\big(\epsilon_{jik}\mathcal{M}_i E_k\big)+E_k\bigg(\epsilon_{jik}\frac{\partial \mathcal{M}_i}{\partial x_j}+\epsilon_{klm}\frac{\partial \mathfrak{H}_m}{\partial x_l}\bigg)=\\
&=-\frac{\partial}{\partial x_j}\big(\epsilon_{jik}\mathcal{M}_i E_k\big)+E_k\epsilon_{kji}\frac{\partial H_i}{\partial x_j}=\\
&=-\frac{\partial}{\partial x_j}\big(\epsilon_{jik}\mathcal{M}_i E_k\big)+\frac{\partial}{\partial x_j}\big(E_k\epsilon_{kji}H_i\big)-\epsilon_{kji}\frac{\partial E_k}{\partial x_j}H_i=\\
=\frac{\partial}{\partial x_j}\Big(-\epsilon_{jik}\mathcal{M}_i E_k+\epsilon_{jik}H_i E_k\Big)-\frac{\partial B_i}{\partial t}H_i&=\frac{\partial}{\partial x_j}\big(\epsilon_{jik}\mathfrak{H}_i E_k\big)-\frac{\partial B_i}{\partial t}H_i\,,
\end{aligned}
\tag{3.219}
$$

where we have employed Eqs. (3.197)$_2$, (3.213)$_1$. Hence, by using the latter, Eq. (3.194)$_2$, and the MAXWELL–LORENTZ aether relations, we attain the second term of the production of kinetic energy in its final form:

$$
\begin{aligned}
\mathcal{F}_i v_i=&\frac{\partial}{\partial x_j}\Big((-P_jE_i+\mathcal{M}_iB_j)v_i+\big(\mathfrak{H}\times\boldsymbol{E}\big)_j\Big)+\\
&+\frac{\partial}{\partial t}\bigg(B_i\mathcal{M}_i-P_jE_j-\frac{1}{2}D_jE_j-\frac{1}{2}B_iH_i\bigg)-(-P_jE_i+\mathcal{M}_iB_j)\frac{\partial v_i}{\partial x_j}-\\
&-\mathcal{E}_i\mathcal{J}_i^{\text{fr.}}+P_i\frac{\mathrm{d}E_i}{\mathrm{d}t}-B_i\frac{\mathrm{d}\mathcal{M}_i}{\mathrm{d}t}\,.
\end{aligned}
\tag{3.220}
$$

The balance of kinetic energy can be rewritten by using the latter and by employing the MAXWELL–LORENTZ aether relations as follows

$$
\begin{aligned}
&\frac{\partial}{\partial t}\Big(\frac{1}{2}\rho v_iv_i-B_i\mathcal{M}_i+P_jE_j+\frac{1}{2}D_jE_j+\frac{1}{2}B_iH_i\Big)-\\
&-\frac{\partial}{\partial x_j}\Big(-v_j\frac{1}{2}\rho v_iv_i+\big(\sigma_{ji}-P_jE_i+\mathcal{M}_iB_j\big)v_i-\big(\boldsymbol{E}\times\mathfrak{H}\big)_j\Big)-\\
&-\rho f_iv_i=-\big(\sigma_{ji}-P_jE_i+\mathcal{M}_iB_j\big)\frac{\partial v_i}{\partial x_j}-\mathcal{E}_i\mathcal{J}_i^{\text{fr.}}+P_i\frac{\mathrm{d}E_i}{\mathrm{d}t}-B_i\frac{\mathrm{d}\mathcal{M}_i}{\mathrm{d}t}\,.
\end{aligned}
\tag{3.221}
$$

This balance equation is well-known for unpolarized materials. For a polarized material it depends on the choice of electromagnetic momentum; we have used $\mathcal{G}_i^{\text{M}}$ leading to Eq. (3.221). Especially $\boldsymbol{E}\times\mathfrak{H}$ is the term responsible for the electromagnetic radiation. It is the energy flux over the boundary due to the electromagnetic fields. We can simplify the balance of kinetic energy by using a similar notation as in Sect. 3.3. The kinetic energy density consists of three terms, $\rho e^{\text{kin.}}=\rho e^{\text{m.}}+e^{\text{p.}}+e^{\text{f.}}$, first term is due to the mass, $\rho e^{\text{m.}}$, second term is caused by the polarized material, $e^{\text{p.}}$, and the third term is the energy density of the field, $e^{\text{f.}}$, as follows

$$
\rho e^{\text{m.}}=\frac{1}{2}\rho v_iv_i\,,\quad e^{\text{p.}}=-B_i\mathcal{M}_i+P_jE_j\,,\quad e^{\text{f.}}=\frac{1}{2}\big(D_jE_j+B_iH_i\big)\,.
\tag{3.222}
$$

By using the above definitions we just rewrite the balance of kinetic energy:

$$\begin{aligned}\frac{\partial \rho e^{\text{kin.}}}{\partial t} - \frac{\partial}{\partial x_j}\Big(- v_j(\rho e^{\text{kin.}} - e^{\text{p.}} - e^{\text{f.}}) - (\boldsymbol{E} \times \mathfrak{H})_j + \\ +(\sigma_{ji} - P_j E_i + \mathcal{M}_i B_j) v_i \Big) - \rho f_i v_i = \\ = -(\sigma_{ji} - P_j E_i + \mathcal{M}_i B_j) \frac{\partial v_i}{\partial x_j} - \mathcal{E}_i \mathcal{J}_i^{\text{fr.}} + P_i \frac{\mathrm{d} E_i}{\mathrm{d}t} - B_i \frac{\mathrm{d}\mathcal{M}_i}{\mathrm{d}t} \ .\end{aligned} \tag{3.223}$$

We aim for the GIBBS equation in polarized material. Thus, we need to acquire the balance of internal energy. Since we have found the balance of kinetic energy we can subtract it from the balance of total energy and acquire the balance of internal energy. The specific total energy, e, is the sum of specific kinetic energy, $e^{\text{kin.}}$, and specific internal energy, u, as, $e = e^{\text{kin.}} + u$. We start off with the balance of (total) energy:

$$\frac{\partial \rho e}{\partial t} - \frac{\partial}{\partial x_j}\big(- v_j \rho e + F_j \big) - \rho s = 0 \ , \tag{3.224}$$

where the flux of energy, F_j, and the energy supply per mass, s, are not defined, yet. Total energy is a conserved quantity implying that the balance of energy lacks a production term. By subtracting the balance of kinetic energy from the balance of total energy we obtain

$$\begin{aligned}\frac{\partial(\rho e - \rho e^{\text{kin.}})}{\partial t} - \frac{\partial}{\partial x_j}\big(- v_j(\rho e - \rho e^{\text{kin.}} + e^{\text{p.}} + e^{\text{f.}}) + F_j + (\boldsymbol{E} \times \mathfrak{H})_j - \\ -(\sigma_{ji} - P_j E_i + \mathcal{M}_i B_j) v_i \big) - \rho(s - f_i v_i) = \\ = (\sigma_{ji} - P_j E_i + \mathcal{M}_i B_j) \frac{\partial v_i}{\partial x_j} + \mathcal{E}_i \mathcal{J}_i^{\text{fr.}} - P_i \frac{\mathrm{d} E_i}{\mathrm{d}t} + B_i \frac{\mathrm{d}\mathcal{M}_i}{\mathrm{d}t} \ .\end{aligned} \tag{3.225}$$

The latter is the balance of internal energy:

$$\frac{\partial \rho u}{\partial t} - \frac{\partial}{\partial x_j}\big(- v_j \rho u - q_j \big) - \rho r = \Gamma \ , \tag{3.226}$$

with the specific internal energy, $u = e - e^{\text{kin.}}$, with the flux of internal energy or heat flux:

$$-q_j = -v_j(e^{\text{p.}} + e^{\text{f.}}) + F_j + (\boldsymbol{E} \times \mathfrak{H})_j - (\sigma_{ji} - P_j E_i + \mathcal{M}_i B_j) v_i \ , \tag{3.227}$$

with the supply of internal energy or internal heating:

$$r = s - f_i v_i \ , \tag{3.228}$$

and with the production of internal energy:

$$\Gamma = \left(\sigma_{ji} - P_j E_i + \mathcal{M}_i B_j\right)\frac{\partial v_i}{\partial x_j} + \mathcal{E}_i \mathcal{J}_i^{\text{fr.}} - P_i \frac{\mathrm{d}E_i}{\mathrm{d}t} + B_i \frac{\mathrm{d}\mathcal{M}_i}{\mathrm{d}t} \,. \tag{3.229}$$

After inserting the balance of mass we obtain the following balance of internal energy in the current frame:

$$\rho\frac{\mathrm{d}u}{\mathrm{d}t} + \frac{\partial q_j}{\partial x_j} - \rho r = \left(\sigma_{ji} - P_j E_i + \mathcal{M}_i B_j\right)\frac{\partial v_i}{\partial x_j} + \mathcal{E}_i \mathcal{J}_i^{\text{fr.}} - P_i \frac{\mathrm{d}E_i}{\mathrm{d}t} + B_i \frac{\mathrm{d}\mathcal{M}_i}{\mathrm{d}t} \,. \tag{3.230}$$

We will use thermodynamics to obtain the necessary constitutive equations, viz., u, q_i, σ_{ji}, P_i, $\mathcal{M}_i$, and $\mathcal{J}_i^{\text{fr.}}$. Then the aforementioned definitions lead to e and F_i.

In this section we search for field equations for deformable solids. Thus, we need to transform the balance of internal energy from the current frame to the LAGRANGEan frame. For the sake of brevity we neglect the geometric nonlinearities. Then the transformation onto the LAGRANGEan frame is an ease

$$\rho_0\frac{\mathrm{d}u}{\mathrm{d}t} + \frac{\partial q_j}{\partial X_j} - \rho_0 r = \left(\sigma_{ji} - P_j E_i + \mathcal{M}_i B_j\right)\frac{\partial v_i}{\partial X_j} + \mathcal{E}_i \mathcal{J}_i^{\text{fr.}} - P_i \frac{\mathrm{d}E_i}{\mathrm{d}t} + B_i \frac{\mathrm{d}\mathcal{M}_i}{\mathrm{d}t} \,. \tag{3.231}$$

As we have decomposed the stress, σ_{ji}, we also split the polarization tensors, P_i, $\mathcal{M}_i$, into reversible and dissipative terms:

$$\sigma_{ji} = {}^{\mathrm{r}}\sigma_{ji} + {}^{\mathrm{d}}\sigma_{ji} \,, \quad P_i = {}^{\mathrm{r}}P_i + {}^{\mathrm{d}}P_i \,, \quad \mathcal{M}_i = {}^{\mathrm{r}}\mathcal{M}_i + {}^{\mathrm{d}}\mathcal{M}_i \,. \tag{3.232}$$

We analyze first the equilibrium state. At the equilibrium state all dissipative terms vanish. The supply terms shall vanish at the equilibrium, too. For a mechanical equilibrium the velocity gradient deduces to rate of strains:

$$\frac{\partial v_{(i}}{\partial X_{j)}} = \frac{\partial}{\partial X_{j)}}\frac{\mathrm{d}u_{(i}}{\mathrm{d}t} = \frac{\mathrm{d}}{\mathrm{d}t}\frac{\partial u_{(i}}{\partial X_{j)}} = \frac{\mathrm{d}\varepsilon_{ij}}{\mathrm{d}t} \,. \tag{3.233}$$

For a symmetric reversible stress tensor, ${}^{\mathrm{r}}\sigma_{ij} = {}^{\mathrm{r}}\sigma_{ji}$, we can write

$${}^{\mathrm{r}}\sigma_{ij}\frac{\partial v_{(i}}{\partial X_{j)}} = {}^{\mathrm{r}}\sigma_{ij}\frac{\mathrm{d}\varepsilon_{ij}}{\mathrm{d}t} \,. \tag{3.234}$$

For an electromagnetic equilibrium the electric current vanishes. For a thermal equilibrium we again introduce[68] as exactly in Sect. 3.3 a new quantity called entropy as follows

$$-\frac{\partial q_i}{\partial X_i}\bigg|_{\text{eq.}} = \frac{\mathrm{d}q}{\mathrm{d}t}\bigg|_{\text{eq.}} \,, \quad \frac{1}{T}\frac{\mathrm{d}q}{\mathrm{d}t}\bigg|_{\text{eq.}} = \rho_0\frac{\mathrm{d}\eta}{\mathrm{d}t} \,, \tag{3.235}$$

[68] See [15, Chap. XIV, Sect. 2].

where the specific entropy, η, is responsible for a reversible exchange from mechanical or electromagnetic energies to thermal energy. Hence, at equilibrium, the balance of internal energy reads[69]

$$\rho_0 \frac{\mathrm{d}u}{\mathrm{d}t} = \rho_0 T \frac{\mathrm{d}\eta}{\mathrm{d}t} + {}^{\mathrm{r}}\sigma_{ji} \frac{\mathrm{d}\varepsilon_{ij}}{\mathrm{d}t} - {}^{\mathrm{r}}P_i \frac{\mathrm{d}E_i}{\mathrm{d}t} + B_i \frac{\mathrm{d}\,{}^{\mathrm{r}}\mathcal{M}_i}{\mathrm{d}t} . \tag{3.236}$$

According to the 1st law of thermodynamics we can rewrite the latter in terms of differentials:

$$\mathrm{d}u = T\mathrm{d}\eta + v\,{}^{\mathrm{r}}\sigma_{ji}\mathrm{d}\varepsilon_{ij} - v\,{}^{\mathrm{r}}P_i\mathrm{d}E_i + vB_i\mathrm{d}\,{}^{\mathrm{r}}\mathcal{M}_i , \tag{3.237}$$

where we have utilized the specific volume, $v = 1/\rho_0$. The latter equation can be called the GIBBS equation for polarized materials. Next, we will define the dual variables η, ${}^{\mathrm{r}}\sigma_{ji}$, ${}^{\mathrm{r}}P_i$, and ${}^{\mathrm{r}}\mathcal{M}_i$ with respect to the primary (or state) variables T, ε_{ij}, E_i, and B_i. We choose the primary variables as T, ε_{ij}, E_i, B_i, since we know their relations to the primitive variables, T, u_i, ϕ, A_i. In order to obtain an energy depending on the primary variables, we introduce a free energy:

$$\psi = u - T\eta - B_i v\,{}^{\mathrm{r}}\mathcal{M}_i , \tag{3.238}$$

and by taking its total differential we obtain

$$\begin{aligned} \mathrm{d}\psi &= \mathrm{d}u - \eta\mathrm{d}T - T\mathrm{d}\eta - B_i v\mathrm{d}\,{}^{\mathrm{r}}\mathcal{M}_i - {}^{\mathrm{r}}\mathcal{M}_i v\mathrm{d}B_i = \\ &= -\eta\mathrm{d}T + {}^{\mathrm{r}}\sigma_{ji} v\mathrm{d}\varepsilon_{ij} - {}^{\mathrm{r}}P_i v\mathrm{d}E_i - {}^{\mathrm{r}}\mathcal{M}_i v\mathrm{d}B_i . \end{aligned} \tag{3.239}$$

The free energy depends on the primary variables, $\psi = \psi(T, \varepsilon_{ij}, E_i, B_i)$. Obviously, each dual variable, η, ${}^{\mathrm{r}}\sigma_{ij}$, ${}^{\mathrm{r}}P_i$, ${}^{\mathrm{r}}\mathcal{M}_i$, may depend on the primary variables, T, ε_{ij}, E_i, B_i, such that we acquire

$$\begin{aligned} \mathrm{d}\eta &= \tilde{c}\mathrm{d}T + \tilde{p}_{ij}\mathrm{d}\varepsilon_{ij} + \tilde{\ell}_i\mathrm{d}E_i + \tilde{o}_i\mathrm{d}B_i , \\ \mathrm{d}\,{}^{\mathrm{r}}\sigma_{ij} &= \tilde{P}_{ij}\mathrm{d}T + C_{ijkl}\mathrm{d}\varepsilon_{kl} + \tilde{t}_{kij}\mathrm{d}E_k + \tilde{s}_{ijk}\mathrm{d}B_k , \\ \mathrm{d}\,{}^{\mathrm{r}}P_i &= \tilde{L}_i\mathrm{d}T + \tilde{T}_{ijk}\mathrm{d}\varepsilon_{jk} + \tilde{\chi}^{\mathrm{el.}}_{ij}\mathrm{d}E_j + \tilde{r}_{ij}\mathrm{d}B_j , \\ \mathrm{d}\,{}^{\mathrm{r}}\mathcal{M}_i &= \tilde{O}_i\mathrm{d}T + \tilde{S}_{ijk}\mathrm{d}\varepsilon_{jk} + \tilde{R}_{ij}\mathrm{d}E_j + \tilde{\chi}^{\mathrm{mag.}}_{ij}\mathrm{d}B_j . \end{aligned} \tag{3.240}$$

[69] An alternative is to use the following formulation:

$$\rho_0 \frac{\mathrm{d}u}{\mathrm{d}t} = \rho_0 T \frac{\mathrm{d}\eta}{\mathrm{d}t} + ({}^{\mathrm{r}}\sigma_{ji} - {}^{\mathrm{r}}P_j E_i + {}^{\mathrm{r}}\mathcal{M}_i B_j) \frac{\mathrm{d}\varepsilon_{ij}}{\mathrm{d}t} - {}^{\mathrm{r}}P_i \frac{\mathrm{d}E_i}{\mathrm{d}t} + B_i \frac{\mathrm{d}\,{}^{\mathrm{r}}\mathcal{M}_i}{\mathrm{d}t} .$$

This formulation leads to the same constitutive equations and is also admissible. The argumentation for explaining the measurements in the dual variables becomes more difficult by using the alternative formulation.

All material parameters, $\tilde{c}$, $\tilde{p}_{ij}$, $\tilde{\ell}_i$, $\tilde{o}_i$, $\tilde{P}_{ij}$, C_{ijkl}, $\tilde{t}_{kij}$, $\tilde{s}_{ijk}$, $\tilde{L}_i$, $\tilde{T}_{ijk}$, $\chi^{\text{el.}}_{ij}$, $\tilde{r}_{ij}$, $\tilde{O}_i$, $\tilde{S}_{ijk}$, $\tilde{R}_{ij}$, $\tilde{\chi}^{\text{mag.}}_{ij}$ shall be determined experimentally. By using the MAXWELL symmetry or reciprocal relations:

$$
\begin{aligned}
\tilde{P}_{ij} &= \frac{\partial\,{}^{\mathrm{r}}\sigma_{ij}}{\partial T} = \rho_0 \frac{\partial\,{}^{\mathrm{r}}\sigma_{ij} v}{\partial T} = \rho_0 \frac{\partial^2 \psi}{\partial T \partial \varepsilon_{ij}} = \rho_0 \frac{\partial^2 \psi}{\partial \varepsilon_{ij} \partial T} = -\rho_0 \frac{\partial \eta}{\partial \varepsilon_{ij}} = -\rho_0 \tilde{p}_{ij} \\
&\Rightarrow \tilde{p}_{ij} = -v \tilde{P}_{ij} \ , \\
\tilde{L}_i &= \frac{\partial\,{}^{\mathrm{r}}P_i}{\partial T} = \rho_0 \frac{\partial\,{}^{\mathrm{r}}P_i v}{\partial T} = -\rho_0 \frac{\partial^2 \psi}{\partial T \partial E_i} = -\rho_0 \frac{\partial^2 \psi}{\partial E_i \partial T} = \rho_0 \frac{\partial \eta}{\partial E_i} = \rho_0 \tilde{\ell}_i \\
&\Rightarrow \tilde{\ell}_i = v \tilde{L}_i \ , \\
\tilde{O}_i &= \frac{\partial\,{}^{\mathrm{r}}\mathcal{M}_i}{\partial T} = \rho_0 \frac{\partial\,{}^{\mathrm{r}}\mathcal{M}_i v}{\partial T} = -\rho_0 \frac{\partial^2 \psi}{\partial T \partial B_i} = -\rho_0 \frac{\partial^2 \psi}{\partial B_i \partial T} = \rho_0 \frac{\partial \eta}{\partial B_i} = \rho_0 \tilde{o}_i \\
&\Rightarrow \tilde{o}_i = v \tilde{O}_i \ ,
\end{aligned}
\tag{3.241}
$$

as well as

$$
\begin{aligned}
\tilde{T}_{ijk} &= \frac{\partial\,{}^{\mathrm{r}}P_i}{\partial \varepsilon_{jk}} = \rho_0 \frac{\partial\,{}^{\mathrm{r}}P_i v}{\partial \varepsilon_{jk}} = -\rho_0 \frac{\partial^2 \psi}{\partial \varepsilon_{jk} \partial E_i} = -\rho_0 \frac{\partial^2 \psi}{\partial E_i \partial \varepsilon_{jk}} = -\rho_0 \frac{\partial\,{}^{\mathrm{r}}\sigma_{jk} v}{\partial E_i} = -\tilde{t}_{ijk} \\
&\Rightarrow \tilde{t}_{kij} = -\tilde{T}_{kij} \ , \\
\tilde{S}_{ijk} &= \frac{\partial\,{}^{\mathrm{r}}\mathcal{M}_i}{\partial \varepsilon_{jk}} = \rho_0 \frac{\partial\,{}^{\mathrm{r}}\mathcal{M}_i v}{\partial \varepsilon_{jk}} = -\rho_0 \frac{\partial^2 \psi}{\partial \varepsilon_{jk} \partial B_i} = -\rho_0 \frac{\partial^2 \psi}{\partial B_i \partial \varepsilon_{jk}} = -\frac{\partial\,{}^{\mathrm{r}}\sigma_{kj}}{\partial B_i} = -\tilde{s}_{kji} \\
&\Rightarrow \tilde{s}_{ijk} = \tilde{S}_{kji} \ , \\
\tilde{R}_{ij} &= \frac{\partial\,{}^{\mathrm{r}}\mathcal{M}_i}{\partial E_j} = \rho_0 \frac{\partial\,{}^{\mathrm{r}}\mathcal{M}_i v}{\partial E_j} = -\rho_0 \frac{\partial^2 \psi}{\partial E_j \partial B_i} = -\rho_0 \frac{\partial^2 \psi}{\partial B_i \partial E_j} = \frac{\partial\,{}^{\mathrm{r}}P_j}{\partial B_i} = \tilde{r}_{ji} \\
&\Rightarrow \tilde{r}_{ij} = \tilde{R}_{ji} ,
\end{aligned}
\tag{3.242}
$$

we can reduce the necessary amount of measurements for determining materials parameters. The experiments are established by varying one of the primary variable as holding the other primary variables fixed and measuring the corresponding dual variable.

The stiffness tensor, C_{ijkl}, is measured for a specific (fixed) temperature, $\mathrm{d}T = 0$, for a fixed electric field, $\mathrm{d}E_i = 0$, for a fixed magnetic intensity, $\mathrm{d}B_i = 0$, by varying strain, $\mathrm{d}\varepsilon_{ij}$, and measuring stress, $\mathrm{d}\sigma_{ij}$. The electric susceptibility, $\chi^{\text{el.}}_{ij}$, constituting $\tilde{\chi}^{\text{el.}}_{ij} = \varepsilon_0 \chi^{\text{el.}}_{ij}$ is measured analogously by varying $\mathrm{d}E_i$ and measuring $\mathrm{d}\,{}^{\mathrm{r}}P_i$. The magnetic susceptibility, $\chi^{\text{mag.}}_{ij}$, in $\tilde{\chi}^{\text{mag.}}_{ij} = \mu^{-1} \chi^{\text{mag.}}_{ij}$ is measured by varying $\mathrm{d}B_i$ and measuring $\mathrm{d}\,{}^{\mathrm{r}}\mathcal{M}_i$. We cannot measure entropy directly but the specific heat energy, $\delta Q = T \mathrm{d}\eta = T \tilde{c} \mathrm{d}T$, can be measured by varying temperature, $\delta Q = c \mathrm{d}T$, where the specific heat capacity, $c = T \tilde{c}$, has been introduced. All direct relations, namely stiffness, susceptibility, heat capacity, can easily be found in the literature for various materials.

Moreover, there are the coupling terms, viz., the thermal pressure, $\tilde{P}_{ij}$, the pyroelectric tensor, $\tilde{L}_i$, the electromagnetic coupling, $\tilde{R}_{ij}$, the magnetothermal coupling, $\tilde{O}_i$, the piezoelectric tensor, $\tilde{T}_{ijk}$, and the piezomagnetic tensor, $\tilde{S}_{ijk}$. Especially for the thermal pressure, pyroelectric tensor, and the magnetothermal coupling the material parameters are difficult to find in the literature. However, we can rewrite them by using coefficients of thermal expansion, α_{ij}, which is determined by varying temperature and measuring strain

$$\mathrm{d}\varepsilon_{ij} = \alpha_{ij}\mathrm{d}T \ . \tag{3.243}$$

The coefficients of thermal expansion are measured by holding the other variables fixed, hence, from Eq. (3.240)$_2$ we obtain for fixed ${}^{\mathrm{r}}\sigma_{ij}$, E_i, B_i,

$$\begin{aligned} 0 = \tilde{P}_{ij}\mathrm{d}T + C_{ijkl}\mathrm{d}\varepsilon_{kl} = \tilde{P}_{ij}\mathrm{d}T + C_{ijkl}\alpha_{kl}\mathrm{d}T \\ \Rightarrow \tilde{P}_{ij} = -C_{ijkl}\alpha_{kl} \ . \end{aligned} \tag{3.244}$$

Analogously we acquire

$$\tilde{L}_i = -\tilde{T}_{ijk}\alpha_{jk} \ , \quad \tilde{O}_i = -\tilde{S}_{ijk}\alpha_{jk} \ . \tag{3.245}$$

Now by using the thermal expansion coefficients and employing the MAXWELL relations we obtain

$$\begin{aligned} \mathrm{d}\eta &= \frac{c}{T}\mathrm{d}T + vC_{ijkl}\alpha_{kl}\mathrm{d}\varepsilon_{ij} - v\tilde{T}_{ijk}\alpha_{kl}\mathrm{d}E_i - v\tilde{S}_{ijk}\alpha_{kl}\mathrm{d}B_i \ , \\ \mathrm{d}{}^{\mathrm{r}}\sigma_{ij} &= -C_{ijkl}\alpha_{kl}\mathrm{d}T + C_{ijkl}\mathrm{d}\varepsilon_{kl} - \tilde{T}_{kij}\mathrm{d}E_k + \tilde{S}_{kji}\mathrm{d}B_k \ , \\ \mathrm{d}{}^{\mathrm{r}}\mathcal{P}_i &= -\tilde{T}_{ijk}\alpha_{jk}\mathrm{d}T + \tilde{T}_{ijk}\mathrm{d}\varepsilon_{jk} + \varepsilon_0\chi_{ij}^{\mathrm{el.}}\mathrm{d}E_j + \tilde{R}_{ji}\mathrm{d}B_j \ , \\ \mathrm{d}{}^{\mathrm{r}}\mathcal{M}_i &= -\tilde{S}_{ijk}\alpha_{jk}\mathrm{d}T + \tilde{S}_{ijk}\mathrm{d}\varepsilon_{jk} + \tilde{R}_{ij}\mathrm{d}E_j + \mu_0^{-1}\chi_{ij}^{\mathrm{mag.}}\mathrm{d}B_j \ . \end{aligned} \tag{3.246}$$

Instead of the piezoelectric tensor, $\tilde{T}_{ijk}$, it is more common to find the so called piezoelectric coefficients, $\tilde{d}_{ijk}$, in the literature.[70] Variation of the electric field and measurement of the induced strain are used to determine the piezoelectric coefficients with $\mathrm{d}\varepsilon_{kl} = \tilde{d}_{mkl}\mathrm{d}E_m$ by fixed stress, $\mathrm{d}{}^{\mathrm{r}}\sigma_{ij} = 0$, at a constant temperature, $\mathrm{d}T = 0$, and for a constant magnetic flux, $\mathrm{d}B_i = 0$, such that we observe from Eq. (3.246)$_2$

$$\begin{aligned} 0 = C_{ijkl}\mathrm{d}\varepsilon_{kl} - \tilde{T}_{mij}\mathrm{d}E_m = C_{ijkl}\tilde{d}_{mkl}\mathrm{d}E_m - \tilde{T}_{mij}\mathrm{d}E_m \ , \\ \tilde{T}_{mij} = C_{ijkl}\tilde{d}_{mkl} \ . \end{aligned} \tag{3.247}$$

The notation $\tilde{d}_{mkl}$ is a standardized notation (without tilde) for the piezoelectric coefficients.

[70] See the measurements described in [27].

All material parameters, namely $\tilde{c}$, C_{ijkl}, $\chi^{\text{el.}}_{ij}$, $\chi^{\text{mag.}}_{ij}$, α_{ij}, $\tilde{T}_{ijk}$, $\tilde{S}_{ijk}$, and $\tilde{R}_{ij}$ may depend on the primary variables, T, ε_{ij}, E_i, and B_i. Some materials show an *electro-magneto-striction* meaning that the susceptibilities depend on electric field and magnetic flux. For a simple and linear material, where material parameters are constants in the primary variables, we can obtain the dual variables simply by integrating from the ground state, $T = T_{\text{ref.}}$, $\varepsilon_{ij} = 0$, $E_i = 0$, $B_i = 0$, to the present state, T, ε_{ij}, E_i, B_i, as follows

$$\begin{aligned} \eta &= c \ln\left(\frac{T}{T_{\text{ref.}}}\right) + v C_{ijkl}\alpha_{kl}\varepsilon_{ij} - v\tilde{T}_{ijk}\alpha_{jk}E_i - v\tilde{S}_{ijk}\alpha_{jk}B_i \ , \\ {}^{\text{r}}\sigma_{ij} &= -C_{ijkl}\alpha_{kl}(T - T_{\text{ref.}}) + C_{ijkl}\varepsilon_{kl} - \tilde{T}_{kij}E_k + \tilde{S}_{kji}B_k \ , \\ {}^{\text{r}}P_i &= -\tilde{T}_{ijk}\alpha_{jk}(T - T_{\text{ref.}}) + \tilde{T}_{ijk}\varepsilon_{jk} + \varepsilon_0\chi^{\text{el.}}_{ij}E_j + \tilde{R}_{ji}B_j \ , \\ {}^{\text{r}}\mathcal{M}_i &= -\tilde{S}_{ijk}\alpha_{jk}(T - T_{\text{ref.}}) + \tilde{S}_{ijk}\varepsilon_{jk} + \tilde{R}_{ij}E_j + \mu_0^{-1}\chi^{\text{mag.}}_{ij}B_j \ . \end{aligned} \tag{3.248}$$

We have defined all of the reversible parts of σ_{ij}, P_i, and $\mathcal{M}_i$ by using the GIBBS equation at the equilibrium state. We assume that the rate of internal energy holds at the non-equilibrium state, too. In other words, we assume that the internal energy is recoverable. Now the balance of internal energy in Eq. (3.231) can be rewritten by inserting the GIBBS equation

$$\begin{aligned} \rho_0 T\frac{\mathrm{d}\eta}{\mathrm{d}t} + \frac{\partial q_j}{\partial X_j} - \rho_0 r = \big({}^{\text{d}}\sigma_{ji} - P_j E_i + \mathcal{M}_i B_j\big)\frac{\partial v_i}{\partial X_j} + \\ + \mathcal{E}_i \mathcal{J}^{\text{fr.}}_i - {}^{\text{d}}P_i\frac{\mathrm{d}E_i}{\mathrm{d}t} + B_i\frac{\mathrm{d}\,{}^{\text{d}}\mathcal{M}_i}{\mathrm{d}t} \ , \end{aligned} \tag{3.249}$$

or even as follows

$$\begin{aligned} \rho_0 T\frac{\partial\eta}{\partial t} + \frac{\partial q_j}{\partial X_j} - \rho_0 r = \big({}^{\text{d}}\sigma_{ji} - P_j E_i + \mathcal{M}_i B_j\big)\frac{\partial v_i}{\partial X_j} + \\ + \mathcal{E}_i \mathcal{J}^{\text{fr.}}_i - {}^{\text{d}}P_i\frac{\partial E_i}{\partial t} + B_i\frac{\partial\,{}^{\text{d}}\mathcal{M}_i}{\partial t} \ , \end{aligned} \tag{3.250}$$

since the formulation is in the LAGRANGEan frame. We can rewrite the latter in order to obtain a balance equation:

$$\begin{aligned} \rho_0\frac{\partial\eta}{\partial t} + \frac{\partial}{\partial X_j}\left(\frac{q_j}{T}\right) - \rho_0\frac{r}{T} = q_j\frac{\partial}{\partial X_j}\left(\frac{1}{T}\right) + \frac{1}{T}\big({}^{\text{d}}\sigma_{ji} - P_j E_i + \mathcal{M}_i B_j\big)\frac{\partial v_i}{\partial X_j} + \\ + \frac{1}{T}\mathcal{E}_i \mathcal{J}^{\text{fr.}}_i - \frac{1}{T}{}^{\text{d}}P_i\frac{\partial E_i}{\partial t} + \frac{1}{T}B_i\frac{\partial\,{}^{\text{d}}\mathcal{M}_i}{\partial t} \ . \end{aligned} \tag{3.251}$$

This balance equation is the balance of entropy in the LAGRANGEan frame:

$$\rho_0\frac{\partial\eta}{\partial t} + \frac{\partial\Phi_j}{\partial X_j} - \rho_0\frac{r}{T} = \Sigma \ , \tag{3.252}$$

with a flux term, Φ_i, and a production term, Σ, as

$$\Phi_j = \frac{q_j}{T} ,$$
$$\Sigma = -\frac{q_j}{T^2}\frac{\partial T}{\partial X_j} + \frac{1}{T}\big({}^{\mathrm{d}}\sigma_{ji} - P_j E_i + \mathcal{M}_i B_j\big)\frac{\partial v_i}{\partial X_j} + \tag{3.253}$$
$$+\frac{1}{T}\mathcal{E}_i \mathcal{J}_i^{\mathrm{fr.}} - \frac{1}{T}{}^{\mathrm{d}}P_i \frac{\partial E_i}{\partial t} + \frac{1}{T} B_i \frac{\partial\, {}^{\mathrm{d}}\mathcal{M}_i}{\partial t} .$$

The entropy flux, Φ_i, might become differently by redefining the energy flux in Eq. (3.227). We could include the radiation term $\boldsymbol{E} \times \boldsymbol{\mathfrak{H}}$ into the balance of internal energy by leaving out the radiation term in Eq. (3.227). Then the entropy flux and the entropy production would have an additional term in the heat flux similar to a HALL effect.[71] For its interpretation see [28, Sect. 9.9.4]. We propose a choice, where the radiation is included in the heat flux. Both choices are reasonable for different experimental setups. In other words, the appropriate choice depends on measuring the heat flux with or without the radiation term.

There are various heat flux gauges, they mostly rely on a heat flux definition, which we will derive in the following. As we know from FOURIER's law, the heat flux depends on the temperature, thus, the temperature measurement plays a crucial role. Although temperature is a primitive variable—we accept its existence without questioning—measuring the temperature is always succeeded in an indirect way. In case of a non-contact temperature measurement with an infrared camera, the detector outputs an electric signal denoting the temperature (after a calibration with a known temperature). The detector in the camera is a sensor. If a quantum detector is used, where the sensor is a photodiode transforming the electromagnetic radiation into an electric signal, then we have to include the radiation term in the heat flux. However, if a thermal detector with a pyroelectric material as sensor is employed, then we shall separate radiation and heat flux.

We continue by using the above definition of heat flux including radiation. Moreover, we restrict our formalism to the case that polarization causes a reversible evolution. In other words, we assume

$${}^{\mathrm{d}}P_i = 0 , \quad {}^{\mathrm{d}}\mathcal{M}_i = 0 \Rightarrow P_i = {}^{\mathrm{r}}P_i , \quad \mathcal{M}_i = {}^{\mathrm{r}}\mathcal{M}_i . \tag{3.254}$$

Then the entropy production reads by introducing the comma notation for derivatives in X_i, as follows

$$\Sigma = -\frac{q_j}{T^2}T_{,j} + \frac{1}{T}\big({}^{\mathrm{d}}\sigma_{ji} - P_j E_i + \mathcal{M}_i B_j\big)v_{i,j} + \frac{1}{T}\mathcal{E}_i \mathcal{J}_i^{\mathrm{fr.}} . \tag{3.255}$$

According to the 2nd law of thermodynamics, the entropy production vanishes, $\Sigma = 0$, for reversible processes (at the equilibrium state); and it is positive, $\Sigma > 0$, on

[71] It is named after Edwin Herbert Hall.

irreversible processes (at non-equilibrium). Moreover, tensors of rank two can be decomposed into the spherical, symmetric deviatoric, and antisymmetric terms, for example, in the case of a generic tensor of rank two, A_{ij}, this would be

$$A_{ji} = A_{(ji)} + A_{[ji]} = A_{|(ji)|} + \frac{1}{3}A_{kk}\delta_{ji} + A_{[ji]} \,,$$
$$A_{(ji)} = \frac{1}{2}\big(A_{ji} + A_{ij}\big) \,, \quad A_{|(ji)|} = A_{(ji)} - \frac{1}{3}A_{kk}\delta_{ji} \,, \quad A_{[ji]} = \frac{1}{2}\big(A_{ji} - A_{ij}\big) \,, \tag{3.256}$$

leading to

$$\begin{aligned} &{}^{\mathrm{d}}\sigma_{ji} - P_j E_i + \mathcal{M}_i B_j = \Xi_{ji} \,, \\ &\frac{1}{T}\Xi_{ji} v_{i,j} = \frac{1}{T}\Xi_{|(ji)|} v_{|(i,j)|} + \frac{1}{3T}\Xi_{ii} v_{j,j} + \frac{1}{T}\Xi_{[ji]} v_{[i,j]} \,. \end{aligned} \tag{3.257}$$

Hence we can rewrite the production of entropy by introducing two lists:

$$\begin{aligned} \mathcal{K}^\alpha &= \left\{ \frac{1}{T^2}T_{,j} \,,\ \frac{1}{T}v_{|(i,j)|} \,,\ \frac{1}{T}v_{j,j} \,,\ \frac{1}{T}v_{[i,j]} \,,\ \frac{1}{T}\mathcal{E}_i \right\} , \\ \mathcal{F}^\alpha &= \left\{ -q_j \,,\ \Xi_{|(ji)|} \,,\ \Xi_{ii} \,,\ \Xi_{[ji]} \,,\ \mathcal{J}_i^{\text{fr.}} \right\} , \end{aligned} \tag{3.258}$$

where the *thermodynamical forces*, $\mathcal{K}^\alpha$, are derived from the primitive variables and *thermodynamical fluxes*, $\mathcal{F}^\alpha$, are going to be defined in the following. The production of entropy reads

$$\Sigma = \mathcal{K}^\alpha \cdot \mathcal{F}^\alpha \geq 0 \,, \quad \alpha = 1, 2, \ldots, 5 \,, \tag{3.259}$$

where the summation convention is used for α. The thermodynamical fluxes depend only on the thermodynamical forces of the same rank. For example, $\mathcal{F}^1$ is a thermodynamical flux of rank one and depends only on the thermodynamical forces of rank one, namely on $\mathcal{K}^1$ and $\mathcal{K}^5$. This restriction is due to the different transformation properties of tensors of different ranks and referred to as the CURIE principle. The thermodynamical flux of rank zero reads

$$\Xi_{ii} = \bar{A}\frac{1}{T}v_{j,j} \,. \tag{3.260}$$

Analogously the thermodynamical fluxes of rank one become

$$-q_j = \bar{a}\frac{1}{T^2}T_{,j} + \bar{b}\frac{1}{T}\mathcal{E}_j \,, \quad \mathcal{J}_i^{\text{fr.}} = \bar{B}\frac{1}{T^2}T_{,j} + \bar{c}\frac{1}{T}\mathcal{E}_j \,. \tag{3.261}$$

The thermodynamical fluxes of rank two are

$$\Xi_{|(ji)|} = \bar{d}\frac{1}{T}v_{|(i,j)|} \,, \quad \Xi_{[ji]} = \bar{e}\frac{1}{T}v_{[i,j]} \,. \tag{3.262}$$

In order to fulfill the 2nd law of thermodynamics we acquire

$$\begin{aligned} \Sigma = &\frac{\bar{A}}{T^2} v_{i,i} v_{j,j} + \frac{\bar{a}}{T^4} T_{,j} T_{,j} + \frac{1}{T^3} T_{,j} \mathcal{E}_j (\bar{b} + \bar{B}) + \\ &+ \bar{c} \frac{1}{T^2} \mathcal{E}_i \mathcal{E}_i + \frac{\bar{d}}{T^2} v_{|(i,j)|} v_{|(i,j)|} + \frac{\bar{e}}{T^2} v_{[i,j]} v_{[i,j]} \geq 0 , \end{aligned} \tag{3.263}$$

such that $\bar{A} > 0, \bar{a} > 0, \bar{c} > 0, \bar{d} > 0$, and $\bar{e} > 0$ have to hold. For the coupling constants we have to obtain $\bar{b} + \bar{B} = 0$ in order to inhibit a negative entropy production, since we cannot assure for all processes that the temperature gradient and the electric field shall have the same direction.

Piezoceramics are insulators such that free conducting current shall vanish, $\mathcal{J}_i^{\text{fr.}} = 0$, i.e., $\bar{b} = 0, \bar{c} = 0$. Hence we obtain for the piezoceramic material:

$$q_j = -\kappa T_{,j} , \quad \kappa = \frac{\bar{a}}{T^2} > 0 , \tag{3.264}$$

this material equation is referred to as FOURIER's law for the case of $\kappa = \text{const}$. We also suppose that an irreversible motion fails to exist, in other words, the material is elastic. In order to achieve that we set $\bar{A} = 0, \bar{d} = 0$, and $\bar{e} = 0$. Hence the dissipative stress tensor becomes

$$\begin{gathered} \Xi_{ii} = 0 , \quad \Xi_{|(ij)|} = 0 , \quad \Xi_{[ij]} = 0 , \\ {}^{\mathrm{d}}\sigma_{ii} = P_i E_i - \mathcal{M}_i B_i , \quad {}^{\mathrm{d}}\sigma_{|(ji)|} = P_{|(j} E_{i)|} - \mathcal{M}_{|(i} B_{j)|} , \quad {}^{\mathrm{d}}\sigma_{[ji]} = P_{[j} E_{i]} - \mathcal{M}_{[i} B_{j]} , \\ {}^{\mathrm{d}}\sigma_{ji} = P_j E_i - \mathcal{M}_i B_j . \end{gathered} \tag{3.265}$$

Moreover, we suppose that the piezoceramic material lacks a magnetic polarization, $\mathcal{M}_i = {}^{\mathrm{r}}\mathcal{M}_i = 0$. In order to succeed this the corresponding materials parameters have to vanish, $\tilde{S}_{ijk} = 0$, $\tilde{R}_{ij} = 0$, $\chi_{ij}^{\text{mag.}} = 0$. Hence the following constitutive equations are valid for piezoceramics:

$$\begin{aligned} \eta &= c \ln\left(\frac{T}{T_{\text{ref.}}}\right) + v C_{ijkl} \alpha_{kl} \varepsilon_{ij} - v \tilde{T}_{ijk} \alpha_{jk} E_i , \quad q_i = -\kappa T_{,i} , \\ \sigma_{ji} &= {}^{\mathrm{d}}\sigma_{ji} + {}^{\mathrm{r}}\sigma_{ji} = P_j E_i - C_{jikl} \alpha_{kl} (T - T_{\text{ref.}}) + C_{jikl} \varepsilon_{kl} - \tilde{T}_{kji} E_k , \\ P_i &= {}^{\mathrm{r}}P_i = -\tilde{T}_{ijk} \alpha_{jk} (T - T_{\text{ref.}}) + \tilde{T}_{ijk} \varepsilon_{jk} + \varepsilon_0 \chi_{ij}^{\text{el.}} E_j , \end{aligned} \tag{3.266}$$

with the following primary variables:

$$\varepsilon_{ij} = u_{(i,j)} , \quad E_i = -\phi_{,i} - \frac{\partial A_i}{\partial t} , \quad B_i = \epsilon_{ijk} A_{k,j} . \tag{3.267}$$

We have deduced all necessary constitutive equations by using thermodynamics. The governing equations are now closed and can be solved. Our aim is to compute the electric potential, ϕ, the magnetic potential, A_i, the displacement, u_i, and the

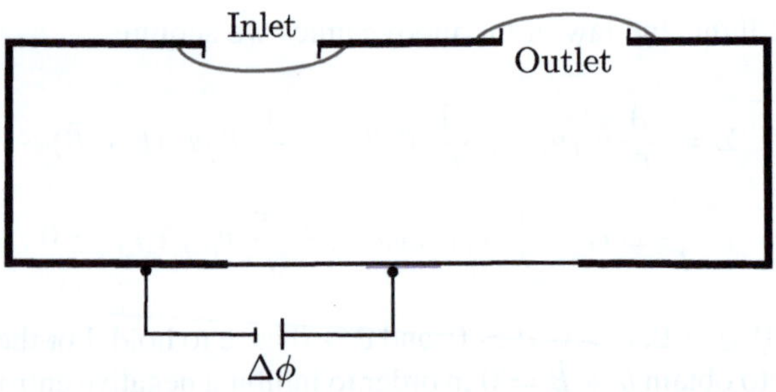

Fig. 3.14 Drawing of a micropump with two valves on *top* and a thin membrane on its *bottom*. The piezo actuator moves the membrane up- and downward such that the volume changes and fluid is pumped from inlet to outlet

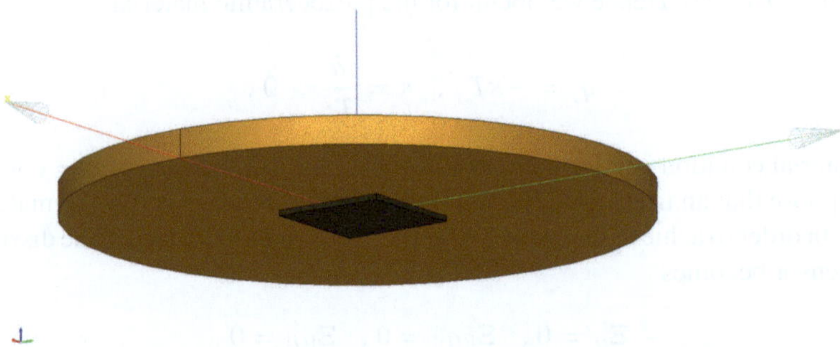

Fig. 3.15 The membrane of the micropump, a PZT-5H piezoceramic rectangular sheet (*green*) is mounted on a thin circular plate out of brass (*yellow*)

temperature, T, as functions in space, X_i, and time, t, in the LAGRANGEan frame for a solid body.

Consider a micropump as drawn in Fig. 3.14. A diaphragm or membrane is actuated with the mounted piezoceramic on the outer side. By applying a cyclic electric potential difference on the piezoceramic plate, the membrane bends down and up such that the chamber sucks in a medium through inlet and pumps out through outlet. The deformation of the membrane is controlled over the given potential difference. Hence, the volume rate out of the pump is dosed very accurately. Such micropumps with piezo drivers are used in medical devices, they are called *lab-on-a-chip*. We want to simulate the motion of the membrane of such a micropump. The membrane comprises a circular plate out of brass (copper and zinc alloy) and a $PbZr_{0.53}Ti_{0.47}O_3$ (PZT-5H) piezoceramic sheet on the plate—it is shown in Fig. 3.15. Since the brass plate is a conductor we can connect the electric supply on any point on it and the piezo sheet has an electric potential on its top. Although it is not modeled, there is a thin sheet of a conductor (like brass) on piezo sheet's bottom utilizing an electric potential on the whole bottom-side.

The piezoelectric properties are naturally found in some single-crystals, however nowadays, the synthetic (polycrystalline) piezoceramics are used in the industry. The piezoelectric ceramic is *poled* in a certain direction.[72] This process creates the piezoelectric properties. The poling direction is also called the piezoelectric axis and it is always chosen along x_3 in the literature. Therefore, we have modeled the geometry such that x_3 is directed along the thickness. The potential difference is set along the poling direction. Therefore, the material is isotropic but during (piezoelectric) poling the thermal and mechanical properties along x_3-axis alter. In a data sheet for PZT-5H we find the following values:[73]

$$\begin{gathered}\rho_0 = 7500\,\mathrm{kg/m^3}\,, \quad S_{33} = 20 \cdot 10^{-12}\,\mathrm{m^2/N}\,,\\ S_{11} = 15.6 \cdot 10^{-12}\,\mathrm{m^2/N}\,, \quad \nu = 0.31\,,\\ \tilde{d}_{33} = 585 \cdot 10^{-12}\,\mathrm{m/V}\,, \quad \tilde{d}_{31} = -265 \cdot 10^{-12}\,\mathrm{m/V}\,,\\ \tilde{d}_{15} = 730 \cdot 10^{-12}\,\mathrm{m/V}\,, \quad \bar{\varepsilon}^{\mathrm{el.}}_{33} = 3400\,, \quad \bar{\varepsilon}^{\mathrm{el.}}_{11} = 3130\,,\\ c = 350\,\mathrm{J/(kg\,K)}\,, \quad \kappa = 1.1\,\mathrm{W/(m\,K)}\,,\\ \alpha_{33} = -4 \cdot 10^{-6}\,\mathrm{1/K}\,, \quad \alpha_{11} = 6 \cdot 10^{-6}\,\mathrm{1/K}\,.\end{gathered} \tag{3.268}$$

The compliance matrix, S_{IJ}, in the VOIGT notation reads

$$S_{IJ} = \begin{pmatrix} S_{11} & -\nu S_{11} & -\nu S_{11} & 0 & 0 & 0\\ -\nu S_{11} & S_{11} & -\nu S_{11} & 0 & 0 & 0\\ -\nu S_{11} & -\nu S_{11} & S_{33} & 0 & 0 & 0\\ 0 & 0 & 0 & (1+\nu)S_{11} & 0 & 0\\ 0 & 0 & 0 & 0 & (1+\nu)S_{11} & 0\\ 0 & 0 & 0 & 0 & 0 & (1+\nu)S_{11} \end{pmatrix}. \tag{3.269}$$

The inverse of the compliance matrix is the stiffness matrix in the VOIGT notation:

$$(S_{JI})^{-1} = C_{IJ} = \begin{pmatrix} C_{1111} & C_{1122} & C_{1133} & C_{1123} & C_{1113} & C_{1112}\\ C_{2211} & C_{2222} & C_{2233} & C_{2223} & C_{2213} & C_{2212}\\ C_{3311} & C_{3322} & C_{3333} & C_{3323} & C_{3313} & C_{3312}\\ C_{2311} & C_{2322} & C_{2333} & C_{2323} & C_{2313} & C_{2312}\\ C_{1311} & C_{1322} & C_{1333} & C_{1323} & C_{1313} & C_{1312}\\ C_{1211} & C_{1222} & C_{1233} & C_{1223} & C_{1213} & C_{1212} \end{pmatrix}. \tag{3.270}$$

Analogously, we have the piezoelectric constants, $\tilde{d}_{iJ}$, by using the VOIGT notation on the indices belonging to the strain

---

[72] Some materials have different stable states where the configuration of molecules are different. By applying a high electric field, i.e., poling, the configuration is changed to the new stable state. The lattice is distorted in this new state so that the electric field and deformation are connected. An electric field (smaller than the poling field) induces a strain (distortion in the lattice). This shape change is reversible as long as the applied field remains smaller than the poling field.

[73] All parameters of PZT-5H are compiled from http://www.piceramic.com/piezo-technology/fundamentals.html and http://bostonpiezooptics.com/ceramic-materials-pzt

$$\tilde{d}_{iJ} = \begin{pmatrix} \tilde{d}_{111} & \tilde{d}_{122} & \tilde{d}_{133} & \tilde{d}_{123} & \tilde{d}_{131} & \tilde{d}_{112} \\ \tilde{d}_{211} & \tilde{d}_{222} & \tilde{d}_{233} & \tilde{d}_{223} & \tilde{d}_{231} & \tilde{d}_{212} \\ \tilde{d}_{311} & \tilde{d}_{322} & \tilde{d}_{333} & \tilde{d}_{323} & \tilde{d}_{331} & \tilde{d}_{312} \end{pmatrix} = \begin{pmatrix} 0 & 0 & 0 & 0 & \tilde{d}_{15} & 0 \\ 0 & 0 & 0 & \tilde{d}_{15} & 0 & 0 \\ \tilde{d}_{31} & \tilde{d}_{31} & \tilde{d}_{33} & 0 & 0 & 0 \end{pmatrix}, \tag{3.271}$$

where the necessary coefficients are calculated with the aforementioned relation:

$$\tilde{T}_{mij} = C_{ijkl}\tilde{d}_{mkl} \,. \tag{3.272}$$

The susceptibility is given by the relative permittivity values:

$$\chi_{ij}^{\text{el.}} = \begin{pmatrix} \varepsilon_{11}^{\text{el.}} & 0 & 0 \\ 0 & \varepsilon_{11}^{\text{el.}} & 0 \\ 0 & 0 & \varepsilon_{33}^{\text{el.}} \end{pmatrix} - \delta_{ij} \,. \tag{3.273}$$

The thermal expansion coefficients read

$$\alpha_{ij} = \begin{pmatrix} \alpha_{11} & 0 & 0 \\ 0 & \alpha_{11} & 0 \\ 0 & 0 & \alpha_{33} \end{pmatrix} . \tag{3.274}$$

Piezoceramic is an insulator and brass is a conductor. For the computational domain where the piezo sheet is, we use $\Omega_{\text{p.}}$, and for the brass plate we employ $\Omega_{\text{b.}}$, so the whole domain reads $\Omega = \Omega_{\text{p.}} \cup \Omega_{\text{b.}}$. Between brass and piezoceramic materials there exists a singular surface, $\partial\Omega_{\text{s.}} = \partial\Omega_{\text{p.}} \cap \partial\Omega_{\text{b.}}$, on which the balance equations on singular surfaces need to be fulfilled. We simply employ the weak forms in Eqs. (3.80), (3.92) derived in Sect. 3.2. For simplicity we skip modeling the air surrounding the membrane and piezo sheet. On the boundaries to air, $\partial\Omega_{\text{a.}} = \partial\Omega \setminus \partial\Omega_{\text{s.}}$, either the electric potential is given as a DIRICHLET boundary condition or the electric current vanishes in the plane normal direction. Similarly we assume that the gradient of magnetic potential is directed along the plane normal. We recall that the piezoceramic material lacks magnetic polarization, $\mathcal{M}_i = 0$. The weak forms for electromagnetic potentials read in the energy unit

$$\mathrm{F}_\phi = \int_\Omega \big(-(\mathfrak{D}_i - \mathfrak{D}_i^0)\,\delta\phi_{,i} - \Delta t J_i^{\text{fr.}}\,\delta\phi_{,i} \big)\mathrm{d}v + \int_{\partial\Omega^I} \Big(n_i \Delta t [\![J_i^{\text{fr.}}]\!]\,\delta\phi \Big)\mathrm{d}a \,.$$

$$\mathrm{F}_{A} = \int_\Omega \Big(\varepsilon_0 \frac{A_i - 2A_i^0 + A_i^{00}}{\Delta t \Delta t}\,\delta A_i + \frac{1}{\mu_0} A_{i,j}\,\delta A_{i,j} - J_i^{\text{fr.}}\,\delta A_i - \frac{P_i - P_i^0}{\Delta t}\,\delta A_i \Big)\mathrm{d}v \,. \tag{3.275}$$

The free electric current is

$$J_i^{\text{fr.}} = \mathcal{J}_i^{\text{fr.}} + \rho z^{\text{fr.}} v_i = \mathcal{J}_i^{\text{fr.}} + \mathfrak{D}_{i,i} \frac{u_i - u_i^0}{\Delta t} \,, \tag{3.276}$$

by using the usual time discretization. For the piezo sheet the free objective current vanishes, $\mathcal{J}_i^{\text{fr.}} = 0$, whereas for the brass it is modeled by using OHM's law and neglecting the thermoelectric coupling for brass, since the PELTIER coefficient is small.

In order to compute the displacement we employ the balance of linear momentum in the LAGRANGEan frame after having neglected the geometric nonlinearities

$$\rho_0 \frac{\partial v_i}{\partial t} - \sigma_{ji,j} - \rho_0 f_i = \mathcal{F}_i \,. \tag{3.277}$$

By inserting the electromagnetic supply from Eq. (3.190) in the LAGRANGEan frame for small deformations we acquire

$$\rho_0 \frac{\partial v_i}{\partial t} + \frac{\partial \mathcal{G}_i}{\partial t} - \sigma_{ji,j}^{\text{tot.}} - \rho_0 f_i = 0 \,, \quad \sigma_{ji}^{\text{tot.}} = \sigma_{ji} + m_{ji} \,. \tag{3.278}$$

The electromagnetic momentum and stress are

$$\mathcal{G}_i = \epsilon_{ijk} \mathfrak{D}_j B_k \,, \quad m_{ji} = -\frac{1}{2} \delta_{ij} (H_k B_k + D_k E_k) + H_i B_j + D_j E_i \,. \tag{3.279}$$

For obtaining the weak form we need to integrate by parts solely the terms already including a derivative. Thus, we separate the stress into τ_{ij} including terms with derivatives of the primitive variables and into $\bar{\sigma}_{ij}$ with the primitive variables without derivatives as follows

$$\begin{aligned} \sigma_{ji}^{\text{tot.}} &= \bar{\sigma}_{ji} + \tau_{ji} \,, \\ \tau_{ji} &= m_{ji} + P_j E_i + C_{jikl} \varepsilon_{kl} - \tilde{T}_{kji} E_k \,, \\ \bar{\sigma}_{ji} &= -C_{jikl} \alpha_{kl} (T - T_{\text{ref.}}) \,. \end{aligned} \tag{3.280}$$

From the balance of linear momentum we acquire the following weak form in the unit of energy:

$$\begin{aligned} \mathrm{F}_{\boldsymbol{u}} = \int_\Omega \bigg(& \rho_0 \frac{u_i - 2u_i^0 + u_i^{00}}{\Delta t \Delta t} \delta u_i + \frac{\epsilon_{ijk} (\mathfrak{D}_j B_k - \mathfrak{D}_j^0 B_k^0)}{\Delta t} \delta u_i - \\ & - \bar{\sigma}_{ji,j} \, \delta u_i + \tau_{ji} \, \delta u_{i,j} - \rho_0 f_i \, \delta u_i \bigg) \mathrm{d}V - \int_{\partial \Omega} (\hat{t}_i - N_j \bar{\sigma}_{ji}) \, \delta u_i \mathrm{d}A \,. \end{aligned} \tag{3.281}$$

The traction vector, $\hat{t}_i = N_j \hat{\sigma}_{ji}^{\text{tot.}}$, is given on the boundaries. Either it is a free surface such that no traction occurs or there is a loading measured in Pa and stated by the given traction vector, $\hat{t}_i$. It is of importance to recall that the traction on the surface is undertaken by the mechanical and electromagnetic stress, together. Therefore, under the same traction vector different combinations might occur where a high mechanical stress is compensated by an electromagnetic stress, or otherwise. The electromagnetic stress is generated by the electric and magnetic potentials computed by their corresponding weak forms.

For computing the temperature distribution we acquire the weak form from the balance of entropy in the unit of energy:

$$\begin{aligned} \mathrm{F}_T = \int_\Omega \Big(\rho_0(\eta - \eta^0)\,\delta T - \Delta t \Phi_i\,\delta T_{,i} - \Delta t \rho_0 \frac{r}{T}\,\delta T - \Delta t \Sigma\,\delta T \Big) \mathrm{d}V + \\ + \int_{\partial\Omega} \Delta t \hat{\Phi}\,\delta T \mathrm{d}A \,, \end{aligned} \tag{3.282}$$

where flux and production of entropy are given in Eq. (3.253). After setting the assumptions, ${}^{\mathrm{d}}P_i = 0$, ${}^{\mathrm{d}}\mathcal{M}_i = 0$, ${}^{\mathrm{r}}\mathcal{M}_i = 0$, we obtain

$$\Phi_i = \frac{q_i}{T} \,, \quad \hat{\Phi} = \frac{\hat{q}_i N_i}{T} \,, \quad \Sigma = -\frac{q_i}{T^2} T_{,i} + \frac{1}{T} \mathcal{E}_i \mathcal{J}_i^{\mathrm{fr.}} \,. \tag{3.283}$$

For the boundaries we implement the usual ROBIN boundaries:

$$\hat{q}_i N_i = h(T - T_{\mathrm{amb}}) \,, \tag{3.284}$$

with the ambient temperature chosen equal to the reference temperature.

Since all forms are in the same unit we can sum them up and obtain a weak form:

$$\mathrm{Form} = \mathrm{F}_\phi + \mathrm{F}_A + \mathrm{F}_u + \mathrm{F}_T \,, \tag{3.285}$$

in order to compute the primitive variables, viz., ϕ, A_i, u_i, T. The nonlinear and coupled form will be computed as usual monolithically after a NEWTON–RAPHSON linearization at the level of the partial differential equations.

The aforementioned micropump's membrane is simulated by setting a harmonic potential difference:

$$\hat{\phi} = A \sin(2\pi\nu t) \,, \tag{3.286}$$

where the amplitude is chosen as $A = 100\,\mathrm{V}$ and the frequency as $\nu = 10\,\mathrm{Hz}$. The diaphragm out of brass is fixed on the pump's body at its circumference. We simulate the circumference as clamped and grounded. The piezoceramic sheet deforms due to the potential difference. Electric potential at the bottom of the piezoceramic sheet is the input and the motion of diaphragm's top middle is the output. The response, i.e., input versus output is plotted in Fig. 3.16. The response is nearly instantaneous, which enables an accurate controlling of the micropump's response. All primitive variables are calculated at once. The maximum deflections occur roughly at the first and third quarter of the cycle, see Fig. 3.17. The change in temperature is not significant and the distribution of the magnetic potential is such that no magnetic flux arises. The deformation is small since we have modeled one piezo sheet. In reality the actuator consists of many layers—multilayer piezoceramics—such that the piezoelectric coefficient in the poling direction increases. The geometry can be obtained in [1] and the code used for the simulation is given below.

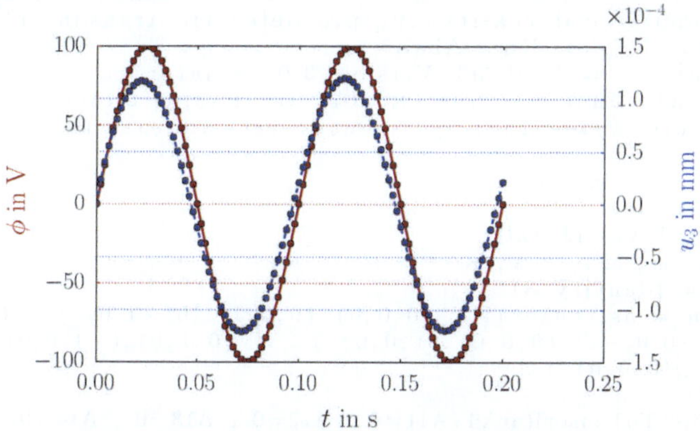

Fig. 3.16 The input (*continuous, red*) and output (*dashed, blue*) of the piezo membrane in the micropump for one cycle in 10 Hz

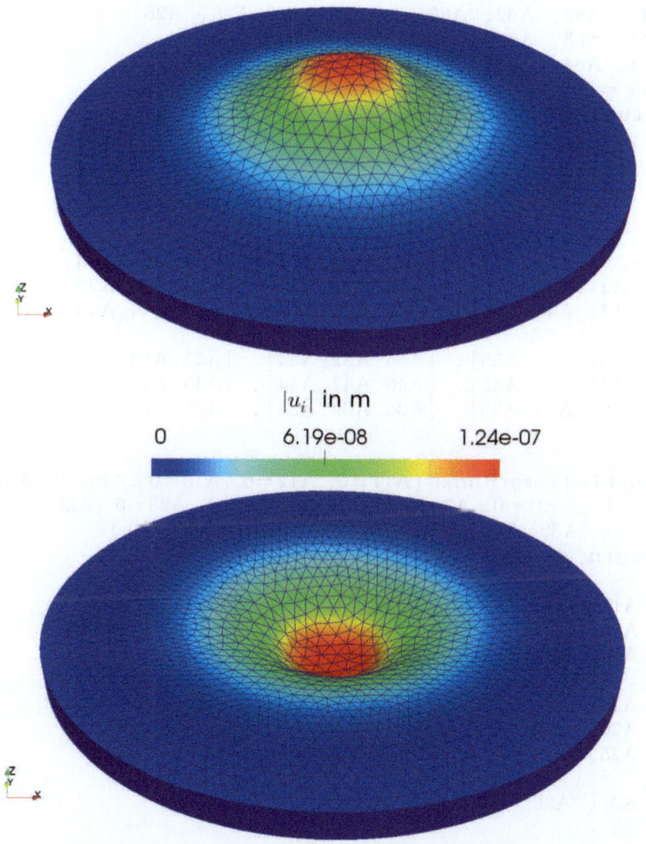

Fig. 3.17 The deformation of the piezo membrane at the first (*top*) and third (*bottom*) quarter of the first cycle with a scale factor of 10 000. The membrane moves upward and downward

```python
"""Computational reality 19, piezoelectric transducer"""
__author__ = "B. Emek Abali"
__license__ = "GNU GPL Version 3.0 or later"
#This code underlies the GNU General Public License, http://www.gnu.org/licenses/gpl-3.0.en.html

from fenics import *
import numpy
set_log_level(ERROR)
#units: m, kg, s, V, K
delta = Identity(3)
epsilon = as_tensor([ ( (0,0,0),(0,0,1),(0,-1,0) ) , ( (0,0,-1),(0,0,0),(1,0,0) ) , ( (0,1,0),(-1,0,0),(0,0,0) ) ])

def VoigtToTensorRank4(A11=0., A12=0., A13=0., A14=0., A15=0., A16=0., A22=0., A23=0., A24=0., A25=0., A26=0., A33=0., A34=0., A35=0., A36=0., A44=0., A45=0., A46=0., A55=0., A56=0., A66=0.):
    A21, A31, A41, A51, A61 = A12, A13, A14, A15, A16
    A32, A42, A52, A62 = A23, A24, A25, A26
    A43, A53, A63 = A34, A35, A36
    A54, A64 = A45, A46
    A65 = A56
    return as_tensor([ \
    [ \
    [ [A11,A16,A15], [A16,A12,A14], [A15,A14,A13]] , \
    [ [A61,A66,A65], [A66,A62,A64], [A65,A64,A63]] , \
    [ [A51,A56,A55], [A56,A52,A54], [A55,A54,A53]] \
    ] , [ \
    [ [A61,A66,A65], [A66,A62,A64], [A65,A64,A63]] , \
    [ [A21,A26,A25], [A26,A22,A24], [A25,A24,A23]] , \
    [ [A41,A46,A45], [A46,A42,A44], [A45,A44,A43]] \
    ] , [ \
    [ [A51,A56,A55], [A56,A52,A54], [A55,A54,A53]] , \
    [ [A41,A46,A45], [A46,A42,A44], [A45,A44,A43]] , \
    [ [A31,A36,A35], [A36,A32,A34], [A35,A34,A33]] ] \
    ])

def VoigtToTensorRank3(A11=0.,A12=0.,A13=0.,A14=0.,A15=0.,A16=0., A21=0.,A22=0.,A23=0.,A24=0.,A25=0.,A26=0., A31=0.,A32=0.,A33=0.,A34=0.,A35=0.,A36=0.):
    return as_tensor([ \
    [ \
    [ A11, A16, A15 ] , \
    [ A16, A12, A14 ] , \
    [ A15, A14, A13 ] \
    ] , [ \
    [ A21, A26, A25 ] , \
    [ A26, A22, A24 ] , \
    [ A25, A24, A23 ] \
    ] , [ \
    [ A31, A36, A35 ] , \
```

```
    [ A36, A32, A34 ] , \
    [ A35, A34, A33 ] ] \
    ])

def ToTensorRank2(A11=0.,A12=0.,A13=0., A21=0.,A22=0.,A23=0., A31=0.,A32=0.,A33=0.):
    A21, A31, A32 = A12, A13, A23
    return as_tensor([ \
    [ A11, A12, A13 ] , \
    [ A21, A22, A23 ] , \
    [ A31, A32, A33 ] \
    ] )

def material_coefficient(target_mesh, cells_list, coeffs):
    coeff_func = Function(FunctionSpace(target_mesh, 'DG', 0))
    markers = numpy.asarray(cells_list.array(), dtype=numpy.int32)
    coeff_func.vector()[:] = numpy.choose(markers-1, coeffs)
    return coeff_func

'''
2D 1 "singularity"
2D 2 "boundary_air"
2D 3 "clamp"
2D 4 "loading"
3D 1 "brass"
3D 2 "piezo"
'''
mesh = Mesh('geo/CR19_piezo.xml')
cells = MeshFunction('size_t',mesh,'geo/CR19_piezo_physical_region.xml')
facets = MeshFunction('size_t',mesh,'geo/CR19_piezo_facet_region.xml')
N = FacetNormal(mesh)

Scalar = FunctionSpace(mesh, 'P', 1)
Vector = VectorFunctionSpace(mesh, 'P', 1)
#phi, A, u, T
Space = MixedFunctionSpace([Scalar, Vector, Vector, Scalar])
dI = Measure('dS', domain=mesh, subdomain_data=facets)
dA = Measure('ds', domain=mesh, subdomain_data=facets)
dV = Measure('dx', domain=mesh, subdomain_data=cells)

nu = 10. #in Hz
t = 0.0
tMax = 2./nu
Dt = tMax/100.

i, j, k, l, n = indices(5)
f = Constant((0., 0., 0.)) #N/kg
r = Constant(0.0)
Tref = 300. #in K
Tamb = Tref
```

```python
eps_0 = 8.85E-12 #in A s/(V m)
mu_0 = 12.6E-7 #in V s/(A m)
null=1E-20 #for numerical reasons it is not zero
h = 10. #in J / (s m^2 K)

# brass
E_b = 105E9 #in Pa
G_b = 36E9 #in Pa
la_b = (E_b-2.*G_b)*G_b/(3.*G_b-E_b)
mu_b = G_b
C_voigt_b = numpy.array([ \
[la_b+2.*mu_b, la_b, la_b, 0, 0, 0],\
[la_b, la_b+2.*mu_b, la_b, 0, 0, 0],\
[la_b, la_b, la_b+2.*mu_b, 0, 0, 0],\
[0, 0, 0, mu_b, 0, 0],\
[0, 0, 0, 0, mu_b, 0],\
[0, 0, 0, 0, 0, mu_b]  ])
alpha_b = 19E-6 #in 1/K
varsigma_b = 0.6 #in S/m or in 1/(Ohm m)
kappa_b = 109. #in W/(K m)
c_b = 380. #in J/(kg K)
rho_0_b = 8400. #in kg/m3

# PZT-5H material
S11_p = 15.6E-12 #1/Pa
S33_p = 20E-12 #1/Pa
nu_p = 0.31
S_voigt_p = numpy.array([ \
[S11_p, -nu_p*S11_p, -nu_p*S11_p, 0, 0, 0],\
[-nu_p*S11_p, S11_p, -nu_p*S11_p, 0, 0, 0],\
[-nu_p*S11_p, -nu_p*S11_p, S33_p, 0, 0, 0],\
[0, 0, 0, (1.+nu_p)*S11_p, 0, 0],\
[0, 0, 0, 0, (1.+nu_p)*S11_p, 0],\
[0, 0, 0, 0, 0, (1.+nu_p)*S11_p]  ])
C_voigt_p = numpy.linalg.inv(S_voigt_p)
dtilde_p_31 = -265E-12 #in m/V
dtilde_p_33 = 585E-12 #in m/V
dtilde_p_15 = 730E-12 #in m/V
eps_rel_el_p_11 = 3130.
eps_rel_el_p_33 = 3400.
alpha_p_11 = 6E-6 #in 1/K
alpha_p_33 =-4E-6 #in 1/K
varsigma_p = null
kappa_p = 1.1 #in W/(K m)
c_p = 350. #in J/(kg K)
rho_0_p = 7500. #in kg/m3

C11 = material_coefficient(mesh, cells, [C_voigt_b[0,0], C_voigt_p[0,0]])
C12 = material_coefficient(mesh, cells, [C_voigt_b[0,1], C_voigt_p[0,1]])
C13 = material_coefficient(mesh, cells, [C_voigt_b[0,2], C_voigt_p[0,2]])
C22 = material_coefficient(mesh, cells, [C_voigt_b[1,1],
```

```
    C_voigt_p[1,1]])
C33 = material_coefficient(mesh, cells, [C_voigt_b[2,2], C_voigt_p[2,2]])
C44 = material_coefficient(mesh, cells, [C_voigt_b[3,3], C_voigt_p[3,3]])
C = VoigtToTensorRank4(A11=C11,A12=C12,A13=C13,A22=C22,A23=C13,A33=C33,A44=C44,A55=C44,A66=C44)

dtilde31 = material_coefficient(mesh, cells, [0., dtilde_p_31])
dtilde33 = material_coefficient(mesh, cells, [0., dtilde_p_33])
dtilde15 = material_coefficient(mesh, cells, [0., dtilde_p_15])
dtilde = VoigtToTensorRank3(A31=dtilde31, A32=dtilde31, A33=dtilde33, A15=dtilde15, A24=dtilde15)
Ttilde = as_tensor( dtilde[i,j,k]*C[n,l,j,k], (i,n,l))

eps_rel_el11 = material_coefficient(mesh, cells, [1., eps_rel_el_p_11])
eps_rel_el33 = material_coefficient(mesh, cells, [1., eps_rel_el_p_33])
eps_rel_el = ToTensorRank2(A11=eps_rel_el11, A22=eps_rel_el11, A33=eps_rel_el33)
chi_el = eps_rel_el - delta

alpha11 = material_coefficient(mesh, cells, [alpha_b, alpha_p_11])
alpha33 = material_coefficient(mesh, cells, [alpha_b, alpha_p_33])
alpha = ToTensorRank2(A11=alpha11, A22=alpha11, A33=alpha33)

varsigma = material_coefficient(mesh, cells, [varsigma_b, varsigma_p])
kappa = material_coefficient(mesh, cells, [kappa_b, kappa_p])
c = material_coefficient(mesh, cells, [c_b, c_p])
rho_0 = material_coefficient(mesh, cells, [rho_0_b, rho_0_p])
v_0 = 1./rho_0

actuator = Expression("100.0*sin(2.0*pi*freq*t)", freq=nu, t=0)
bc1 = DirichletBC(Space.sub(0), 0.0, facets, 3)
bc2 = DirichletBC(Space.sub(0), actuator, facets, 4)
bc3 = DirichletBC(Space.sub(2), Constant((0.0, 0.0, 0.0)), facets, 3)
bc = [bc1,bc2,bc3]

dunkn = TrialFunction(Space)
test = TestFunction(Space)
unkn = Function(Space)
unkn0 = Function(Space)
unkn00 = Function(Space)

#initial values for phi, A, u, T
```

```
unkn_init = Expression(('0.', '0.','0.','0.', '0.','0.','0.',
    'T_r'), T_r=Tref)
unkn00 = interpolate(unkn_init,Space)
unkn0.assign(unkn00)
unkn.assign(unkn0)

del_phi,del_A,del_u,del_T = split(test)
phi,A,u,T = split(unkn)
phi0,A0,u0,T0 = split(unkn0)
phi00,A00,u00,T00 = split(unkn00)

eps = sym(grad(u))
eps0 = sym(grad(u0))
E = -grad(phi) - (A-A0)/Dt
E0 = -grad(phi0) - (A0-A00)/Dt
B = as_tensor(epsilon[i,j,k]*A[k].dx(j) , (i,))
B0 = as_tensor(epsilon[i,j,k]*A0[k].dx(j) , (i,))
EE = as_tensor(E[i] + epsilon[i,j,k]*(u-u0)[j]/Dt*B[k], (i,))

#constitutive equations
D = eps_0*E
D0 = eps_0*E0
H = 1./mu_0*B
m = as_tensor( -1./2.*delta[i,j]*(H[k]*B[k]+D[k]*E[k]) + H[i
    ]*B[j] + D[j]*E[i] , (j,i))
P = as_tensor(-Ttilde[i,j,k]*alpha[j,k]*(T-Tref) + Ttilde[i,j
    ,k]*eps[j,k] + eps_0*chi_el[i,j]*E[j] , (i,))
P0 = as_tensor(-Ttilde[i,j,k]*alpha[j,k]*(T0-Tref) + Ttilde[i
    ,j,k]*eps0[j,k] + eps_0*chi_el[i,j]*E0[j] , (i,))
mD = D + P
mD0 = D0 + P0
 JJ_fr = varsigma*EE
 J_fr = as_tensor( JJ_fr[i] + mD[j].dx(j)*(u-u0)[i]/Dt , (i,))
 sigmaBar = as_tensor( -C[j,i,k,l]*alpha[k,l]*(T-Tref) , (j,i)
    )
 tau = as_tensor(m[j,i] + P[j]*E[i] + C[j,i,k,l]*eps[k,l] -
    Ttilde[k,j,i]*E[k] , (j,i))
 eta = as_tensor( c*ln(T/Tref) + v_0*C[i,j,k,l]*alpha[k,l]*eps
    [i,j] - v_0*Ttilde[i,j,k]*alpha[j,k]*E[i] , ())
 eta0 = as_tensor( c*ln(T0/Tref) + v_0*C[i,j,k,l]*alpha[k,l]*
    eps0[i,j] - v_0*Ttilde[i,j,k]*alpha[j,k]*E0[i] , ())
 q = as_tensor(-kappa*T.dx(i), (i,))
 Phi = q/T
 Sigma = as_tensor(-q[i]/T/T*T.dx(i) + 1./T*EE[i]*JJ_fr[i], ())

 qHat = h*(T-Tamb)
 PhiHat = qHat/T

 #weak forms
 F_phi = ( -(mD-mD0)[i]*del_phi.dx(i) - Dt*J_fr[i]*del_phi.dx(
    i) )*(dV(1)+dV(2)) + N('+')[i]*Dt*(J_fr('+') - J_fr('-
    '))[i]*del_phi('+')*dI(1)

```

```
F_A = (eps_0*(A-2.*A0+A00)[i]/Dt/Dt*del_A[i] + 1./mu_0*A[i].dx(j)*del_A[i].dx(j) -J_fr[i]*del_A[i] - (P-P0)[i]/Dt*del_A[i] ) * (dV(1)+dV(2))

F_u = (rho_0*(u-2.*u0+u00)[i]/Dt/Dt*del_u[i] - sigmaBar[j,i].dx(j)*del_u[i] + tau[j,i]*del_u[i].dx(j) - rho_0*f[i]*del_u[i] )*(dV(1)+dV(2)) + N[j]*sigmaBar[j,i]*del_u[i]*(dA(2)+dA(4))

F_T = ( rho_0*(eta-eta0)*del_T - Dt*Phi[i]*del_T.dx(i) -Dt*rho_0*r/T*del_T - Dt*Sigma*del_T )*(dV(1)+dV(2)) + Dt*PhiHat*del_T*(dA(2)+dA(3)+dA(4))

Form = F_phi + F_A + F_u + F_T
Gain = derivative(Form, unkn, dunkn)

pwd = '/calcul/CR19/'
file_phi = File(pwd + 'phi.pvd')
file_A = File(pwd + 'A.pvd')
file_u = File(pwd + 'u.pvd')
file_T = File(pwd + 'T.pvd')

import matplotlib as mpl
mpl.use('Agg')
import matplotlib.pyplot as pylab
from mpl_toolkits.axes_grid1 import host_subplot
import mpl_toolkits.axisartist as AA

pylab.rc('text', usetex=True )
pylab.rc('font', family='serif', serif='cm', size=30 )
pylab.rc('legend', fontsize=30)
pylab.rc(('xtick.major','ytick.major'), pad=15)
c1, c2 = '#990000', '#0033FF'
fig = pylab.figure(1, figsize=(14,8))
pylab.subplots_adjust(top=0.85)
pylab.subplots_adjust(bottom=0.15)
pylab.subplots_adjust(left=0.18)
pylab.subplots_adjust(right=0.82)
fig.clf()
ax1 = host_subplot(111, axes_class=AA.Axes)
ax1.ticklabel_format(style='sci',scilimits=(-3,+3),axis='y')
ax2 = ax1.twinx()
ax2.ticklabel_format(style='sci',scilimits=(-3,+3),axis='y')
ax1.grid(True, axis='x')
ax1.set_xlabel(r'$t$ in s')
ax1.set_ylabel(r'$\phi$ in V', color=c1)
ax1.tick_params(axis='y', colors=c1)
ax1.grid(True, axis='y', color=c1)
ax2.set_ylabel(r'$u_3$ in mm', color=c2)
ax2.tick_params(axis='y', colors=c2)
ax2.grid(True, axis='y', color=c2)

time_plot, phi_plot, u_plot = [0],[0],[0]
tic()
```

```
while t < tMax:
    t += Dt
    actuator.t = t
    print 'time: ',t, ' after ',toc(),' seconds'
    tic()
    solve(Form==0, unkn, bc, J=Gain, \
        solver_parameters={"newton_solver":{"linear_solver": "mumps", "relative_tolerance": 1e-5} }, \
        form_compiler_parameters={"cpp_optimize": True, "representation": "quadrature", "quadrature_degree": 2} )


    time_plot.append(t)
    phi_plot.append(unkn.split()[0](0.,0.,-0.0001))
    u_plot.append(unkn.split()[2](0.,0.,0.0005)[2]*1000.)

    ax1.plot(time_plot, phi_plot, color=c1, linewidth=3, linestyle='-', marker='o', markersize=8)
    ax2.plot(time_plot, u_plot, color=c2, linewidth=3, linestyle='--', marker='o', markersize=8)

    pylab.savefig(pwd + 'CompReal19_input_output.pdf')
    file_phi << (unkn.split()[0], t)
    file_A << (unkn.split()[1], t)
    file_u << (unkn.split()[2], t)
    file_T << (unkn.split()[3], t)

    unkn00.assign(unkn0)
    unkn0.assign(unkn)
```

To-do

Piezo drives are used in many measurement, medical, and high-precision devices. Make a web based search for different systems using piezoelectricity, for example:

- Quartz oscillators,
- ultrasonic motors (or engines) for autofocusing in camera lenses,
- energy scavengers (piezoelectric generators),
- piezo gyros (gyroscopes).

We have implemented a uniform piezoceramic sheet, however, usually multilayer piezoceramics are used. Search for a piezoceramic bimorph to comprehend the idea of a multilayered piezoceramic sheet.

3.6 Magnetohydrodynamics in Metal Smelting

Compounds like minerals are found in Earth's crust. These minerals are nothing else than rocks on streets. Some of them contain a high amount of metals like iron, copper, aluminum, or even gold, and silver. These useful minerals are called *ores*. Gold and silver have a low chemical activity such that they are found in the ore in their pure state. Aluminum is an *active* metal such that it needs to be extracted from the ore by using electrolysis. Copper and iron are *reactive* materials such that they can be extracted by heating and adding carbon.[74]

One way of extracting metal from the ore is called *smelting*. By increasing temperature the ore melts and due to the different mass densities, metal and the impurities separate. Either the impurities are burned off such that they leave the molten metal as gases or they form a molten slag on top of the metal. The molten metal is stirred for helping the separation. A type of smelting is the electric smelting, where a magnetic flux is applied on the molten metal. This field induces an electric current. Another type of smelting is an electrolytic reduction process. This electrochemical process is based on the fact that a molten metal is an *ionic* conductor called *electrolyte*. Anode and cathode are immersed into the electrolyte and pass an electric current directly. Anode supplies positive charges such that the positive ions are "pushed" from anode to cathode. In the ore the metal is a positive ion. In other words, the metal is separated from the chemical solution by using a powerful electric current supplied directly into the molten metal. The pure metal is extracted from its ore and collected on the cathode. First, a process called *electrowinning* is applied, then, another process called *electrorefining* is used. Metals as pure as 99.999 % are possible to be produced by using these methods.

Electrowinning is the primary extraction from the ore. For reactive materials like copper, the electrolyte is a concentrated acidic solution with a very low pH value such that the metal ions electrodeposit more efficiently. For an active material like aluminum a fused-salt electrolysis is used. For aluminum extraction the ore, mainly bauxite, is converted by the Bayer process to alumina (aluminum oxide). In the Bayer process bauxite is mixed with sodium hydroxide under high temperatures and pressures creating dissolved aluminum oxide. Unfortunately, this solution cannot be used as the electrolyte in the electrowinning process. The alumina is dissolved in molten cryolite in order to get a proper electrolyte. Historically, the discovery of using the cyrolite as an electrolyte took a long time. Although bauxite is easy to find in the nature, high production costs led to an aluminum consumption as a precious metal until the end of 19th century. The discovery did decrease the aluminum production costs extremely. Nowadays, aluminum is used in a wide variety of products. The electrowinning using cyrolite is called the HALL–HEROULT process.[75] Suppose we want to simulate the primary extraction process of aluminum by using the

---

[74] Iron with carbon is named as steel.

[75] In 1886 Charles Martin Hall and (his sister) Julia Brainerd Hall developed the process. In the same year Paul Héroult did develope the same process, independently to them. Therefore, the process is called after all of them.

HALL–HEROULT process. The electrolyte is conducting a high electric current, inducing a magnetic flux, thus, producing a body force—electromagnetic supply term. Since the electrolyte is a viscous fluid, this supply term alters the velocity. In this section we are interested in the hydrodynamics of a viscous conducting fluid.

In order to comprehend the interaction of the electromagnetic supply term with hydrodynamics we need to obtain the constitutive equations in magnetohydrodynamics. We start with the balance of mass and (linear) momentum:

$$\frac{\partial \rho}{\partial t} + \frac{\partial \rho v_i}{\partial x_i} = 0 \,, \quad \frac{\partial \rho v_i}{\partial t} - \frac{\partial}{\partial x_j}\big(- v_j \rho v_i + \sigma_{ji}\big) - \rho f_i = \mathcal{F}_i \,. \tag{3.287}$$

Since the metal will be polarized we choose the following electromagnetic supply term as motivated in the last section

$$\mathcal{F}_i = \rho z E_i + \epsilon_{ijk} J_j B_k - \epsilon_{ijk} \frac{\partial P_j}{\partial t} B_k - \epsilon_{ijk} P_j \frac{\partial B_k}{\partial t} \,. \tag{3.288}$$

By repeating the same steps from the last section we obtain the balance of internal energy:

$$\rho \frac{\mathrm{d}u}{\mathrm{d}t} + \frac{\partial q_j}{\partial x_j} - \rho r = \big(\sigma_{ji} - P_j E_i + \mathcal{M}_i B_j\big) \frac{\partial v_i}{\partial x_j} + \mathcal{E}_i \mathcal{J}_i^{\text{fr.}} - P_i \frac{\mathrm{d}E_i}{\mathrm{d}t} + B_i \frac{\mathrm{d}\mathcal{M}_i}{\mathrm{d}t} \,. \tag{3.289}$$

The molten salt is a viscous fluid such that the reversible part of the stress is given by the hydrostatic pressure, p, as follows

$$\sigma_{ji} = -p\delta_{ji} + {}^{\mathrm{d}}\sigma_{ji} \,. \tag{3.290}$$

By rewriting the balance of mass:

$$\begin{aligned} \frac{\mathrm{d}\rho}{\mathrm{d}t} + \rho \frac{\partial v_i}{\partial x_i} &= 0 \,, \\ -\frac{\partial v_i}{\partial x_i} &= \frac{1}{\rho} \frac{\mathrm{d}\rho}{\mathrm{d}t} \,, \end{aligned} \tag{3.291}$$

and then using it in the balance of internal energy with the stress tensor in Eq. (3.290), we obtain

$$\begin{aligned} \rho \frac{\mathrm{d}u}{\mathrm{d}t} + \frac{\partial q_j}{\partial x_j} - \rho r = \frac{p}{\rho} \frac{\mathrm{d}\rho}{\mathrm{d}t} + \big({}^{\mathrm{d}}\sigma_{ji} - P_j E_i + \mathcal{M}_i B_j\big) \frac{\partial v_i}{\partial x_j} + \\ + \mathcal{E}_i \mathcal{J}_i^{\text{fr.}} - P_i \frac{\mathrm{d}E_i}{\mathrm{d}t} + B_i \frac{\mathrm{d}\mathcal{M}_i}{\mathrm{d}t} \,. \end{aligned} \tag{3.292}$$

We introduce the specific volume in the EULERian frame, $v = 1/\rho$, such that the first term on the right-hand side becomes

$$\frac{p}{\rho}\frac{\mathrm{d}\rho}{\mathrm{d}t} = pv\frac{\mathrm{d}v^{-1}}{\mathrm{d}t} = -pvv^{-2}\frac{\mathrm{d}v}{\mathrm{d}t} = -p\rho\frac{\mathrm{d}v}{\mathrm{d}t}\ . \tag{3.293}$$

The first simplification relies on the assumption that polarization is reversible. Hence, we neglect the dissipative electric and magnetic polarizations, ${}^{\mathrm{d}}\mathcal{P}_i = 0$, ${}^{\mathrm{d}}\mathcal{M}_i = 0$. At a mechanical equilibrium the velocity gradient vanishes; at a thermodynamical equilibrium the heat flux is expressed via entropy as well as the supply term, r, disappears; and at an electromagnetic equilibrium a conducting current is vanishing. Thus, at an equilibrium we acquire from the balance of internal energy the following GIBBS equation:

$$\mathrm{d}u = T\mathrm{d}\eta - p\mathrm{d}v - vP_i\mathrm{d}E_i + vB_i\mathrm{d}\mathcal{M}_i\ . \tag{3.294}$$

By inserting GIBBS's equation into the balance of internal energy we obtain the balance of entropy without dissipative polarization terms:

$$\rho\frac{\mathrm{d}\eta}{\mathrm{d}t} + \frac{\partial}{\partial x_j}\Big(\frac{q_j}{T}\Big) - \rho\frac{r}{T} = -\frac{q_j}{T^2}\frac{\partial T}{\partial x_j} + \frac{1}{T}\big({}^{\mathrm{d}}\sigma_{ji} - P_jE_i + \mathcal{M}_iB_j\big)\frac{\partial v_i}{\partial x_j} + \frac{1}{T}\mathcal{E}_i\mathcal{J}_i^{\mathrm{fr.}}\ . \tag{3.295}$$

We have already introduced the entropy flux, $\Phi_i = q_i/T$, which will be used in the following. The entropy production:

$$\Sigma = -q_j\frac{1}{T^2}T_{,j} + \Xi_{ji}\frac{1}{T}v_{i,j} + \mathcal{J}_i^{\mathrm{fr.}}\frac{1}{T}\mathcal{E}_i\ , \tag{3.296}$$

has to be positive, where we have used again the comma notation $_{,i}$ as the partial derivative with respect to x_i in the EULERian frame, and where we have introduced the abbreviation:

$$\Xi_{ji} = {}^{\mathrm{d}}\sigma_{ji} - P_jE_i + \mathcal{M}_iB_j\ . \tag{3.297}$$

Although Ξ_{ji} is a tensor of rank two, it is not symmetric. Again by decomposing the tensor of rank two into symmetric-deviatoric, spherical, and antisymmetric terms and by writing the thermodynamical forces:

$$\mathcal{K}^\alpha = \left\{\frac{1}{T^2}T_{,j}\ ,\ \frac{1}{T}v_{|(i,j)|}\ ,\ \frac{1}{T}v_{j,j}\ ,\ \frac{1}{T}v_{[i,j]}\ ,\ \frac{1}{T}\mathcal{E}_i\right\}, \tag{3.298}$$

as well as the thermodynamical fluxes:

$$\mathcal{F}^\alpha = \left\{-q_j\ ,\ \Xi_{|(ji)|}\ ,\ \Xi_{ii}\ ,\ \Xi_{[ji]}\ ,\ \mathcal{J}_i^{\mathrm{fr.}}\right\}, \tag{3.299}$$

we write the 2 law of thermodynamics,

$$\Sigma = \mathcal{K}^\alpha\cdot\mathcal{F}^\alpha \geq 0\ ,\quad \alpha = 1, 2, \ldots, 5\ . \tag{3.300}$$

With the help of the CURIE principle such that each thermodynamical flux depends on all thermodynamical forces of the same rank,

$$\Xi_{ii} = \Xi_{ii}\Big(\frac{1}{T}v_{j,j}\Big) \,, \quad -q_i = -q_i\Big(\frac{1}{T^2}T_{,j}, \frac{1}{T}\mathcal{E}_i\Big) \,, \quad \mathcal{J}_i^{\text{fr.}} = \mathcal{J}_i^{\text{fr.}}\Big(\frac{1}{T^2}T_{,j}, \frac{1}{T}\mathcal{E}_i\Big) \,,$$
$$\Xi_{|(ji)|} = \Xi_{|(ji)|}\Big(\frac{1}{T}v_{|(i,j)|}\Big) \,, \quad \Xi_{[ji]} = \Xi_{[ji]}\Big(\frac{1}{T}v_{[i,j]}\Big) \,, \tag{3.301}$$

we propose the following constitutive equations:

$$\Xi_{ii} = \bar{c}_1\frac{1}{T}v_{j,j} \,, \quad -q_i = \bar{c}_2\frac{1}{T^2}T_{,i} + \bar{c}_3\frac{1}{T}\mathcal{E}_i \,,$$
$$\mathcal{J}_i^{\text{fr.}} = \bar{c}_4\frac{1}{T^2}T_{,i} + \bar{c}_5\frac{1}{T}\mathcal{E}_i \,, \quad \Xi_{|(ji)|} = \bar{c}_6\frac{1}{T}v_{|(i,j)|} \,, \quad \Xi_{[ji]} = \bar{c}_7\frac{1}{T}v_{[i,j]} \,, \tag{3.302}$$

where all coefficients, $\bar{c}_\times$, are functions of the corresponding thermodynamical forces. By inserting them into Eq. (3.300) we result in the following conditions:

$$\bar{c}_1 \geq 0 \,, \quad \bar{c}_2 \geq 0 \,, \quad \bar{c}_3 + \bar{c}_4 = 0 \,, \quad \bar{c}_6 \geq 0 \,, \quad \bar{c}_7 \geq 0 \,, \tag{3.303}$$

in order to guarantee the 2nd law of thermodynamics, $\Sigma \geq 0$. Now by renaming,

$$\frac{1}{T^2}\bar{c}_2 = \kappa \,, \quad \frac{1}{T}\bar{c}_5 = \varsigma \,, \quad \frac{1}{T^2\varsigma}\bar{c}_4 = \pi \,, \tag{3.304}$$

we obtain the heat flux and the electric current:

$$q_i = -\kappa T_{,i} + \varsigma\pi\mathcal{E}_i \,, \quad \mathcal{J}_i^{\text{fr.}} = \varsigma\pi T_{,i} + \varsigma\mathcal{E}_i \,, \tag{3.305}$$

with the thermoelectric coupling, π, identical to the case of unpolarized materials introduced in Sect. 3.3. Often, the simplification of linearity is utilized and κ, ς, π are assumed to be constant values. Moreover, by renaming,

$$\bar{c}_1\frac{1}{T} = 3\lambda + 2\mu \,, \quad \bar{c}_6\frac{1}{T} = 2\mu \,, \quad \bar{c}_7\frac{1}{T} = \nu \,, \tag{3.306}$$

we acquire

$$\Xi_{ji} = \frac{1}{3}\Xi_{kk}\delta_{ji} + \Xi_{|(ji)|} + \Xi_{[ji]} = \Big(\lambda + \frac{2}{3}\mu\Big)v_{k,k}\delta_{ji} + 2\mu v_{|(i,j)|} + \nu v_{[i,j]} \,,$$
$$\Xi_{ji} = \lambda v_{k,k}\delta_{ji} + 2\mu v_{(i,j)} + \nu v_{[i,j]} \,,$$
$${}^{\text{d}}\sigma_{ji} = P_j E_i - \mathcal{M}_i B_j + \lambda v_{k,k}\delta_{ji} + 2\mu v_{(i,j)} + \nu v_{[i,j]} \,. \tag{3.307}$$

Here again we might assume a linear model, such that λ, μ, ν are constants. The volumetric viscosity, λ, and shear viscosity, μ, have been introduced in Sect. 1.7 in the NAVIER–STOKES equation. Their values can be found for a molten salt in the literature. However, the viscosity ν causing an antisymmetric velocity gradients is difficult to measure. As the second simplification we neglect this term and acquire

$$\sigma_{ji} = -p\delta_{ji} + {}^{\mathrm{d}}\sigma_{ji} = P_j E_i - \mathcal{M}_i B_j + (-p + \lambda v_{k,k})\delta_{ji} + 2\mu v_{(i,j)} \,. \quad (3.308)$$

The mechanical pressure in the molten salt is simply the volumetric part of the CAUCHY stress:

$$-\frac{1}{3}\sigma_{ii} = -\frac{1}{3}P_i E_i + \frac{1}{3}\mathcal{M}_i B_i + p - \Big(\lambda + \frac{2}{3}\Big)v_{i,i} = p + \frac{1}{3}\mathfrak{p} \,, \quad (3.309)$$

where p is the hydrostatic pressure and $\mathfrak{p}$ is called the dynamic pressure. At equilibrium the velocity gradient vanishes such that the dynamic pressure becomes

$$\mathfrak{p} = -P_i E_i + \mathcal{M}_i B_i \,. \quad (3.310)$$

The fluid flows with the velocity v_i such that a part of the motion of electric charge is due to convection. The dielectric displacement caused by the electric polarization is in fact quite small with respect to the electric charge motion due to the convection. Hence, we neglect this term, $P_i = 0$, and obtain at equilibrium

$$\begin{aligned} du &= T\mathrm{d}\eta - p\mathrm{d}v + vB_i\mathrm{d}\mathcal{M}_i \,, \\ du &= T\mathrm{d}\eta - (p + B_i\mathcal{M}_i)\mathrm{d}v + B_i\mathrm{d}(v\mathcal{M}_i) \,. \end{aligned} \quad (3.311)$$

By introducing a specific magnetic polarization, $m_i = v\mathcal{M}_i$, and total pressure, $\bar{p} = p + \mathfrak{p}$, we acquire

$$du = T\mathrm{d}\eta - \bar{p}\mathrm{d}v + B_i\mathrm{d}m_i \,. \quad (3.312)$$

For the extraction metallurgy, where the ore is melted such that the metal is extracted from the ore, setting the correct temperature is of paramount importance. There are many measurements of the GIBBS free energy, g, indicating the necessary energy for a reaction to occur. For example, in order to extract aluminum from its ore[76] first the ore is crushed and washed in sodium hydroxide, NaOH. This so-called Bayer process leaches aluminum from bauxite in form of aluminum hydroxide, $Al(OH)_3$, which is calcined into alumina, Al_2O_3. From the alumina by using the HALL–HEROULT process aluminum is smelted. Alumina is dissolved in the molten cryolite, Na_3AlF_6. Cryolite is just another mineral of aluminum found in nature. Nowadays, cyrolite is synthetically produced to decrease the production costs and optimize its physical

[76]There are many different ores containing aluminum. Mainly bauxite is used. It is a mixture of aluminum minerals, clay minerals, and insoluble materials. The main aluminum minerals in bauxite are gibbsite $Al(OH)_3$, boehmite γ–AlO(OH), and diaspore α–AlO(OH).

properties.[77] The positive ions of alumina are attracted to the cathode such that they move to the cathode. By touching the cathode they are reduced to pure aluminum by embodying electrons supplied from the cathode. This reduction (of charge) process is called smelting or electrowinning and it uses high amount of electric current. Basically, the aluminum reduction is done in areas with low electricity costs.

We want to simulate the motion of molten cryolite in an electrowinning process by computing the electromagnetic potentials, pressure, velocity, and temperature. Specific volume and specific magnetic polarization are derived from these primitive variables. Hence, the internal energy, $u = u(\eta, v, m_i)$, depending on specific entropy, specific volume, and specific magnetic polarization is unpractical. We introduce the specific GIBBS free energy at the equilibrium:

$$g = u - T\eta + \bar{p}v - B_i m_i \ . \tag{3.313}$$

Its differential:

$$\mathrm{d}g = \mathrm{d}u - \mathrm{d}T\eta - T\mathrm{d}\eta + \mathrm{d}\bar{p}v + \bar{p}\mathrm{d}v - \mathrm{d}B_i m_i - B_i \mathrm{d}m_i \ , \tag{3.314}$$

leads to the following relation:

$$\mathrm{d}g = -\eta \mathrm{d}T + v\mathrm{d}\bar{p} - m_i \mathrm{d}B_i \ , \tag{3.315}$$

after inserting the GIBBS equation. From the latter relation we observe

$$g = g(T, \bar{p}, B_i) \ , \tag{3.316}$$

the primary or state variables are now T, $\bar{p}$, B_i. Their dual variables read η, v, m_i with the following relations:

$$\eta = -\frac{\partial g}{\partial T} \ , \quad v = \frac{\partial g}{\partial \bar{p}} \ , \quad m_i = -\frac{\partial g}{\partial B_i} \ . \tag{3.317}$$

Since the dual variables may depend on the primary variables we obtain

$$\begin{aligned} \mathrm{d}\eta &= c^1 \mathrm{d}T + c^2 \mathrm{d}\bar{p} + c_i^3 \mathrm{d}B_i \ , \\ \mathrm{d}v &= c^4 \mathrm{d}T + c^5 \mathrm{d}\bar{p} + c_i^6 \mathrm{d}B_i \ , \\ \mathrm{d}m_i &= c_i^7 \mathrm{d}T + c_i^8 \mathrm{d}\bar{p} + c^9 \mathrm{d}B_i \ . \end{aligned} \tag{3.318}$$

By using the MAXWELL symmetry or reciprocal relations:

[77] There are many investigations for optimizing the extraction conditions, for example, see [5, 36].

$$c^2 = \frac{\partial \eta}{\partial \bar{p}} = -\frac{\partial^2 g}{\partial \bar{p} \partial T} = -\frac{\partial^2 g}{\partial T \partial \bar{p}} = -\frac{\partial v}{\partial T} = -c^4 \,,$$
$$c^3 = \frac{\partial \eta}{\partial B_i} = -\frac{\partial^2 g}{\partial B_i \partial T} = -\frac{\partial^2 g}{\partial T \partial B_i} = -\frac{\partial m_i}{\partial T} = -c^7 \,, \tag{3.319}$$
$$c^6 = \frac{\partial v}{\partial B_i} = \frac{\partial^2 g}{\partial B_i \partial \bar{p}} = \frac{\partial^2 g}{\partial \bar{p} \partial B_i} = \frac{\partial m_i}{\partial \bar{p}} = c_i^8 \,,$$

we conclude

$$\begin{aligned} d\eta &= c^1 dT - c^4 d\bar{p} - c_i^7 dB_i \,, \\ dv &= c^4 dT + c^5 d\bar{p} + c_i^8 dB_i \,, \\ dm_i &= c_i^7 dT + c_i^8 d\bar{p} + c^9 dB_i \,. \end{aligned} \tag{3.320}$$

In principle, we can measure all material coefficients, $c^\times$, as functions depending on the primary variables. Instead of this method one might measure the GIBBS free energy and finds out the material coefficients as derivatives of the free energy. For example, the specific heat capacity is

$$c^1 = \frac{\partial \eta}{\partial T} = -\frac{\partial^2 g}{\partial T^2} \,. \tag{3.321}$$

In chemical engineering one important measure is *enthalpy*. We introduce the specific enthalpy, $\hbar$, as follows

$$\begin{aligned} \hbar &= u + \bar{p} v - B_i m_i \,, \\ \hbar &= g + T\eta \,. \end{aligned} \tag{3.322}$$

Since $T\eta = Q$ is the heat (per mass) added to the system we can obtain the following relation:

$$g = \hbar - Q \,, \tag{3.323}$$

where the total energy of a chemical reaction, $\hbar$, minus the heat dissipating within the system, Q, can be seen as a driving energy necessary for a chemical process to occur. In chemical engineering the GIBBS free energy is interpreted as the necessary amount of energy to start a reaction or the excess energy generated in the process. Therefore, the enthalpy is often used by measuring the material coefficients.

Cyrolite's[78] specific heat is well-documented over temperature. In its solid phase it depends on temperature linearly. A liquid cyrolite has a constant specific heat capacity as in [4, Table 8], which is measured by holding the total pressure, $\bar{p}$, constant, $d\bar{p} = 0$,

$$c_{\bar{p}} = 394.7 \text{ J/(mol K)} \,, \quad 1 \text{ mol } Na_3AlF_6 \,\hat{=}\, 209.9413 \cdot 10^{-3} \text{ kg} \,, \quad c^1 = \frac{c_{\bar{p}}}{T} \,. \tag{3.324}$$

[78]The chemical composition of cyrolite is Na_3AlF_6.

Since the volume variation with respect to a pressure change is difficult to measure for liquid metals, the coefficient c^5 will be approximated.[79] In order to determine c^4 we exploit the data from [23, Table 1], where the mass density over temperature is given. The measured dependency is linear in the temperature, so we need only two values in order to define the linear temperature dependency of the specific volume. We obtain the following specific volume (under constant total pressure, $\mathrm{d}\bar{p} = 0$):

$$v = v_{\text{ref.}} + c^4 T \ , \quad v_{\text{ref.}} = 1.891 \cdot 10^{-4}\,\text{m}^3/(\text{kg}) \ , \quad c^4 = 2.259 \cdot 10^{-7}\,\text{m}^3/(\text{kg K}) \ . \tag{3.325}$$

The magnetocaloric coefficient, c_i^7, occurs only in special alloys, for the molten cyrolite we can neglect this effect by setting $c_i^7 = 0$. Furthermore, the parameter c_i^8 claims a magnetization owing to pressure. This effect is so small that it gets important in thin films. For a bulk of molten salt we exclude such an effect and implement a magnetization occurring only as a consequence of electromagnetism. For molten cyrolite we use the same constant magnetic susceptibility as solid aluminum.

Since all material parameters are constants, the underlying material is a linear material and we obtain the dual variables by integrating from the reference state, $T = T_{\text{ref.}}$, $p = p_{\text{ref.}}$, $B_i = 0$, to the current state, by using $\bar{p} = p + \mathcal{M}_i B_i$,

$$\begin{aligned} \eta &= c_{\bar{p}} \ln\left(\frac{T}{T_{\text{ref.}}}\right) - c^4 (p - p_{\text{ref.}} + \mathcal{M}_i B_i) \ , \\ v &= v_0 + c^4 (T - T_{\text{ref.}}) + c^5 (p - p_{\text{ref.}} + \mathcal{M}_i B_i) \ , \\ m_i &= c^9 B_i \Rightarrow \mathcal{M}_i = \rho c^9 B_i \ . \end{aligned} \tag{3.326}$$

The coefficient c^9 can be rewritten, $(\mu_0 \mu^{\text{mag.}})^{-1} \chi^{\text{mag.}} = \rho c^9$, where $\chi^{\text{mag.}}$ is the magnetic susceptibility of cyrolite, which is approximately the same constant value in solid and liquid state, $\chi^{\text{mag.}} = 2.2 \cdot 10^{-5}$.

In the following we continue using the comma notation for the space derivative. The objective is to solve the electric potential, ϕ, the magnetic potential, A_i, the pressure, p, the velocity, v_i, and the temperature, T, in space, x_i, and time, t. The electromagnetic potentials ϕ and A_i are introduced as solutions of the following MAXWELL equations:

$$B_{i,i} = 0 \ , \quad \frac{\partial B_i}{\partial t} + \epsilon_{ijk} E_{k,j} = 0 \ , \tag{3.327}$$

[79] In solid bodies the so-called *equation of state* (EOS) is well-documented for many materials. For example, MIE- GRUNEISEN approximation, named after Gustav Adolf Feodor Wilhelm Ludwig Mie and Eduard Grüneisen, is used for a relation between energy and pressure, in our notation $\partial g/\partial \bar{p} = v/\Gamma$, where Γ denotes the GRUNEISEN parameter. See for values of such a parameter for Cu in [19] or for Cu, Al, Pb in [17, Table 1].

with adequate ansatz functions:

$$B_i = \epsilon_{ijk} A_{k,j} \ , \quad E_i = -\phi_{,i} - \frac{\partial A_i}{\partial t} \ . \tag{3.328}$$

After implementing LORENZ's gauge, we have obtained the weak forms for solving ϕ and A_i in Sect. 3.2. The weak form for the computation of the electric potential as in Eq. (3.80) reads in the unit of energy

$$\begin{aligned} \mathrm{F}_\phi = & \int_\Omega \big(-(\mathfrak{D}_i - \mathfrak{D}_i^0)\, \delta\phi_{,i} - \Delta t J_i^{\text{fr.}}\, \delta\phi_{,i} - \Delta t \epsilon_{ijk} \mathcal{M}_{k,j}\, \delta\phi_{,i} \big) \mathrm{d}v + \\ & + \int_{\partial\Omega^I} \Big(n_i \Delta t \big[J_i^{\text{fr.}} \big] \delta\phi + n_i \Delta t \epsilon_{ijk} \big[\mathcal{M}_{k,j} \big] \delta\phi \Big) \mathrm{d}a \ . \end{aligned} \tag{3.329}$$

We recall that we employ $P_i = 0$ in this section such that $\mathfrak{D}_i = D_i = \varepsilon_0 E_i$. The free electric current is

$$J_i^{\text{fr.}} = \mathcal{J}_i^{\text{fr.}} + \rho z^{\text{fr.}} v_i = \mathcal{J}_i^{\text{fr.}} + \mathfrak{D}_{j,j} v_i \ , \tag{3.330}$$

for $\mathcal{J}_i^{\text{fr.}}$ we have derived the necessary constitutive equation in Eq. (3.305) with the objective electric field:

$$\mathcal{E}_i = E_i + \epsilon_{ijk} v_j B_k \ . \tag{3.331}$$

For the computation of the magnetic potential we use Eq. (3.92) in the unit of energy:

$$\mathrm{F}_A = \int_\Omega \Big(\varepsilon_0 \frac{A_i - 2A_i^0 + A_i^{00}}{\Delta t \Delta t}\, \delta A_i + \frac{1}{\mu_0} A_{i,j}\, \delta A_{i,j} - J_i^{\text{fr.}}\, \delta A_i + \epsilon_{ijk} \mathcal{M}_k\, \delta A_{i,j} \Big) \mathrm{d}v \ , \tag{3.332}$$

where again the electric polarization is set to zero. In order to determine the hydrostatic pressure, p, we use the balance of mass in Eq. $(3.287)_1$. The weak form becomes in the unit of energy after multiplying with the time step and dividing by the mass density

$$\mathrm{F}_p = \int_\Omega \Big((\rho - \rho^0) + \Delta t \rho_{,i} v_i + \Delta t \rho v_{i,i} \Big) \frac{\delta p}{\rho} \mathrm{d}v \ , \tag{3.333}$$

where the mass density is given by $\rho = 1/v$ and for the specific volume, v, we have derived a constitutive equation in Eq. $(3.326)_2$. For computing velocity, v_i, we rewrite the balance of linear momentum in Eq. $(3.287)_2$ by using the balance of electromagnetic momentum:

$$\frac{\partial \mathcal{G}_i}{\partial t} = m_{ji,j} - \mathcal{F}_i \ , \tag{3.334}$$

with the electromagnetic momentum for polarized systems and with the electromagnetic stress:

$$\mathcal{G}_i = \epsilon_{ijk}\mathfrak{D}_j B_k \ , \quad m_{ji} = -\frac{1}{2}\delta_{ji}\big(H_k B_k + D_k E_k\big) + H_i B_j + D_j E_i \ , \tag{3.335}$$

such that we acquire

$$\frac{\partial}{\partial t}\big(\rho v_i + \mathcal{G}_i\big) - \frac{\partial}{\partial x_j}\big(- \rho v_i v_j + \sigma_{ji}^{\text{tot.}}\big) - \rho f_i = 0 \ . \tag{3.336}$$

The total stress,

$$\sigma_{ji}^{\text{tot.}} = \sigma_{ji} + m_{ji} = -p\delta_{ji} + m_{ji} + {}^{\text{d}}\sigma_{ji} \ , \tag{3.337}$$

is fully defined since the mechanical stress is derived in Eq. (3.308) and the electromagnetic stress is given by Eq. $(3.335)_2$. By integrating by parts on the terms including a gradient of the primitive variables we acquire the following weak form in the unit of energy:

$$\begin{aligned} \mathrm{F}_{\boldsymbol{v}} = \int_\Omega \Big(\big(\rho v_i - \rho^0 v_i^0 + \mathcal{G}_i - \mathcal{G}_i^0\big)\, \delta v_i + \Delta t \rho_{,j} v_i v_j \, \delta v_i + \\ + \Delta t \rho v_{i,j} v_j \, \delta v_i + \Delta t \rho v_i v_{j,j} \, \delta v_i + \Delta t p_{,i} \, \delta v_i + \Delta t (m_{ji} + {}^{\text{d}}\sigma_{ji})\, \delta v_{i,j} - \\ - \Delta t \rho f_i \, \delta v_i \Big) \mathrm{d}v - \int_{\partial\Omega} \Delta t \big(\hat{t}_i + p n_i\big)\, \delta v_i \mathrm{d}a \ , \end{aligned} \tag{3.338}$$

after multiplying by the time step. The traction belongs to the total stress

$$\hat{t}_i = n_j \sigma_{ji}^{\text{tot.}} \ . \tag{3.339}$$

In other words, if a force is applied on the boundary then the mechanical and the electromagnetic stress react together against this force. We will give the velocities on the domain boundary such that the boundary integrals vanish. Equation (3.295) is the balance of entropy in the EULERian frame:

$$\rho \frac{\partial \eta}{\partial t} + \rho v_i \eta_{,i} + \Phi_{i,i} - \rho \frac{r}{T} = \Sigma \ , \tag{3.340}$$

with the entropy flux, $\Phi_i = q_i / T$, and entropy production:

$$\Sigma = -\frac{q_i}{T^2} T_{,i} + \frac{1}{T}({}^{\text{d}}\sigma_{ji} + \mathcal{M}_i B_j) v_{i,j} + \frac{1}{T} \mathcal{J}_i^{\text{fr.}} \mathcal{E}_i \ , \tag{3.341}$$

as given in Eq. (3.296). All terms including derivatives are integrated by parts and we acquire the following weak form in the unit of energy:

$$
\begin{aligned}
\mathrm{F}_T = \int_\Omega \Big(& \rho(\eta-\eta^0)\,\delta T - \Delta t\eta(\rho v_i\,\delta T)_{,i} - \Delta t\frac{q_i}{T}\,\delta T_{,i} - \Delta t\rho\frac{r}{T}\,\delta T - \\
& -\Delta t\Sigma\,\delta T\Big)\mathrm{d}v + \int_{\partial\Omega^I} \Delta t\Big([\rho\eta]v_i + \frac{1}{T}[q_i]\Big)\,\delta T n_i \mathrm{d}a + \\
& + \int_{\partial\Omega} \Delta t\Big(\rho v_i\eta + \frac{q_i}{T}\Big)\,\delta T n_i \mathrm{d}a\ .
\end{aligned}
\tag{3.342}
$$

The weak form is the sum of all weak forms in the same unit

$$
\mathrm{Form} = \mathrm{F}_\phi + \mathrm{F}_A + \mathrm{F}_p + \mathrm{F}_v + \mathrm{F}_T\ , \tag{3.343}
$$

which is a coupled and nonlinear integral form.

We compile all material coefficients of the molten cryolite from the literature. The electrical conductivity depends on the temperature. We use values for a low-melting point electrolyte produced synthetically,

$$
\varsigma = 3.95 \cdot 10^2 \exp\Big(-\frac{1192}{T}\Big)\ \text{S/m}\ , \tag{3.344}
$$

taken from [18, Table 1]. From the investigations in [22] we realize that cryolite shows a thermoelectric coupling, we set $\pi = -1.28 \cdot 10^{-3}$ V/K based on [12, Fig. 3]. Shear viscosity's temperature dependency might be neglected[80] and we use $\mu = 2.5 \cdot 10^{-2}$ Pa s based on [31, Table 1]. As we fail to find experiments on the volume change due to pressure variation, as expected, measurements of volume viscosity seems to be missing, too. Often, the liquid metals are computed as incompressible. An experimental validation of this assumption seems to be missing. We will approximate λ for the simulation and simulate the molten salt as compressible.

There is a considerable amount of computational effort and attempts to analyze the complex phenomenon of metal smelting. Beyond the complicated coupling between electromagnetism with hydrodynamics, there are chemical reactions occurring in the real process. Even by neglecting the chemical processes in an aluminum cell filled with an assumed to be a one-phase, homogeneous cyrolite as a linear viscous fluid, it is still difficult to obtain numerical results.

Magnetohydrodynamics is used for simulations of molten salts (electrolytes) in smelting processes.[81] There are mainly two drawbacks. First, the system of equations in magnetohydrodynamics varies in different works, especially if polarization is incorporated. We have seen herein a thermodynamically consistent formulation, however, equally well-justified works arrive at slightly different formulations.[82] Since the

[80] See [23, Fig. 5].

[81] For a copper cell simulation see [34] and for an aluminum cell simulation see [13]. A review of such simulations can be found in [39].

[82] See [32, 35].

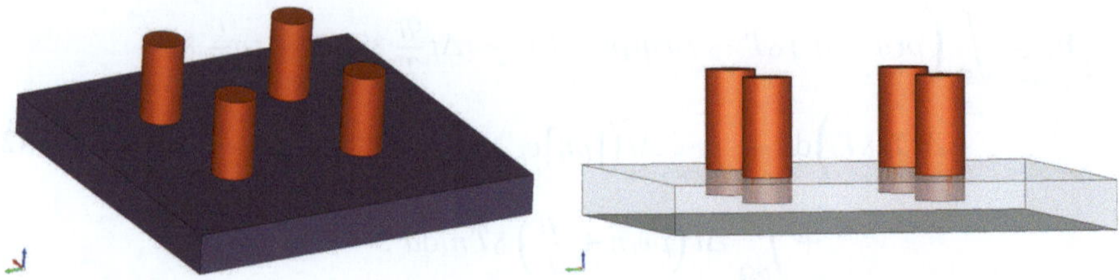

Fig. 3.18 A simplified drawing of HALL–HEROULT process. *Left* Orange anode rods are immersed into the purple cyrolite. *Right* The anodes in the cyrolite are near to the green cathode at the *bottom* of the (transparent) cyrolite

system is rather complex, it is difficult to verify the different formulations by using experiments. Secondly, there occur instabilities in the real process. The high amount of electric current may lead to shortcuts due to the motion of the electrolyte. Hence, a simulation of the motion is of interest. Especially for a case, which likely results in an instability. Unfortunately, there exist many numerical instabilities in fluid flow simulations, especially for the incompressible, steady motion of fluids[83] as mainly done in the literature. Therefore, the subject attracts many interests and there are even highly complicated simulations with attempts to include effects of gases in the process.[84] Such simulations of real processes help us to comprehend the interaction between different physical phenomena based on electromagnetism, hydrodynamics, even on chemistry.

Herein, we simulate a simplified HALL–HEROULT process in order to see the electromagnetism induced hydrodynamics in the molten salt in an aluminum cell. Four conductive rods are immersed into the molten salt cyrolite bath of dimensions $1.8 \times 1.8 \times 0.2$ m, see Fig. 3.18. The rods work as anodes under a given electric potential set as DIRICHLET boundary conditions. The bottom of the cyrolite is in touch with the molten aluminum, which grows on top of the cathode. We exclude the aluminum from the simulation. Since the aluminum is highly conductive, we simulate the cathode as being on top of aluminum. In a real process 4 V potential difference is used between anodes and cathode. We set the cathode zero and change the the electric potential in anode linearly in time such that in 1000 s the realistic difference of 4 V is attained. The potential difference generates an electric field that induces an electric current. The current creates a magnetic flux. The effect of magnetism in the total stress generates a linear momentum change such that the viscous fluid is set in motion. This motion is very slow, therefore, by using a transient solution and aforementioned constitutive equations for a compressible fluid flow we manage to obtain simulation results, see Fig. 3.19. After 1000 s the anode rods attain the optimum level

[83]See [16].

[84]For example in [9–11, 38].

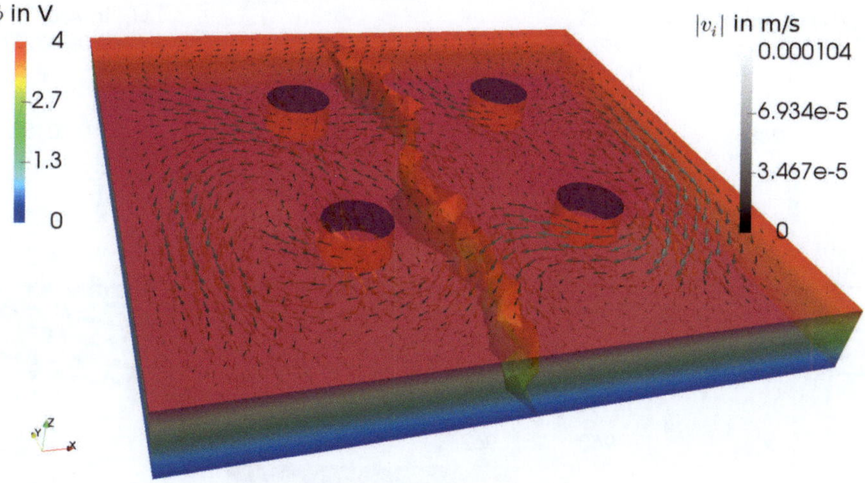

Fig. 3.19 Within electrolyte the electric potential is presented as a color distribution and the velocity is shown as *scaled arrows* at 1000 s just after reaching 4 V potential difference between anode rods and cathode (*the bottom*). Electric potential is shown as transparent for a better visualization of the motion. We have simulated in two processors such that the computational domain is calculated separately in two processors and the results are stitched afterward leading to a fictitious separation in the middle of the whole domain

of 4 V. Since the cyrolite is a good conductor there occurs a high amount of electric current from anodes to the bottom layer of the electrolyte, which is the cathode (grounded in the simulation). The high electric current is presented in Fig. 3.19. The electric current is applied directly by exerting a potential difference between anode and cathode. This current, $\mathcal{J}_i^{\text{fr.}}$, induces a magnetic flux, B_i, leading to a magnetic polarization, $\mathcal{M}_i$, in the electrolyte. Magnetic flux and polarization are parallel due to the constitutive equation. We see that the current and magnetic polarization (or flux) are perpendicular to each other. Hence, there is a contribution of a body force in the electromagnetic supply term, $\mathcal{F}_i$, in Eq. (3.288) leading to a mechanical momentum change in Eq. (3.287)$_2$. Moreover, the electric current produces the JOULE heating leading to a temperature increase. There occurs a significant increase in the temperature, especially near to the anodes. We have simulated the walls of the cyrolite bath as efficient insulators to present this effect. In reality, the cyrolite bath is cooled down outside the walls in order to hold the electrolyte at the optimum temperature. The hydrostatic pressure remains the same throughout the simulation indicating that the motion of the fluid is caused only by the electromagnetic interaction (Fig. 3.20).

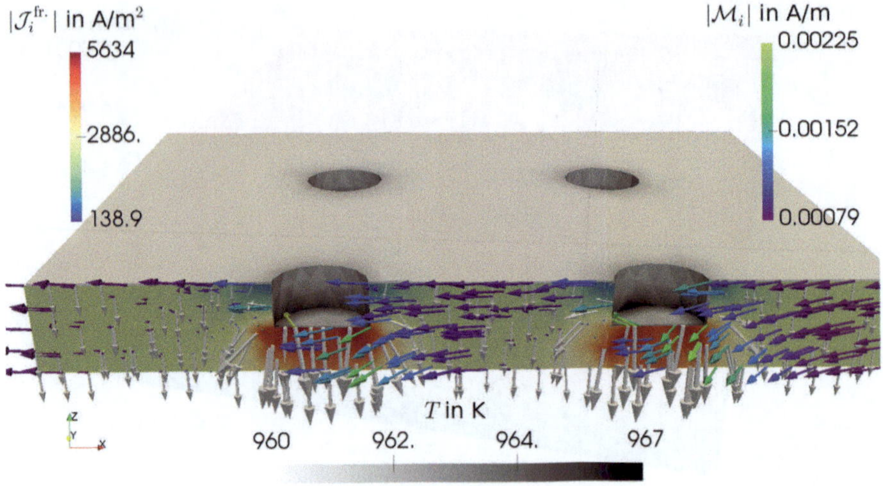

Fig. 3.20 The electric current as a color distribution and as *white arrows* are presented on the cut plane. On the same cut plane the magnetic polarization is shown as *colored arrows*. The magnetization and current are perpendicular to each other. The temperature distribution is shown in *grayscale*. All quantities are presented at 1000 s

The geometry can be received from [1]. In order to solve in multiple processors we converted the mesh into another format by using the following code:

```
from fenics import *
mesh = Mesh('geo/CR20_smelting_aluminum.xml')
cells = MeshFunction('size_t',mesh,'geo/CR20_smelting_aluminum_physical_region.xml')
facets = MeshFunction('size_t',mesh,'geo/CR20_smelting_aluminum_facet_region.xml')
hdf = HDF5File(mesh.mpi_comm(),'geo/CR20_smelting_aluminum.h5','w')
hdf.write(mesh, '/mesh')
hdf.write(cells, '/cells')
hdf.write(facets, '/facets')
```

Afterward by using

```
$mpirun -n 2 python CompReal20_smelting.py
```

we started a parallel computation in two processors with the code below:

```python
"""Computational reality 20, electric smelting"""
__author__ = "B. Emek Abali"
__license__ = "GNU GPL Version 3.0 or later"
#This code underlies the GNU General Public License, http://www.gnu.org/licenses/gpl-3.0.en.html

from fenics import *
import numpy
processID = MPI.rank(mpi_comm_world())
#units: m, kg, s, V, K
delta = Identity(3)
epsilon = as_tensor([ ( (0,0,0),(0,0,1),(0,-1,0) ), ( (0,0,-1),(0,0,0),(1,0,0) ), ( (0,1,0),(-1,0,0),(0,0,0) ) ])

t = 0.0
tMax = 1000.
Dt = 50.

'''
2D 1 "anode"
2D 2 "cathode"
2D 3 "walls"
2D 4 "top"

3D 1 "cyrolite"
'''

mesh = Mesh()
hdf = HDF5File(mesh.mpi_comm(), 'geo/CR20_smelting_aluminum.h5', 'r')
hdf.read(mesh, '/mesh', False)
cells = CellFunction('size_t', mesh)
hdf.read(cells, '/cells')
facets = FacetFunction('size_t', mesh)
hdf.read(facets, '/facets')

Scalar = FunctionSpace(mesh, 'P', 1)
Vector = VectorFunctionSpace(mesh, 'P', 1)
Space = MixedFunctionSpace([Scalar, Vector, Scalar, Vector, Scalar]) #phi, A, p, v, T
da = Measure('ds')[facets]
dv = Measure('dx')[cells]

n = FacetNormal(mesh)

dunkn = TrialFunction(Space)
test = TestFunction(Space)
unkn = Function(Space)
unkn0 = Function(Space)
unkn00 = Function(Space)

del_phi,del_A,del_p,del_v,del_T = split(test)
phi,A,p,v,T = split(unkn)
```

```

f = Constant((0.0, 0.0, 0.0))
r = Constant(0.0)
Tref = 960. #in K
pref = 1E5 #in Pa
Tamb = Tref
eps_0 = 8.85E-12 #in A s/(V m)
mu_0 = 12.6E-7 #in V s/(A m)
h = 0.1 #in J / (s m^2 K)

varsigma = 3.95E2*exp(-1192./T)
pi = -1.28E-3 #V/K
kappa = 0.8 #in W/(m K)
chi_mag = 2.2E-5
mu_mag_rel = chi_mag + 1.
cp = 394.7/(209.9413E-3) #in J/(kg K)
c4 = 2.259E-7 #in m3/(kg K)
vol_ref = 1.891E-4 #in m3/kg
c5 = 1E2
mu = 2.5E-2 #in Pa s
la = mu*1E3 #in Pa s

#boundary conditions--------------------------------------
phi_in = Expression('4.0E-3*t', t=0. )
phi_out = Expression('0.0')
bc = []
#electric potential on anode and cathode
bc.append( DirichletBC(Space.sub(0), phi_in, facets,1) )
bc.append( DirichletBC(Space.sub(0), phi_out, facets,2) )
#reference p on the top opening
bc.append( DirichletBC(Space.sub(2), pref, facets,4) )
#zero velocity on the walls of the bath
bc.append( DirichletBC(Space.sub(3).sub(2), Constant(0.0), facets,4) )
bc.append( DirichletBC(Space.sub(3), Constant((0.0,0.0,0.0)), facets,3) )
#zero velocity on anode, cathode
bc.append( DirichletBC(Space.sub(3), Constant((0.0,0.0,0.0)), facets,1) )
bc.append( DirichletBC(Space.sub(3), Constant((0.0,0.0,0.0)), facets,2) )

#initial values for phi, A, p, v, T ------------------------
unkn_init = Expression(('0.', '0.','0.','0.', 'p_r', '0.','0.','0.', 'T_r'), T_r=Tref, p_r = pref)
unkn00 = interpolate(unkn_init,Space)
unkn0.assign(unkn00)
unkn.assign(unkn0)
phi0,A0,p0,v0,T0 = split(unkn0)
phi00,A00,p00,v00,T00 = split(unkn00)

i, j, k, l = indices(4)

#electromagnetic fields -------------------------
```

```
E = -grad(phi) - (A-A0)/Dt
E0 = -grad(phi0) - (A0-A00)/Dt
B = as_tensor(epsilon[i,j,k]*A[k].dx(j) , (i,))
B0 = as_tensor(epsilon[i,j,k]*A0[k].dx(j) , (i,))
EE = as_tensor( E[i] + epsilon[i,j,k]*v[j]*B[k] , (i,))

#constitutive equations ------------------------------
D = eps_0*E
D0 = eps_0*E0
H = 1./mu_0*B
H0 = 1./mu_0*B0
m = as_tensor( -1./2.*delta[j,i]*(H[k]*B[k]+D[k]*E[k]) + H[i]*B[j] + D[j]*E[i] , (j,i))

mD = D
mD0 = D0
GG = as_tensor( epsilon[i,j,k]*mD[j]*B[k] , (i,))
GG0 = as_tensor( epsilon[i,j,k]*mD0[j]*B0[k] , (i,))
MM = chi_mag/mu_0/mu_mag_rel*B
MM0 = chi_mag/mu_0/mu_mag_rel*B0
dsigma = as_tensor( -MM[i]*B[j] + la*v[k].dx(k)*delta[j,i] + 2.*mu*sym(grad(v))[i,j] , (j,i))
eta = as_tensor(cp*ln(T/Tref) - c4*(p-pref+MM[i]*B[i]) , ())
eta0 = as_tensor(cp*ln(T0/Tref) - c4*(p0-pref+MM0[i]*B0[i]) , ())
vol = as_tensor(vol_ref + c4*(T-Tref) + c5*(p-pref + MM[i]*B[i]) , ())
vol0 = as_tensor(vol_ref + c4*(T0-Tref) + c5*(p0-pref + MM0[i]*B0[i]) , ())
rho = 1./vol
rho0 = 1./vol0
JJ_fr = as_tensor(varsigma*pi*T.dx(i) + varsigma*EE[i] , (i,))
J_fr = as_tensor( JJ_fr[i] + mD[j].dx(j)*v[i] , (i,))
q = as_tensor(-kappa*T.dx(i) + varsigma*pi*EE[i] , (i,))
Phi = q/T
Sigma = as_tensor(-q[i]/T/T*T.dx(i) + 1./T*(dsigma[j,i]+MM[i]*B[j])*v[i].dx(j) + 1./T*JJ_fr[i]*EE[i] , ())

#weak forms -------------------------------------------

F_phi = (-(mD-mD0)[i]*del_phi.dx(i) - Dt*J_fr[i]*del_phi.dx(i) )*dv(1) \
+ n[i]*Dt*( J_fr[i] + epsilon[i,j,k]*MM[k].dx(j) )*del_phi*(da(3)+da(4)) #walls and top

F_A = (eps_0*(A-2.*A0+A00)[i]/Dt/Dt*del_A[i] + 1./mu_0*A[i].dx(j)*del_A[i].dx(j) -J_fr[i]*del_A[i] + epsilon[i,j,k]*MM[k]*del_A[i].dx(j) )*dv(1)

F_p = ( (rho-rho0) + Dt*rho.dx(i)*v[i] + Dt*rho*v[i].dx(i) )*del_p/rho*dv(1)

F_v = ( (rho*v[i] - rho0*v0[i] + GG[i] - GG0[i] )*del_v[i] +
```

```
    Dt*rho.dx(j)*v[i]*v[j]*del_v[i] + Dt*rho*v[i].dx(j)*v[j]*del_v[i] + Dt*rho*v[i]*v[j].dx(j)*del_v[i] + Dt*p.dx(i)*del_v[i] + Dt*(m[j,i]+dsigma[j,i])*del_v[i].dx(j) ) - Dt*rho*f[i]*del_v[i] )*dv(1)

F_T = ( rho*(eta-eta0)*del_T - Dt*eta*(rho*v[i]*del_T).dx(i) - Dt*Phi[i]*del_T.dx(i) - Dt*rho*r/T*del_T - Dt*Sigma*del_T )*dv(1) \
+ Dt/T*h*(T-Tamb) *del_T*(da(1)+da(2)+da(3)+da(4)) #all

Form = F_phi + F_A + F_p + F_v + F_T
Gain = derivative(Form, unkn, dunkn)

pwd = '/calcul/CR20/'
file_phi = File(pwd + 'phi.pvd')
file_A = File(pwd + 'A.pvd')
file_p = File(pwd + 'p.pvd')
file_v = File(pwd + 'v.pvd')
file_T = File(pwd + 'T.pvd')
file_J = File(pwd + 'J.pvd')
file_B = File(pwd + 'B.pvd')

tic()
while t < tMax:
    t += Dt
    phi_in.t = t
    if processID==0: print 'time: ',t, 'after ',toc(),' seconds'
    tic()

    solve(Form==0, unkn, bc, J=Gain, \
        solver_parameters={"newton_solver":{"linear_solver": "mumps", "relative_tolerance": 1e-3} }, \
        form_compiler_parameters={"cpp_optimize": True, "representation": "quadrature", "quadrature_degree": 2} )
    unkn00.assign(unkn0)
    unkn0.assign(unkn)

    file_phi << (unkn.split()[0], t)
    file_A << (unkn.split()[1], t)
    file_p << (unkn.split()[2], t)
    file_v << (unkn.split()[3], t)
    file_T << (unkn.split()[4], t)
    file_J << (project(JJ_fr, Vector), t)
    file_B << (project(B, Vector), t)
```

To-do

Metal smelting is a highly complex phenomenon attracting much interest in the literature. Search for simulations of other smelting processes such as:

- Electric smelting,
- electrorefining,
- flash smelting.

References

1. Abali, B.E.: Supply material. Computational Reality. http://www.lkm.tu-berlin.de/ComputationalReality/ (2016)
2. Abali, B.E., Lofink, P., Müller, W.H.: Investigations regarding variations of material properties and their impact on lifetime prediction for Cu-vias in circuit boards. In: IEEE Electronics System-Integration Technology Conference (2014)
3. Abali, B.E., Reich, F.A., Müller, W.H.: Fatigue analysis of anisotropic copper-vias in a circuit board. In: GMM, Mikro-Nano-Integration, pp. 92–95. VDE Verlag, Berlin (2014)
4. Anovitz, L.M., Hemingway, B.S., Westrum, E.F., Metz, G.W., Essene, E.J.: Heat capacity measurements for cryolite (Na3AlF6) and reactions in the system Na-Fe-Al-Si-O-F. Geochimica et Cosmochimica Acta **51**(12), 3087–3103 (1987)
5. Apisarov, A., Dedyukhin, A., Nikolaeva, E., Tinghaev, P., Tkacheva, O., Redkin, A., Zaikov, Y.: Liquidus temperatures of cryolite melts with low cryolite ratio. Metallur. Mater. Trans. B **42**(1), 236–242 (2011)
6. Bethune-Waddell, M., Chau, K.J.: Simulations of radiation pressure experiments narrow down the energy and momentum of light in matter. Reports Progr. Phys. **78**(12), 122, 401 (2015)
7. Bobbio, S.: Electrodynamics of materials. In: Forces, Stresses, and Energies in Solids and Fluids. Academic Press, Orlando, Florida (2000)
8. Brevik, I.: Experiments in phenomenological electrodynamics and the electromagnetic energy-momentum tensor. Phys. Reports **52**(3), 133–201 (1979)
9. Caboussat, A.: Numerical simulation of two-phase free surface flows. Arch. Comput. Meth. Eng. **12**(2), 165–224 (2005)
10. Doheim, M., El-Kersh, A., Ali, M.: Computational modeling of flow in aluminum reduction cells due to gas bubbles and electromagnetic forces. Metallur. Mater. Trans. B **38**(1), 113–119 (2007)
11. Einarsrud, K.E.: The effect of detaching bubbles on aluminum-cryolite interfaces: an experimental and numerical investigation. Metall. Mat. Trans. B **41**(3), 560–573 (2010)
12. Flem, B.E., Xu, Q., Kjelstrup, S., Sterten, Å.: Thermoelectric powers of cells with NaF-AlF3-Al2O3 melts. J. Non-Equilibrium Thermodyn. **26**(2), 125–151 (2001)
13. Gerbeau, J.F., Lelièvre, T., Le Bris, C.: Simulations of MHD flows with moving interfaces. J. Comput. Phys. **184**(1), 163–191 (2003)
14. Griffiths, D.J.: Resource letter em-1: Electromagnetic momentum. Am. J. Phys. **80**(1), 7–18 (2012)
15. de Groot, S.R., Mazur, P.: Non-Equilibrium Thermodynamics. Dover Publications, New York (1984)
16. Gunzburger, M.D., Meir, A.J., Peterson, J.S.: On the existence, uniqueness, and finite element approximation of solutions of the equations of stationary, incompressible magnetohydrodynamics. Math. Comput. **56**(194), 523–563 (1991)
17. Heuzé, O.: General form of the Mie-Grüneisen equation of state. Comptes Rendus Mecanique **340**(10), 679–687 (2012)
18. Hives, J., Thonstad, J.: Electrical conductivity of low-melting electrolytes for aluminium smelting. Electrochimica Acta **49**(28), 5111–5114 (2004)
19. Holzapfel, W., Hartwig, M., Sievers, W.: Equations of state for Cu, Ag, and Au for wide ranges in temperature and pressure up to 500 GPa and above. J. Phys. Chem. Ref. Data **30**(2), 515–529 (2001)
20. Kemp, B.A., Grzegorczyk, T.M.: The observable pressure of light in dielectric fluids. Opt. Lett. **36**(4), 493–495 (2011)
21. Kim, S.M., Kim, H., Nam, Y., Kim, S.: Effects of external surface charges on the enhanced piezoelectric potential of zno and aln nanowires and nanotubes. AIP Adv. **2**(4), 042, 174 (2012)
22. Kjelstrup, S., Qian, J., Haarberg, G.M.: The peltier heating of the aluminium cathode in contact with cryolite-alumina melts. Electrochimica Acta **45**(17), 2707–2717 (2000)

23. Korenko, M., Vasková, Z., Priscák, J., Simko, F., Ambrová, M., Shi, Z.: Density, viscosity and electrical conductivity of the molten cryolite electrolytes (Na3AlF6-SiO2) for solar grade silicon (Si-sog) electrowinning. Silicon **7**(3), 261–267 (2015)
24. Kovetz, A.: Electromagnetic Theory. Oxford University Press, Oxford (2000)
25. Mansuripur, M.: Resolution of the Abraham-Minkowski controversy. Opt. Commun. **283**(10), 1997–2005 (2010)
26. Marcon, P., Ostanina, K.: Overview of methods for magnetic susceptibility measurement. In: PIERS Proceedings, pp. 420–424 (2012)
27. Meitzler, A.H., Tiersten, H.F., Warner, A.W., Berlincourt, D., Couqin, G.A., Welsh III, F.S.: IEEE Standard on Piezoelectricity (1988)
28. Müller, I.: Thermodynamics. Pitman Publishing, London (1985)
29. Obukhov, Y.N.: Electromagnetic energy and momentum in moving media. Annalen der Physik **17**(9–10), 830–851 (2008)
30. Pao, Y.H., Hutter, K.: Electrodynamics for moving elastic solids and viscous fluids. Proc. IEEE **63**(7), 1011–1021 (1975)
31. Robelin, C., Chartrand, P.: A viscosity model for the (NaF+AlF3+ CaF2+Al2O3) electrolyte. J. Chem. Thermodyn. **43**(5), 764–774 (2011)
32. Rosensweig, R.E.: Basic equations for magnetic fluids with internal rotations. In: Ferrofluids, pp. 61–84. Springer (2002)
33. Schwartz, M.: Principles oF Electrodynamics. Dover Publications (1987)
34. Sha, M., Wang, T., Li, J., Li, T., Jin, J.: Numerical simulation of horizontal continuous casting process of round copper billet with electromagnetic stirring. Int. J. Cast Metals Res. **24**(3–4), 197–202 (2011)
35. Steigmann, D.J.: On the formulation of balance laws for electromagnetic continua. Math. Mech. Solids (2007)
36. Thonstad, J., Olsen, E.: Cell operation and metal purity challenges for the use of inert anodes. JOM **53**(5), 36–38 (2001)
37. Trainer, M.: Ferroelectricity: Measurement of the dielectric susceptibility of strontium titanate at low temperatures. A. J. Phys. **69**(9), 966–969 (2001)
38. Xu, Y., Zhang, H., Li, J., Lai, Y.: A nonlinear shallow-water model combined with gas bubble effect for melt flows and interface instability in aluminum reduction cells. JOM **65**(11), 1459–1466 (2013)
39. Zhang, H., Li, J., Wang, Z., Xu, Y., Lai, Y.: The numerical modeling of melt flow and mhd instabilities in an aluminum reduction cell. JOM **62**(11), 26–31 (2010)

Appendix

A.1 Quick Introduction to Programming

The whole book is set up first to discuss the governing equations and then to solve them numerically by programming. As an engineer one needs to fulfill three tasks: Applying theoretical approaches to model the physical phenomenon, solving the model by using appropriate numerical techniques, evaluating and interpreting solutions. The first task is discussed in the underlying book. In every section the theory results in a *weak* form. This integral form can be solved numerically. For the second task we get use of the open-source codes developed under the FEniCS project.[1] We simply utilize FEniCS as a calculator for integral forms. In order to access the functionality of FEniCS we need to write a script either in C++ or in Python. We code in Python.

Installing FEniCS

Follow the instructions in http://fenicsproject.org/ for installing all necessary packages. Please remember that this is a developing code such that there might be small changes in the commands we have used. All presented codes in this book have been tested with the version 1.6, i.e., if a code does not work, take a look at the changes in the methods used in. For the installation under Ubuntu just add the repository and install the latest stable version by running the following commands:

```
sudo add-apt-repository ppa:fenics-packages/fenics
sudo apt-get update
sudo apt-get install fenics
sudo apt-get dist-upgrade
```

[1]The name of the project might be an acronym of Finite Elements of nonlinear iterative Computational Science.

B.E. Abali, *Computational Reality*, Advanced Structured Materials 55,
DOI 10.1007/978-981-10-2444-3

Under Windows and MacOS the codes can be run in a so-called container in the Docker platform. First, the Docker toolbox needs to be installed. Then the latest stable version can be downloaded by running the following line in the Docker toolbox:

```
docker pull quay.io/fenicsproject/stable:latest
```

Docker is a virtual box with a machine running FEniCS on Ubuntu. If we run in the Docker terminal

```
docker run -ti quay.io/fenicsproject/stable
```

then a machine starts and we have a terminal like in a working Ubuntu with FEniCS installed on it. The easiest way to work is to create a directory and bind the directory *shared* in that container. Under Windows for the user *nerd* we create a directory under

```
C:\Users\nerd\compreal
```

and then start a machine in the Docker terminal as follows

```
cd compreal
docker run -ti -v $(pwd):/home/fenics/shared quay.io/
    ↪ fenicsproject/stable
```

where the directory *shared* in the container is connected to the *compreal* in Windows. If the directory compreal contains a code, for example, CompReal01.py then this code can be run as follows

```
fenics@...:~ $ cd shared
fenics@...:~/shared$ python CompReal01.py
```

Python

Basically we use Python for programming in Unified Form Language (UFL) for FEniCS. Some basic Python knowledge might help. Go to the terminal and start a python interface

```
python
```

now we may try the conventional first line

```
>>> print 'Hello world!'
Hello world!
```

or even use a sequence

```
>>> a = 'Hello world!'
>>> print a
Hello world!
>>> print 'I said:', a
I said: Hello world!
>>> print 'I said:', a*2
I said: Hello world!Hello world!
```

which is a list of letters

```
>>> a
'Hello world!'
>>> a[0]
'H'
>>> a[1]
'e'
>>> a[:5]
'Hello'
>>> a[5:]
' world!'
```

like a list of numbers

```
>>> b = [2, 5, 8]
>>> b
[2, 5, 8]
>>> b[2]
8
```

They are all objects and the good thing about programming in Python is that we never need an initializer, constructor, or destructor. We just code anything in the logical way without caring about the memory usage. Python knows the correct type of its arguments without decleration

```
>>> type(a)
<type 'str'>
>>> type(b)
<type 'list'>
```

Every object has its functions (methods) according to its class, ask the possible functions and use one of them as follows

```
>>> dir(b)
['__add__', '__class__', ... , 'remove', 'reverse', 'sort']
>>> b.reverse()
>>> b
[8, 5, 2]
```

Let us start with FEniCS by importing all functions into cache

```
>>> from fenics import *
>>> dir()
```

First, we create a finite element mesh on 2D square and plot it with Viper

```
>>> mesh = UnitSquareMesh(80,20)
>>> plot(mesh, interactive=True)
```

There is also a documentation of the function under the fenics class or namespace (in Python no strict differentiation is made between class and namespace, also between function, method, and definition), which is accessed via (use q to quit)

```
>>> exit()
pydoc fenics.UnitSquareMesh
```

which provides some useful information. The same documentary can be found online under http://fenicsproject.org/documentation/

Gedit

If the code is more than a few lines, it is a good idea to save it in a file, for example code.py and run it by typing

```
python code.py
```

in the shell. One of the simplest and yet powerful text editor is Gnome project text editor. It is in Ubuntu Gnome included and can be installed under Windows or MacOS for free via http://projects.gnome.org/gedit/

You may try under Edit/Preferences/Font&Colors/Color Scheme: Oblivion for a darker background during "long coding nights."

Spyder

A quite nice Scientific PYthon Development EnviRonment (Spyder) is included in Ubuntu packages, see http://pythonhosted.org/spyder/ for instructions to install it under other operating systems. It has nice debugging options and also shows the documentary during coding.

ParaView

For postprocessing (visualizing) Viper may be limited in some features, ParaView is a very powerful tool, get it from http://www.paraview.org

Ready to go?

```
>>> print 'Alright! Start with the first tutorial!'
```

A.2 Solvers in FEniCS

Solvers and its arguments can be listed by using

```
>>>list_linear_solver_methods()
>>> list_linear_algebra_backends()
>>>list_krylov_solver_methods()
>>> list_krylov_solver_preconditioners()
>>> info(NonlinearVariationalSolver.default_parameters(), True)
```

A.3 Complicated Geometries in FEniCS

The geometry and mesh can be automatically acquired by using BoxMesh() in FEniCS. For more advanced geometries we need to use a CAD or CAE (Computer Aided Design or Engineering) program and a preprocessor for meshing. There are different scenarios for importing an advanced geometry into FEniCS. One quite straight-forward way is to use the open-source CAE platform Salome[2] version 7.5 for preprocessing. We will explain these steps in Salome version 7.5.

The geometry can be established by using Salome or can be imported from another program. If it shall be created in Salome, there is a nice feature called *notebook*. In notebook parameters can be declared, like $a = 100$ and $b = 10$ and then these parameters can be used by creating a 3D body, like a box of $a \times b \times b$. If then the parameters have been changed, the geometry can be updated. Slightly different to the most CAD programs; in Salome the geometry is directly created in 3D modeling. There are basic elements like box, cylinder, sphere, which can be created and added together by using boolean operations, like *fuse*, *common*, *cut*. Consider a plate with a hole in it. Instead of a 2D sketch of a rectangle including a hole, box is used for the plate and a cylinder is subtracted from the plate. Any other program like FreeCAD[3] can also be used to create a geometry and import into Salome by using STEP file format.

[2] http://www.salome-platform.org/

[3] http://www.freecadweb.org/

We assume that the model consists of 3 parts out of different materials. Different parts will have unmatched meshes, therefore, the adjacent faces have to be stitched together. This can be done by using *partition* in the *geometry* module. All parts will be collected under one partition named *Partition*. In the *object browser* all parts and partition can be seen separately. Partition will be used further, since it includes all parts as joined together. The difference to *fuse* is that a *partition* preserves the information of different parts. Hence we use *partition* for collecting different materials in one geometry leading to matching meshes.

In order to distinguish between different parts in FEniCS we will use *groups* under Salome to mark them. By *create groups* in *geometry* module we can create 3 volume groups for 3 different parts. Moreover, for boundary conditions we can create surface groups. Later in FEniCS we will use these groups for applying a DIRICHLET condition on a specific surface group or applying a boundary integral only on the chosen surface group for applying a NEUMANN boundary condition by integrating over the marked surface.

After creating groups in the geometry module we switch to *mesh* module and *create mesh* on the *Partition*. A *Mesh* object is created in the *object browser*. It is important to recall that the part *Partition* includes different parts and groups. FEniCS can work with tetrahedron elements such that we choose *mesh type: tetrahedral* and *algorithm: NetGen 1D2D3D*. Other algorithms can be tried, too. NetGen algorithm[4] is quite stable and powerful. Some parameters can be changed for manipulating the total number of nodes. For creating the elements choose *compute*. The mesh has solely the information of nodes and connectivity, but not the topology (geometrical connectivity). In order to pass the knowledge of the created groups in the geometry module we use *create groups from geometry*, choose all groups under *Partition* in the *object browser*, and *apply* it. Then the groups can be seen under the mesh, too.

Select *Mesh* in the *object browser* and export as *.med* file by using *export*. Now the *.med* file shall be opened in Gmsh[5] and saved as a *.msh* file by using *save mesh*. This is an Ascii file such that it can be opened in an editor. In the very beginning the groups with their enumerations can be seen, for example suppose we see

```
2 1 "bottom"
2 2 "top"
3 1 "steel"
3 2 "alu"
3 3 "pvc"
```

[4]https://sourceforge.net/projects/netgen-mesher/

[5]http://geuz.org/gmsh/

for a partition denoted by 3 for 3D with three parts, namely, *steel*, *alu*, *pvc*, marked as 1, 2, 3, respectively; and two faces denoted by 2 for 2D as *bottom* and *top* marked with 1 and 2, respectively. These numbers are the key to use them correctly in FEniCS. The *.msh* file needs to be converted by using the command:

```
dolfin-convert example.msh example.xml
```

into the FEniCS compatible *.xml* file. The *dolfin-convert* produces

- example.xml, including positions of nodes and connectivities of elements,
- example_facet_region.xml, including the information of elements' faces belonging to the chosen faces for boundary conditions,
- example_physical_region.xml, including the information of elements belonging to the parts.

These files can then be applied as follows

```
from fenics import *
mesh = Mesh('example.xml')
V = VectorFunctionSpace(mesh, 'P', 1)
cells=MeshFunction('size_t', mesh, 'example_physical_region.xml')
plot(cells,interactive=True)
facets = MeshFunction('size_t', mesh, 'example_facet_region.xml')
plot(facets,interactive=True)
dA = Measure('ds', domain=mesh, subdomain_data=facets)
dV = Measure('dx', domain=mesh, subdomain_data=cells)
```

Now, in the Form, dA(1) let us integrate over *bottom* since all facets[6] belonging to *bottom* are marked with 1. Integration over a part of boundary is useful for implementing a NEUMANN or ROBIN boundary condition. We can also apply a DIRICHLET boundary condition for example on *top* by writing

```
bc = [DirichletBC(V, null, facets, 2)]
```

Analogously we can integrate over a part of the volume. Consider that we have different material parameters and thus stresses for three materials such that we implement a Form such that:

```
Form = sigma_steel[k,i]*del_u[i].dx(k)*dV(1) + sigma_alu[k,i]*del_u[i].dx(k)*dV(2) + sigma_pvc[k,i]*del_u[i].dx(k)*dV(3) - tr[i]*del_u[i]*dA(2)
```

The primitive variables, in this example the deformation, will be computed for the whole *Partition*. If we want to compute the distribution of equivalent stress according to VON MISES then we need to calculate it for each part separately

[6]Facet is a FEniCS specific terminology meaning a face for a tetrahedron element and a line for a triangle element. Facet is a synonym for element boundaries and they are one dimension less than the element itself.

```
mesh_steel = SubMesh(mesh, cells , 1)
eps1 = project(sym(grad(u)), TensorFunctionSpace(mesh_steel , 'P', 1), solver_type="mumps", form_compiler_parameters ={"cpp_optimize": True, "representation": "quadrature", "quadrature_degree": 2})
sigma1 =as_tensor(C_steel[i,j,k,l]*eps1[k,l] , (i ,j))
sigma_1_dev =as_tensor(sigma1[i,j]-1.0/3.0*sigma1[k,k]*delta[i,j], (i,j))
eqStress_1 =as_tensor((3.0/2.0*sigma_1_dev[i,j]*sigma_1_dev[i,j])**0.5,())
eqS_1=project(eqStress_1, FunctionSpace(mesh_steel, 'P', 1), solver_type="mumps", form_compiler_parameters={"cpp_optimize": True, "representation": "quadrature", "quadrature_degree": 2})

mesh_alu = SubMesh(mesh, cells , 2)
eps2 = project(sym(grad(disp)), TensorFunctionSpace(mesh_alu, 'P', 1), solver_type="mumps", form_compiler_parameters={"cpp_optimize": True, "representation": "quadrature", "quadrature_degree": 2})
sigma2 = as_tensor(C_alu[i,j,k,l]*eps2[k,l] , (i ,j))
sigma_2_dev =as_tensor(sigma2[i,j]-1.0/3.0*sigma2[k,k]*delta[i,j], (i,j))
eqStress_2 = as_tensor((3.0/2.0*sigma_2_dev[i,j]*sigma_2_dev[i,j])**0.5,())
eqS_2 = project(eqStress_2, FunctionSpace(mesh_alu, 'P', 1), solver_type="mumps", form_compiler_parameters={"cpp_optimize": True, "representation": "quadrature", "quadrature_degree": 2})

mesh_pvc = SubMesh(mesh, cells , 3)
eps3 = project(sym(grad(disp)), TensorFunctionSpace(mesh_pvc, 'P', 1), solver_type="mumps", form_compiler_parameters={"cpp_optimize": True, "representation": "quadrature", "quadrature_degree": 2})
sigma3 = as_tensor(C_pvc[i,j,k,l]*eps3[k,l] , (i ,j))
sigma_3_dev=as_tensor(sigma3[i,j]-1.0/3.0*sigma3[k,k]*delta[i,j],(i,j))
eqStress_3=as_tensor((3.0/2.0*sigma_3_dev[i,j]*sigma_3_dev[i,j])**0.5,())
eqS_3 = project(eqStress_3, FunctionSpace(mesh_pvc, 'P', 1), solver_type="mumps", form_compiler_parameters={"cpp_optimize": True, "representation": "quadrature", "quadrature_degree": 2})

file = File('eqS_1.pvd')
file << eqS_1
file = File('eqS_2.pvd')
file << eqS_2
file = File('eqS_3.pvd')
file << eqS_3
```

and open all three files under ParaView to get the whole *Partition*.

Transferring the mesh from Salome to FEniCS can be accelerated by using a bash script

```
#!/bin/bash
fil=$1
strlen=${#filename}
filename=${fil:0:strlen-4}
gmsh $1 -3 -v 0 $fil
dolfin-convert "$filename.msh" "$filename.xml"
```

By saving this script in a file *med2xml* in the same directory as the *.med* file and setting the permission of it as executable, we can run in terminal

```
./med2xml example.med
```

The script produces *.msh* and all *.xml* files.

A.4 Objective Time Rate of Strain Tensor

The generic formulation of the objective time rate can be found in [1]. Here we want to derive a special case applied to the strain. Therefore, we need to introduce a new concept. Consider a continuum body expressed in Cartesian coordinates. We mark on the body the lines of the coordinate system. Since we use a Cartesian coordinate system, the lines on the body compile an orthogonal grid. Subject to a deformation the lines bend such that the grid is curvilinear and oblique. We can visualize this deformation as an evolution of the coordinate system and introduce a *convected* coordinate system deforming with the continuum body.

We introduce two coordinate systems for two instants of time. At the initial time we may use Cartesian coordinates. After deformation at another time instant we have to use curvilinear and oblique coordinates. The transformation between them is the deformation of the underlying body. The intuitive formulation is that the Cartesian coordinates deform and the body in the current frame is in the curvilinear and oblique coordinates. However, evaluation in curvilinear and oblique coordinates can be unpractical. It is much more easier to evaluate in Cartesian coordinates. Therefore, we introduce two coordinate systems, a Cartesian for the current frame, $x_i = x^i$, and a general (a curvilinear and oblique) coordinate system for the initial frame, $Z^i \neq Z_i$. The initial coordinate system denotes again to particles, in other words, the coordinates for a particle remains the same but the metric changes in time. The transformation between Cartesian (current frame) and general (initial frame) represents the deformation.

Despite the expectation, not much will change in the formulation. Two neighboring particles, $\mathrm{d}x^i$, have an initial distance $\mathrm{d}S$ that changes due to the deformation to $\mathrm{d}s$ as

$$(\mathrm{d}s)^2 = \mathrm{d}x^i\,\mathrm{d}x_i = \delta^{ij}\,\mathrm{d}x_i\,\mathrm{d}x_j\;,\;(\mathrm{d}S)^2 = \mathrm{d}Z^i\,\mathrm{d}Z_i = g^{ij}\,\mathrm{d}x_i\,\mathrm{d}x_j \tag{A.1}$$

where g_{ij} represents the *metric* tensor and $g^{ij}g_{jk} = \delta^i_k$. In other words, the deformation is the deviation from the Cartesian system given by the metric tensor:

$$g_{ij} = \frac{\partial x_k}{\partial Z^i}\frac{\partial x^k}{\partial Z^j}\;,\quad g^{ij} = \frac{\partial Z^i}{\partial x_k}\frac{\partial Z^j}{\partial x^k}\;. \tag{A.2}$$

The metric tensor is symmetric, $g_{ij} = g_{ji}$. The neighboring particles in the initial frame, $\mathrm{d}Z^i$, can be transformed by a mapping between the coordinate systems:

$$\begin{aligned}\mathrm{d}x^i &= \frac{\partial x^i}{\partial Z^j}\,\mathrm{d}Z^j\;,\quad \mathrm{d}x_i = \frac{\partial Z^j}{\partial x^i}\,\mathrm{d}Z_j\;,\\ F^i{}_j &= \frac{\partial x^i}{\partial Z^j}\;,\quad (F^{-1})^j{}_i = \frac{\partial Z^j}{\partial x^i}\;,\quad F^i{}_j(F^{-1})^j{}_k = \delta^i_k\;,\end{aligned} \tag{A.3}$$

we call $\boldsymbol{F}$ the deformation gradient. The inverse metric tensor reads

$$g^{ij} = (\boldsymbol{F}^{-1})^{ik}(\boldsymbol{F}^{-1})^j_k\;. \tag{A.4}$$

It can easily be associated with the left CAUCHY–GREEN deformation tensor. Hence, without further ado we observe from Eq. (1.53)

$$g^{ij} = \delta^{ij} - 2e^{ij}\;. \tag{A.5}$$

The time rate of length difference:

$$((\mathrm{d}\ell)^2)^{\boldsymbol{\cdot}} = \big((\mathrm{d}s)^2 - (\mathrm{d}S)^2\big)^{\boldsymbol{\cdot}}\;, \tag{A.6}$$

will result in an identity that we are searching for. We start by using Eq. (A.5) that leads to

$$\begin{aligned}\big((\mathrm{d}s)^2 - (\mathrm{d}S)^2\big)^{\boldsymbol{\cdot}} &= \big((\delta^{ij} - g^{ij})\,\mathrm{d}x_i\,\mathrm{d}x_j\big)^{\boldsymbol{\cdot}} = \big(2e^{ij}\,\mathrm{d}x_i\,\mathrm{d}x_j\big)^{\boldsymbol{\cdot}} =\\ &= 2\big((e^{ij})^{\boldsymbol{\cdot}}\,\mathrm{d}x_i\,\mathrm{d}x_j + e^{ij}\,\mathrm{d}w_i\,\mathrm{d}x_j + e^{ij}\,\mathrm{d}x_i\,\mathrm{d}w_j\big)\;.\end{aligned} \tag{A.7}$$

Generally speaking, the Cartesian coordinates in the current frame may be moving with a velocity of w_i, so we allow $(\mathrm{d}x_i)^{\boldsymbol{\cdot}} = \mathrm{d}w_i$. This velocity is independent of the underlying continuum body. The term, $\mathrm{d}w_i$, needs a special attention. First we show that

$$
\begin{aligned}
\delta^j_k = \frac{\partial Z^j}{\partial Z^k} &= \frac{\partial Z^j}{\partial x^i}\frac{\partial x^i}{\partial Z^k} \,, \\
\Big(\frac{\partial Z^j}{\partial x^i}\frac{\partial x^i}{\partial Z^k}\Big)^{\bullet} &= (\delta^j_k)^{\bullet} = 0 \,, \\
\Big(\frac{\partial Z^j}{\partial x^i}\Big)^{\bullet}\frac{\partial x^i}{\partial Z^k} = -\frac{\partial Z^j}{\partial x^i}\Big(\frac{\partial x^i}{\partial Z^k}\Big)^{\bullet} &= -\frac{\partial Z^j}{\partial x^i}\frac{\partial w^i}{\partial Z^k} \,,
\end{aligned}
\tag{A.8}
$$

since Z^i denotes to particles, thus, $(Z^i)^{\bullet} = 0$. Secondly, we obtain by using the above relation

$$
\begin{aligned}
\mathrm{d}w_i = (\mathrm{d}x_i)^{\bullet} = \Big(\frac{\partial Z^j}{\partial x^i}\,\mathrm{d}Z_j\Big)^{\bullet} &= \Big(\frac{\partial Z^j}{\partial x^i}\Big)^{\bullet}\,\mathrm{d}Z_j \,, \\
\mathrm{d}w_i\frac{\partial x^i}{\partial Z^k} &= -\frac{\partial Z^j}{\partial x^i}\frac{\partial w^i}{\partial Z^k}\,\mathrm{d}Z_j \,, \\
\mathrm{d}w_i\frac{\partial x^i}{\partial Z^k}\frac{\partial Z^k}{\partial x^l} &= -\frac{\partial Z^j}{\partial x^i}\frac{\partial w^i}{\partial Z^k}\,\mathrm{d}Z_j\frac{\partial Z^k}{\partial x^l} \,, \\
\mathrm{d}w_l = -\frac{\partial Z^j}{\partial x^i}\frac{\partial w^i}{\partial x^l}\,\mathrm{d}Z_j &= -\frac{\partial w^i}{\partial x^l}\,\mathrm{d}x_i \,.
\end{aligned}
\tag{A.9}
$$

By introducing comma for the partial derivative, we acquire

$$
\big((\mathrm{d}\ell)^2\big)^{\bullet} = 2\big((e^{ij})^{\bullet} - e^{lj}w^i_{,l} - e^{il}w^j_{,l}\big)\,\mathrm{d}x_i\,\mathrm{d}x_j \,. \tag{A.10}
$$

Moreover, a metric tensor depends only on the coordinates such that it has no partial time derivatives:

$$
g^{\bullet}_{ij} = \frac{\partial g_{ij}}{\partial t} + g_{ij;m}(\boldsymbol{x}^{\bullet})^m = g_{ij;m}w^m \,. \tag{A.11}
$$

According to RICCI's theorem[7] the covariant derivative of the metric tensor vanishes such that

$$
g^{\bullet}_{ij} = g_{ij;m}w^m = 0 \,. \tag{A.12}
$$

We want to calculate the rate of length, $((\mathrm{d}\ell)^2)^{\bullet}$. Although it is counterintuitive, we introduce the same Cartesian coordinate system at each time instant. Hence, the distance in current frame remains the same, $(\mathrm{d}s)^{\bullet} = 0$, leading to

$$
\begin{aligned}
((\mathrm{d}\ell)^2)^{\bullet} &= \big((\mathrm{d}s)^2 - (\mathrm{d}S)^2\big)^{\bullet} = -\big((\mathrm{d}S)^2\big)^{\bullet} = \\
&= -\big(g^{ij}\,\mathrm{d}(u_i + Z_i)\,\mathrm{d}(u_j + Z_j)\big)^{\bullet} = \\
&= -g^{ij}\big((\mathrm{d}u_i)^{\bullet}\,\mathrm{d}x_j + \mathrm{d}x_i(\mathrm{d}u_j)^{\bullet}\big) \,,
\end{aligned}
\tag{A.13}
$$

since the coordinates of the particle remains constant, too. By using the same steps as in Eq. (A.9) we obtain

[7] It is named after Gregorio Ricci-Curbastro.

$$\mathrm{d}v_j = (\mathrm{d}u_j)^{\cdot} = -\frac{\partial v^i}{\partial u^j}\,\mathrm{d}u_i = -\frac{\partial v^i}{\partial (u^j + Z^j)}\,\mathrm{d}(u_i + Z_i) = -\frac{\partial v^i}{\partial x^j}\,\mathrm{d}x_i \ . \tag{A.14}$$

Hence, we acquire

$$\begin{aligned}((\mathrm{d}\ell)^2)^{\cdot} &= g^{ij}\Big(\frac{\partial v^k}{\partial x^i}\,\mathrm{d}x_k\,\mathrm{d}x_j + \mathrm{d}x_i\frac{\partial v^k}{\partial x^j}\,\mathrm{d}x_k\Big) = \frac{\partial v^k}{\partial x_j}\,\mathrm{d}x_k\,\mathrm{d}x_j + \mathrm{d}x_i\frac{\partial v^k}{\partial x_i}\,\mathrm{d}x_k = \\ &= \frac{\partial v^k}{\partial x_j}\,\mathrm{d}x_k\,\mathrm{d}x_j + \frac{\partial v^j}{\partial x_k}\,\mathrm{d}x_j\,\mathrm{d}x_k = 2d^{kj}\,\mathrm{d}x_k\,\mathrm{d}x_j \ ,\end{aligned} \tag{A.15}$$

where the symmetric part of velocity gradient is used

$$d^{ij} = \frac{\partial v^{(i}}{\partial x_{j)}} \ . \tag{A.16}$$

Therefore, finally we acquire by using Eq. (A.10) for the general case:

$$\big((\mathrm{d}\ell)^2\big)^{\cdot} = 2d^{kj}\,\mathrm{d}x_k\,\mathrm{d}x_j = 2\big((e^{ij})^{\cdot} - e^{lj}w^i_{,l} - e^{il}w^j_{,l}\big)\,\mathrm{d}x_i\,\mathrm{d}x_j \ . \tag{A.17}$$

Since $\mathrm{d}x_i$ can be chosen arbitrarily we conclude

$$(e^{ij})^{\cdot} = d^{ij} + e^{lj}w^i_{,l} + e^{il}w^j_{,l} \ . \tag{A.18}$$

This relation is a definition for the objective time rate of a contravariant tensor of rank two, herein, of the nonlinear strain measure. For the specific case without velocity of the domain, $w_i = 0$, in other words, for a fixed frame we obtain

$$(e^{ij})^{\cdot} = d^{ij} \ . \tag{A.19}$$

A.5 Time Discretization

Time is a scalar quantity, thus, it has a simple transformation property and can be approximated with a truncated TAYLOR expansion for an arbitrary variable, A,

$$\begin{aligned}A(x_k, t^0) &= A(x_k, t^0 + \Delta t - \Delta t) = A(x_k, t - \Delta t) = \\ &= A(x_k, t) - \frac{\partial A}{\partial t}(x_k, t)\Delta t + O(\Delta t^2) \ , \\ \frac{\partial A}{\partial t}(x_k, t)\Delta t &\approx A(x_k, t) - A(x_k, t - \Delta t) \ .\end{aligned} \tag{A.20}$$

We can rewrite the latter in a simplified notation

$$\frac{\partial A}{\partial t} = \frac{A - A^0}{\Delta t} \,, \tag{A.21}$$

which is widely known as backward EULER method. Since the derivative in the series in Eq. (A.20) is evaluated at the current time instant, t, the time integration is implicit. This implicit time integration is conditionally stable. The condition vanishes for a *real valued* differential equation. We use only real valued systems, thus it is always stable, not depending on the time step size Δt. Here, it is important to distinguish stability and reliability. If the time steps have been chosen too big, caused by the truncation error, solution may land far from the exact solution. Error will grow in each time step successively, however, the solution will be found, while the time discretization is stable. Therefore, we use only implicit time integration technique and eliminate any discussion about stability conditions[8] for the sake of the time step. If the problem is nonlinear, then a linearization method finds the nearest root (since the form is quadratic, it is the minimum). Hence we need to use small time steps to "land on" the correct solution upon the linearization. The approach for linearization has been discussed in Sect. 1.8 on p. 86.

A.6 Gauge Freedom in Electromagnetic Fields

We can use the ansatz functions:

$$E_i = -\frac{\partial \phi}{\partial x_i} - \frac{\partial A_i}{\partial t} \,, \quad B_i = \epsilon_{ijk} \frac{\partial A_k}{\partial x_j} \,, \tag{A.22}$$

in order to solve the following MAXWELL equations:

$$\frac{\partial B_i}{\partial x_i} = 0 \,, \quad \frac{\partial B_i}{\partial t} + \epsilon_{ijk} \frac{\partial E_k}{\partial x_j} = 0 \,. \tag{A.23}$$

It is obvious that the proposed ansatz functions do solve the latter equations since $\epsilon_{ijk} = -\epsilon_{jik}$ such that from Eq. (A.23)$_1$ we acquire

$$\epsilon_{ijk} \frac{\partial^2 A_k}{\partial x_i \partial x_j} = 0 \,, \tag{A.24}$$

moreover, by using SCHWARZ's rule

[8] Especially in fluid dynamics there are many difficulties due to the time stability measured by the PECLET number or eliminated by satisfying the COURANT- FRIEDRICHS- LEWY condition.

$$\epsilon_{ijk}\frac{\partial^2 A_k}{\partial t \partial x_j} - \epsilon_{ijk}\frac{\partial^2 \phi}{\partial x_j \partial x_k} - \epsilon_{ijk}\frac{\partial^2 A_k}{\partial x_j \partial t} = 0\,. \tag{A.25}$$

Hence, the electric potentials, ϕ, A_i, are formal solutions of the two MAXWELL equations.

In order to see the gauge freedom we need to remember the decomposition of any rank two tensor into a spherical and a deviatoric part:

$$\frac{\partial A_k}{\partial x_j} = \frac{\partial A_{|k}}{\partial x_{j|}} + \frac{1}{3}\frac{\partial A_l}{\partial x_l}\delta_{kj}\,. \tag{A.26}$$

However, by multiplying with the LEVI-CIVITA symbol

$$B_i = \epsilon_{ijk}\frac{\partial A_k}{\partial x_j} = \epsilon_{ijk}\frac{\partial A_{|k}}{\partial x_{j|}} + \frac{1}{3}\epsilon_{ijk}\frac{\partial A_l}{\partial x_l}\delta_{kj}\,, \tag{A.27}$$

we see that the second term vanishes, since

$$\epsilon_{ijk}\delta_{jk} = \epsilon_{ijj} = 0\,, \tag{A.28}$$

no matter what the divergence of A_i is. In other words, we can freely choose $\partial A_i/\partial x_i$.

In order to present the second gauge freedom, we take the time rate of Eq. (A.23)$_2$ and then utilize Eq. (A.22) into it

$$\begin{gathered}
\frac{\partial^2 B_i}{\partial t^2} + \epsilon_{ijk}\frac{\partial^2 E_k}{\partial t \partial x_j} = 0\,, \\
\epsilon_{ijk}\frac{\partial^3 A_k}{\partial t^2 \partial x_j} - \epsilon_{ijk}\frac{\partial^3 \phi}{\partial t \partial x_j \partial x_k} - \epsilon_{ijk}\frac{\partial^3 A_k}{\partial t \partial x_j \partial t} = 0\,, \\
-\epsilon_{ijk}\frac{\partial^2}{\partial x_j \partial x_k}\left(\frac{\partial \phi}{\partial t}\right) = 0\,,
\end{gathered} \tag{A.29}$$

where we have used SCHWARZ's rule twice. Since the second derivation in x_j and x_k is symmetric, its multiplication with the antisymmetric LEVI-CIVITA tensor vanishes for any values of $\partial\phi/\partial t$. In other words we can choose $\partial\phi/\partial t$ as we like, the equations are still fulfilled.

Reference

1. Oldroyd, J.G.: On the formulation of rheological equations of state. Proc. R. Soc. Lond. A. **200**(1063), 523–542 (1950)

Index

B.E. Abali, *Computational Reality*, Advanced Structured Materials 55,
DOI 10.1007/978-981-10-2444-3

MIX
Papier aus verantwortungsvollen Quellen
Paper from responsible sources
FSC® C105338

If you have any concerns about our products,
you can contact us on
ProductSafety@springernature.com

In case Publisher is established outside the EU,
the EU authorized representative is:
Springer Nature Customer Service Center GmbH
Europaplatz 3, 69115 Heidelberg, Germany

Printed by Libri Plureos GmbH
in Hamburg, Germany